METHODS IN MOLECULAR BIOLOGY™

For further volumes:
http://www.springer.com/series/7651

Target Identification and Validation in Drug Discovery

Methods and Protocols

Edited by

Jürgen Moll

Director Pharmacology, Boehringer Ingelheim RCV GmbH & Co KG, Vienna, Austria

Riccardo Colombo

Project and Team Leader, Cell Biology, Nerviano Medical Sciences, Nerviano, MI, Italy

Humana Press

Editors
Jürgen Moll
Director Pharmacology
Boehringer Ingelheim RCV GmbH & Co KG
Vienna, Austria

Riccardo Colombo
Project and Team Leader
Cell Biology
Nerviano Medical Sciences
Nerviano, MI, Italy

ISSN 1064-3745 ISSN 1940-6029 (electronic)
ISBN 978-1-4939-6278-5 ISBN 978-1-62703-311-4 (eBook)
DOI 10.1007/978-1-62703-311-4
Springer New York Heidelberg Dordrecht London

Printed on acid-free paper

Humana Press is a brand of Springer
Springer is part of Springer Science+Business Media (www.springer.com)

Preface

The major reason for the elevated costs of drug development in the pharmaceutical industry is the high attrition rate. Only a minority of molecules entering the clinical phase makes it to the market, and one of the increasing causes for drug failures in recent years is "target failure." Hence investments in better validated targets is the key to reduce attrition rates and certainly also the most cost effective. However, in most indications, validated targets are rare and the highly validated ones are also extremely competitive. This is particularly true for oncology, and therefore we focused mainly on oncology-related methods. The identification of innovative new targets is the real bottleneck of modern drug development and the challenge will be to identify new disease-associated mechanisms and spotting "druggable" targets while minimizing side effects.

In this book state-of-the art methods and approaches to identify and validate new targets are summarized including detailed protocols with a critical assessment of their limitations and strengths. The knowledge of these technologies is essential for any drug developer interested in target identification or validation, and certainly there should be awareness of these technologies for any executive drug development manager. A wide range of technologies relevant in the field are introduced, including in silico and "Omics-" related technologies, RNAi, innovative cell culture technologies as well as tailored animal models and translational applications. The chapters are ordered by major categories covering biochemical, cell-based, in vivo models and translational methods. Besides a complete review of the major technologies, chapters handling selected case reports are included in order to integrate the technologies to real-life experiences and to demonstrate the multiple use of more than one technology to increase knowledge on a specific target. Certainly the degree of target validation correlates with the confidence in a target, which ideally is increasing over the life cycle of a drug development project but reaches a definitive answer only when clinical proof of concept is reached.

This book is aimed at scientists in diverse fields including molecular and cellular biologists, pharmacologists, in silico drug developers, pathologists, geneticists, and clinical investigators. The book contains a comprehensive list of essential methods and clear protocols to follow. The contributions are going beyond typical protocols since the emphasis is on the description of most critical steps and typical pitfalls, written by experienced experts in the field. Last but not the least, this work intends also to help managers in drug development to get a quick overview of available technologies in order to improve attrition rates of their drug development portfolio.

We thank all the authors for their excellent contributions and appreciate their hard work and fruitful interactions with the final aim to make this book a success.

Nerviano, MI, Italy *Jürgen Moll*
Vienna, Austria *Riccardo Colombo*

Acknowledgments

We thank Susan Watts for assisting in editing some of the chapters and our families for their patience during the preparation of this book.

Contents

Part VII Translational Methods to validate Biomarkers

Contributors

ROBERTO ANGIOLI • *Department of Obstetrics and Gynecology, University Campus Bio-Medico of Rome, Rome, Italy*

DAVID O. AZORSA • *Clinical Translational Research Division, Translational Genomics Research Institute (TGen), Scottsdale, AZ, USA*

RICO BARSACCHI • *Max Planck Institute of Molecular Cell Biology and Genetics, Dresden, Germany*

MARC BICKLE • *Max Planck Institute of Molecular Cell Biology and Genetics, Dresden, Germany*

CAROL BOX • *Tumour Biology and Metastasis, Cancer Research UK Cancer Therapeutics Unit, Division of Cancer Therapeutics, McElwain Laboratories, The Institute of Cancer Research, Surrey, UK*

HOLGER BRANDL • *Max Planck Institute of Molecular Cell Biology and Genetics, Dresden, Germany*

MARTA CORTÉS-CROS • *Disease Area Oncology, Novartis Institutes for BioMedical Research, Basel, Switzerland*

KONSTANTINOS DROSOPOULOS • *Division of Breast Cancer Research, Breakthrough Breast Cancer Research Centre, The Institute of Cancer Research, London, UK*

SUZANNE A. ECCLES • *Tumour Biology and Metastasis, Cancer Research UK Cancer Therapeutics Unit, Division of Cancer Therapeutics, McElwain Laboratories, The Institute of Cancer Research, Surrey, UK*

MITCHELL J. FREDERICK • *Department of Head and Neck Surgery, University of Texas M.D. Anderson Cancer Center, Houston, TX, USA*

ROSSELLA GALLI • *Neural Stem Cell Biology Unit, Division of Regenerative Medicine, Stem Cells and Gene Therapy, San Raffaele Scientific Institute, Milan, Italy*

THOMAS M. GROGAN • *Medical Innovation, Ventana Medical Systems, Inc, Tucson, AZ, USA*

MEREDITH C. HENDERSON • *Clinical Translational Research Division, Translational Genomics Research Institute (TGen), Scottsdale, AZ, USA*

FRANCESCO HOFMANN • *Disease Area Oncology, Novartis Institutes for BioMedical Research, Basel, Switzerland*

KIM HOLMSTRØM • *Bioneer A/S, Hørsholm, Denmark*

ALJOSCHA KLEINHAMMER • *Institute for Developmental Genetics, Helmholtz Zentrum München, Munich, Germany*

THOMAS I. KOBLIZEK • *Lonza Cologne GmbH, Köln, Germany*

RALF KÜHN • *Institute for Developmental Genetics, Helmholtz Zentrum München, Munich, Germany*

NIKOLAI V. KUZNETSOV • *InDex Pharmaceuticals, Karolinska Institutet, Stockholm, Sweden*

JENNY K.W. LAM • *Li Ka Shing Faculty of Medicine, Department of Pharmacology and Pharmacy, The University of Hong Kong, Pokfulam, Hong Kong*
WANLING LIANG • *Li Ka Shing Faculty of Medicine, Department of Pharmacology and Pharmacy, The University of Hong Kong, Pokfulam, Hong Kong*
SPIROS LINARDOPOULOS • *Division of Breast Cancer Research, Breakthrough Breast Cancer Research Centre, The Institute of Cancer Research, London, UK*
XUESONG LIU • *Cancer Research, Abbott Laboratories, Abbott Park, IL, USA*
A. JAMES MASON • *Institute of Pharmaceutical Science, King's College London, London, UK*
MARCO MAZZOLETTI • *Pharmacology Department of BU Oncology, Nerviano Medical Sciences, Nerviano, MI, Italy*
GUNTER MEISTER • *Laboratory for RNA Biology, Biochemistry Center Regensburg (BZR), University of Regensburg, Regensburg, Germany*
FELIX MEYENHOFER • *Max Planck Institute of Molecular Cell Biology and Genetics, Dresden, Germany*
ALESSANDRO NATONI • *Centre for Chromosome Biology and National Centre of Biomedical Engineering and Science, School of Natural Sciences, National University of Ireland Galway, Galway, Ireland*
ANTJE NIEDERLEIN • *Max Planck Institute of Molecular Cell Biology and Genetics, Dresden, Germany*
BOYE SCHNACK NIELSEN • *Bioneer A/S, Hørsholm, Denmark*
HIROAKI NITTA • *Medical Innovation, Ventana Medical Systems, Inc., Tucson, AZ, USA*
MICHAEL O'DWYER • *National Centre of Biomedical Engineering and Science, School of Medicine, National University of Ireland Galway, Galway, Ireland*
SEBASTIAN PETRI • *Laboratory for RNA Biology, Biochemistry Center Regensburg (BZR), University of Regensburg, Regensburg, Germany*
EMANUEL F. PETRICOIN III • *Center for Applied Proteomics and Molecular Medicine, George Mason University, Manassas, VA, USA*
MARIAELENA PIEROBON • *Center for Applied Proteomics and Molecular Medicine, George Mason University, Manassas, VA, USA; Department of Obstetrics and Gynecology, University Campus Bio-Medico of Rome, Rome, Italy*
ANTHONY PITT • *Lonza Cologne GmbH, Köln, Germany*
JANOS ROSZIK • *Department of Biophysics and Cell Biology, University of Debrecen, Debrecen, Hungary*
CORRADO SANTOCANALE • *Centre for Chromosome Biology and National Centre of Biomedical Engineering and Science, School of Natural Sciences, National University of Ireland Galway, Galway, Ireland*
TOBIAS SCHMELZLE • *Disease Area Oncology, Novartis Institutes for BioMedical Research, Basel, Switzerland*
MARIA ISABELLA SERENI • *Center for Applied Proteomics and Molecular Medicine, George Mason University, Manassas, VA, USA; Department of Obstetrics and Gynecology, University Campus Bio-Medico of Rome, Rome, Italy*
ANN SIEHOFF • *Lonza Cologne GmbH, Köln, Germany*
MARTIN STÖTER • *Max Planck Institute of Molecular Cell Biology and Genetics, Dresden, Germany*

VOLKER M. STUCKE • *Disease Area Oncology, Novartis Institutes for BioMedical Research, Basel, Switzerland*
JÁNOS SZÖLLŐSI • *Department of Biophysics and Cell Biology, and MTA-DE Cell Biology and Signaling Research Group, University of Debrecen, Debrecen, Hungary*
GEMMA TEXIDÓ • *Pharmacology Department, BU Oncology, Nerviano Medical Sciences, Nerviano, MI, Italy*
GÁBOR TÓTH • *Department of Biophysics and Cell Biology, University of Debrecen, Debrecen, Hungary*
JEAN-YVES TROSSET • *Sup'Biotech, Villejuif, France; Institut National de la Santé et de la Recherche Médicale (INSERM), UMRS 942, Hôpital Lariboisière, Paris, France*
HENNING ULRICH • *Departamento de Bioquímica, Instituto de Química, Universidade de São Paulo, São Paulo, Brazil*
GYÖRGY VEREB • *Department of Biophysics and Cell Biology, and MTA-DE Cell Biology and Signaling Research Group, University of Debrecen, Debrecen, Hungary*
MARIA VINCI • *Division of Cancer Therapeutics, McElwain Laboratories, The Institute of Cancer Research, Surrey, UK*
NICOLAS VODOVAR • *Institut National de la Santé et de la Recherche Médicale (INSERM), UMRS 942, Hôpital Lariboisière, Paris, France*
CARSTEN WRENGER • *Departamento de Parasitologia, Instituto de Ciências Biomédicas, Universidade de São Paulo, São Paulo, Brazil*
WOLFGANG WURST • *Institute for Developmental Genetics, Helmholtz Zentrum München, Munich, Germany*
MIRIAM ZIMMERMANN • *Tumour Biology and Metastasis, Cancer Research UK Cancer Therapeutics Unit, Division of Cancer Therapeutics, McElwain Laboratories, The Institute of Cancer Research, Surrey, UK*

Part I

Background

Chapter 1

The Path to Oncology Drug Target Validation: An Industry Perspective

Marta Cortés-Cros, Tobias Schmelzle, Volker M. Stucke, and Francesco Hofmann

Abstract

The advent of a variety of genomic, proteomic and other system-based scientific approaches has raised the expectations of identifying novel targets for oncology drug discovery. However, the complexity of human genome cancer alterations requires a careful analysis of the function of candidate targets identified by these efforts. The postulation and testing of a hypothesis that modulation of a protein or pathway will result in a therapeutic effect in a preclinical setting is crucial for target validation activities. In this chapter, we provide an overview on target identification and validation approaches to interrogate the functional and therapeutic relevance of a candidate cancer drug target as an essential step towards justifying the subsequent investment in drug discovery efforts.

Key words: Target identification, Target validation, Drug target, Oncology, Cancer genetics

1. Introduction

A variety of information sources in the public domain complemented by proprietary datasets generally serve as a basis for the identification of candidate oncology drug targets. The subsequent industry investments in drug discovery activities aimed at the identification of candidate drugs for clinical testing are substantial from a resourcing, finance, and time perspective (1), for both biologics and small molecule therapeutic modalities. Thus, preceding validation activities on potential targets constitute a necessary and well-justified effort, in order to select those targets that have the highest likelihood to impact disease and benefit the patient.

Marta Cortés-Cros, Tobias Schmelzle, and Volker M. Stucke have contributed equally to this work.

Jürgen Moll and Riccardo Colombo (eds.), *Target Identification and Validation in Drug Discovery: Methods and Protocols*, Methods in Molecular Biology, vol. 986, DOI 10.1007/978-1-62703-311-4_1,

Recent overviews (2), examples (3), and our own experimental efforts (4, 5) raised concerns about the reproducibility of a proportion of published data inferring to novel potential targets for a given disease, further highlighting the need for robust target validation programs in the pharmaceutical industry setting.

While there is no generally applicable paradigm to define oncology target relevance, this chapter aims at providing an overview of the rationales and strategies for tackling the challenges of cancer drug target identification and validation, focusing on three key work streams: (1) linking the target to cancer dependence; (2) assessing the clinical relevance; and (3) defining the approach to drug discovery.

2. Target Identification

The ultimate goal of cancer drug discovery is to identify and develop effective and tolerable therapies. The definition of what constitutes a cancer target has been evolving over time as have the cellular and molecular understanding of cancer pathology. Recently, cancer targets have been classified in four different categories based on the major principles of cancer dependencies: genetics (mutations, translocations, copy number, and epigenetically driven expression changes in the cancer cell), synergy (synthetic lethal genetic interactions of two or more genes), lineage (differentiation status and lineage-specific signaling and markers), and host (tumor stroma, vascularization) (6). Examples of these four different types of cancer dependencies include (1) the BCR-ABL translocation in chronic myeloid leukemia (CML) or the HER2 amplification in breast cancer as genetic drivers; (2) the sensitivity of tumors carrying a BRCA mutation, but not of wild-type tumors, to the inhibition of poly (ADP-ribose) polymerase (PARP) as a synthetic lethal interaction; (3) the specific treatment of breast cancers with estrogen receptor antagonists as a lineage dependency; and (4) the inhibition of the vascular endothelial growth factor (VEGF) directed against the requirement of tumors on angiogenesis as an example of a host track. In this chapter, we focus on the identification and validation of targets belonging to the genetic track, with mention to other tracks as appropriate.

Cancer is mainly a disease of alterations in the genome: gene copy number aberrations, DNA mutations, chromosomal rearrangements, and modification in DNA methylation collectively drive the development and progression of human tumors. With the completion of the human genome sequencing and the recent technological advances that have led to the development of new and increasingly affordable high-throughput methodologies for genomic analysis, it is becoming feasible to comprehensively characterize genomic alterations in cancer (7). From a therapeutic standpoint, this knowledge will contribute to expand upon the concept of targeted therapy and personalized medicine, resulting in the development of disease-relevant drugs for the respective rationally selected patient population, thereby leading to more predictable clinical responses. Several large-scale projects applying genome analysis techniques to catalogue genetic alterations in cancer were started during the last few years. They include The Cancer Genome Atlas (TCGA, http://www.tcga-data.nci.nih.gov/tcga/) project, the International Cancer Genome Consortium (ICGC, http://www.icgc.org), the Cancer Cell Line Encyclopedia (CCLE, http://www.broadinstitute.org/ccle) from the Broad Institute and the Novartis Institutes for BioMedical Research, The Therapeutically Applicable Research to Generate Effective Treatments (TARGET, http://www.target.cancer.gov/dataportal/) Initiative in childhood cancers from the National Cancer Institute, and the Sanger Institute Cancer Cell Line Project (http://www.sanger.ac.uk/genetics/CGP/CellLines), amongst others. They all apply recently developed high-throughput genome analysis techniques and aim at improving our ability to diagnose, treat, and prevent cancer through a better understanding of the molecular basis of this disease. The TCGA pilot project was initially focused on three types of human cancers—glioblastoma multiforme, lung, and ovarian cancer—and it has now been expanded to 20 tumor types. The ICGC, which comprises research organizations around the world, aims at generating comprehensive catalogues of genomic abnormalities in more than 25,000 tumors from 50 different cancer types and/or subtypes. ICGC is a decade-long program, and as of December 2012 it has a portal containing data from 7,358 cancer genomes.

The genomic information generated by these consortia, as well as by other academic and industry research organizations, is the result of combining several techniques that became available over the last years. These include (1) gene expression profiling, where abundance of protein-coding genes and noncoding microRNAs can be measured by transcript-specific and exon-specific probes on multiple platforms; (2) copy number variations determined by SNP arrays; (3) exon sequencing of selected genes and whole exome and whole genome sequencing; and (4) genome-wide DNA methylation profiles.

Once the genomic alterations have been identified, the challenge is to distinguish the "driver" events (which confer the ability to initiate and maintain the neoplastic process) from "passenger or carrier" events (which do not confer any selective advantage and occur randomly throughout tumorigenesis and disease progression) (8). In addition to the experimental target validation approaches that we introduce in the next section, several bioinformatics methods have been described and can be applied upfront to analyze the genomic information with the aim of differentiating driver from passenger alterations. For copy number analysis, these include, but are not limited to, (1) genomic identification of significant targets in cancer (GISTIC), a statistical approach that identifies those regions of the genome that are aberrantly altered more frequently than would be expected by chance, with greater weight given to high-amplitude events (high-level copy-number gains or homozygous deletions) that are less likely to represent random aberrations (9); (2) genome topography scan (GTS), which analyzes a set of copy number profiles to generate scores for each genomic position, summarizing copy number alteration (CNA) recurrence, amplitude, and focality across samples (10); or (3) RAE algorithm, which integrates the noise characteristics of individual tumors producing sample-specific discriminators of single-copy gain, amplification, hemizygous loss, and homozygous deletion (11). For analyzing expression profiles, Tomlins and colleagues developed the "cancer outlier profile analysis" (COPA) method for detecting cancer genes which show increased expression in a subset of disease samples and, more recently, Tibshirani and Hastie proposed the outlier sum (OS) statistic to detect cancer gene outlier expression (12, 13).

In addition to these large-scale genomic efforts to characterize cancer genomes and identify potential cancer drivers, it should be noted that gain- or loss-of-function screens are also widely used for functional target identification. These screens are commonly based on cDNA or RNAi libraries and interrogate large number of genes for their ability to either transform cells or enhance tumor formation/progression or to impair the viability of established cancer cells, respectively. These approaches can be applied to a variety of non-transformed or established cancer cell lines of interest with the aim of discovering cancer-relevant genes in an unbiased manner. These genes can be oncogenic drivers and tumor suppressors or represent cooperating synthetic lethal interactors of a particular preexisting disease-relevant lesion (14).

3. Target Validation

Potential targets emerging from any target identification approach should be subjected to a rigorous validation plan aimed at (1) linking the target to cancer dependence, (2) assessing the clinical relevance, and (3) defining the approach to drug discovery.

3.1. Linking the Target to Cancer Dependence

The first key question in the process of validating a cancer target is whether the target is necessary for the maintenance of a phenotype associated with transformation, i.e.: Can the tumor survive and progress upon modulation of this target? Experiments designed to address this question are frequently referred to as loss-of-function experiments and typically require knockdown of the target by shRNA (constitutive or inducible) delivered to the cells using lentiviruses, or by transfection using siRNA, a faster method for design and delivery, but with the drawback of being transient in nature and not applicable for in vivo experiments. In addition, it is possible to achieve inhibition of the target by dominant-negative constructs, which can be transiently transfected or stably integrated into the cellular genome by generating cell lines with the desired somatic knockout or knockin constructs using technologies such as zinc finger nucleases (15) or homologous recombination (16). Other tools to modulate the target of interest and test dependence in the cells carrying the particular genomic alteration include antibodies, tool compounds, soluble decoy-receptors, and interfering peptides.

The second key question to address when validating a target is whether expression or activation of the target is sufficient to induce a cancer-associated or transformation-associated phenotype. This involves gain-of-function experiments, where the gene is transiently or stably (e.g., inducible cell lines, transduced immortalized human cells) expressed in a model system, or the pathway in which the target is involved is modulated by antibodies, small molecules, or peptides. Experiments that address the above questions are often performed in an in vitro setting and criteria such as, but not limited to, growth rate, immortalization, loss of contact inhibition, two-dimensional clonogenic survival, or anchorage-independent growth in soft agar are tested.

Additionally, in vivo validation by gain- and loss-of-function studies may involve target modulation in tumor cell line xenografts in nude mice and, less frequently although more desirable but technically challenging, genetically engineered mouse models (GEMMs), including inducible, transgenic, or knockout mice (17). These should more accurately recapitulate human disease and predict the outcome of inhibiting a given target in a therapeutic setting. Experimental criteria tested in an in vivo setting comprise tumor volume and pathology, vascularization, metastatic nodule formation, and modulation of target or pathway activity.

3.2. Assessing the Clinical Relevance

Historically, empiric target selection without an in-depth understanding of cancer dependence and of the specific clinical context where such dependency would be dominant has led to large clinical trials without an effective strategy for patient stratification, resulting in a high failure rate for experimental anticancer therapies. More recently, therapeutic agents targeting well-characterized examples of genetic dependencies, accompanied by the respective rational patient selection strategy, have led to the successful development of efficacious therapies targeting BCR-ABL fusions in CML (imatinib), EGFR mutations in lung adenocarcinoma (erlotinib, gefitinib), B-RAF mutations in melanoma (vemurafenib), and ALK translocations in lung cancer (crizotinib) (18).

The actual frequency of the genomic alteration identified to confer dependence in experimental model systems should be confirmed within a given indication, and assessed across a larger panel of histotypes in order to validate the actual clinically relevant population associated with the envisioned target. In particular, this should include validation of the target alteration(s) in independent tumor sample cohorts by alternative methodologies, such as (1) genomic quantitative real-time PCR (qRT-PCR) or fluorescent in situ hybridization (FISH) to assess copy number variations (amplification/deletion), (2) FISH or RNASeq technology to monitor translocations/fusions, (3) standard Sanger (or dideoxy) nucleotide sequencing of PCR products or high-resolution melting analysis (HRMA) variant screening to detect mutations, (4) methylation-specific PCR or HRMA following bisulfite conversion as well as restriction enzyme-based methyl profiling to confirm and validate epigenetic alterations (such as changes in promoter methylation regulating gene expression), or (5) chromatin immunoprecipitation (ChIP)-sequencing to assess changes in histone marks and chromatin remodeling complex components associated to DNA.

To further assess the clinical relevance of a particular cancer target it may be necessary to determine the association of the target protein levels, its localization, and/or its activity with a specific cancer type, stage, or prognosis. This will often require the generation and, most importantly, validation of robust antibodies for immunohistochemistry (IHC) to be used on tissue microarrays (TMAs) encompassing the relevant tumor types and, ideally, matched normal samples.

The epidemiological information obtained from the analysis of clinical samples will then be applied to the overall validation and drug discovery strategy with the aim of closely matching experimental systems in vitro (e.g., cell lines) and in vivo (e.g., GEMMs; primary xenograft) with the envisioned relevant clinical indication. As highlighted in recent reviews (8, 19), the availability of appropriate preclinical models for therapeutic hypothesis testing requires the extensive characterization of cell line panels at the genetic and genomic level. As illustration, the CCLE project, a collaborative effort between the Broad Institute and the Novartis Institutes for BioMedical Research, encompasses circa 1,000 extensively characterized human cancer cell lines (20), i.e., a large, genetically heterogeneous source of models. However, for selected tumor subtypes, the establishment of cell lines has been unsuccessful or has yielded lines lacking key features of the cancers that they should model. In addition, in vitro cell line studies do not recapitulate essential host/stromal interactions, thus somewhat limiting their use to test cell-autonomous dependency hypotheses. More recently, in an effort to circumvent the limitations of established cancer cell lines and increase the diversity of human cancer models, the cancer research community has become increasingly interested in propagating primary human tumors as xenografts in mice without any in vitro manipulation (19). While more limiting in throughput than cell line panels and less amenable to genetic manipulation (e.g., shRNA), these models are becoming increasingly relevant when, for instance, tool compounds or antibodies are available to modulate and interrogate the relevance of a given target (21).

Finally, when envisioning the eventual design of clinical trials around a proposed target and indication, the availability of reasonable patient numbers with an unmet medical need and the feasibility of developing molecular diagnostic tools to detect the relevant alteration predicting dependence also need to be considered at the validation stage, in order to estimate the ability to execute a targeted clinical drug development plan. This may eventually result in considerations for an early effort towards indication expansion to facilitate corresponding proof-of-concept studies.

3.3. Defining the Approach to Drug Discovery

As described in the previous sections, one of the key aims during the initial target validation phase is to establish whether a target of interest confers cancer dependence in a variety of experimental settings. However, to be useful pharmaceutically, there is a need to understand what structures or activities of the target are responsible for conferring cancer dependence. This process will be referred to as "target structure–activity relationship (SAR)" and should be an integral part of the target validation phase.

A careful and thorough validation of a protein's activity towards its cancer function is particularly important in the context of a genetic target validation approach using RNA interference (RNAi) technology. A phenotype caused by RNAi may not always equate to a phenotype resulting from pharmacological inhibition of a target. In many cases, genetic and pharmacological inhibition of a target can perturb its physiological role in different ways, leading to different conclusions about the function and activity of a given protein. It is important to note that, in contrast to gene silencing, small molecules typically do not alter the expression of their target protein. While small molecules target a single function of the protein, such as a kinase activity, silencing of a gene results in complete loss of the target protein, which may disrupt protein complexes or impair protein functional domains that would be otherwise unaffected by a drug. These effects can dominate the phenotype caused by RNAi, such that inhibition by either siRNA/shRNA or a small molecule inhibitor targeting the same protein results in very different outcomes (22). One example to illustrate this is Aurora B, a kinase implicated in mitosis. Knockdown of Aurora B by siRNA leads to a diverse array of strong mitotic defects, while treatment with the Aurora kinase inhibitor ZM-447439 induces a much milder phenotype (23). As Aurora B is part of a centromeric protein complex required for correct execution of mitosis, RNAi targeting of Aurora B leads to protein complex dissociation, thus yielding a more severe phenotype than pharmacological inhibition of its kinase activity (24). There are additional examples where different phenotypes emerge depending on whether a protein is targeted by a small molecule inhibitor or a knockout/siRNA of the gene (22). These discrepancies reveal the diverse biological functions of a single protein, in particular for targets harboring more than one functional domain and/or interacting with several partners.

How to determine which protein activity should be targeted for drug discovery? For proteins with a clearly defined enzymatic activity, the use of catalytically inactive mutants has been proven to be useful. In the field of protein kinase research, kinase-dead mutants frequently behave as dominant-negatives, thus highlighting the importance of the kinase activity. In addition, the Shokat lab has developed a method allowing validation of the specific contribution of a kinase activity in cellular pathways.

Through mutation of the large conserved "gatekeeper residue" in the ATP-binding pocket of kinases to a nonnatural small amino acid, engineered kinases can be sensitized to inhibition by mutant-specific inhibitors, ensuring that only the engineered, but not the wild-type, protein is targeted by the small molecule (25). For a protein for which there is no obvious way for identifying an inhibitor, other alternatives have to be explored. For example, one might have to discover protein-binding partners and map a protein–protein interaction, by defining the minimal peptide sequence in combination with amino acid mutagenesis that is responsible to confer cancer dependence. Ultimately, the most elegant way to elucidate a protein's activity is a siRNA/shRNA rescue experiment. The RNAi-induced phenotype is countered by expression of the target cDNA that is resistant to the silencing reagent. By reexpressing a wild-type, fully functional version of the target cDNA along with rationally designed mutant cDNAs (e.g., catalytically inactive mutants, point mutations in protein–protein interaction motifs) the relative contribution of enzymatic activities or protein domains in reversing the phenotype can be assessed. For example, a growth arrest phenotype of a tumor cell line caused by RNAi of a candidate protein kinase may be rescued with the wild-type, but not with the corresponding kinase-dead, mutant, confirming the contribution of the kinase activity towards establishing cancer dependence. However, this kind of experiment remains technically challenging and has not been carried out on a large scale.

An alternative way to validate the critical function eliciting the disease-relevant phenotype is the use of tool modulators; these can be small molecules, peptides, or antibodies that may not have the properties to be considered a drug, but may display sufficient potency and selectivity to be used to interrogate the specific protein function in the relevant system. Such tools, however, are rarely available with the required characteristics to allow for a robust interpretation of the experiment. The exception is represented by the "pathway-centric" target identification approach based on chemical proteomics. A recent example is the identification of tankyrase and its role in destabilizing axin, a key component of the Wnt pathway (26). A cellular reporter assay designed to monitor Wnt pathway activation was used to screen a large collection of small drug-like molecules leading to the discovery of Wnt signaling inhibitors. To identify the cellular targets of these compounds, a chemical proteomics approach was set up using the identified small molecule as bait to fish for the cellular target in the appropriate cell line models. This approach does not only have the advantage of identifying new targets within key signaling pathways, but it also can directly lead to the identification of the mode of action of a given compound and therefore contributes to our understanding of which activity of the protein is required for conferring cancer dependence.

There has been much debate about what proportion of the human genome is considered as "druggable," with current estimations ranging from 8 to 12% of all genes (27). Key factors needed for such assessments are the availability of structural information on the target or a closely related homologue, information about the enzymology of the target or binding properties of cofactors, interactors, modulators, as well as a variety of computational algorithms to model the above and predict putative binding sites for small drug-like molecules. Thus, a thorough assessment of a target's drugability and drug discovery feasibility (i.e., availability of target-relevant assays, models, readouts) may be warranted as part of the target selection process. However, it should be emphasized that such an assessment is biased by the restricted target space that the pharmaceutical industry has historically tackled and is limited by the static nature of protein structure snapshots, which do not allow appreciating protein plasticity and dynamic. It is therefore important to allow for the drug discovery exploration of unprecedented targets, provided that their disease link is robustly validated, in order to expand the target space. To this end, the establishment of assays to probe the critical function conferring cancer dependence that will allow for the protein plasticity and the dynamic of the system to unfold is essential to enhance the chances of identifying unpredicted binding sites and modulators as starting points for the drug discovery process.

4. Perspectives

Even with the ever increasing amount of genetic and epigenetic information and associated knowledge of the corresponding alterations in cancers, each new therapeutic modality may still face challenges during its clinical development and beyond. Some of the persistent difficulties towards curing cancer, as recently reviewed by

Sellers (8), include a continued lack of complete information about alterations in the tumor genome; the lack of robust predictive preclinical models; "challenging-to-drug" targets; and, finally, the ability of the tumors for rapid genetic evolution and mutation, combined with their genetic heterogeneity, which can result in the rapid emergence of resistance, a setting in which combination therapy is likely required towards achieving therapeutic cure. Thus, a robust target validation process poised at addressing the questions of cancer dependence, clinical relevance, and path to drug discovery will contribute to increase the chances of successfully developing novel cancer therapies and mitigate the intrinsic challenges of this endeavor.

References

1. Paul SM, Mytelka DS, Dunwiddie CT et al (2010) How to improve R&D productivity: the pharmaceutical industry's grand challenge. Nat Rev Drug Discov 9:203–214
2. Prinz F, Schlange T, Asadullah K (2011) Believe it or not: how much can we rely on published data on potential drug targets? Nat Rev Drug Discov 10:712
3. Babij C, Zhang Y, Kurzeja RJ et al (2011) STK33 kinase activity is nonessential in KRAS-dependent cancer cells. Cancer Res 71:5818–5826
4. Stucke VM, Gorses D, Hofmann F (2008) DEAD-box RNA helicase p68 is not required for nuclear translocation of beta-catenin in colon cancer cells. Cell Cycle 7:830–832
5. Cortés-Cros M, Hemmerlin C, Ferretti S et al (2011) The M2 isoform of pyruvate kinase is dispensable for tumor maintenance and growth. In press, Proc Natl Acad Sci USA
6. Benson JD, Chen YN, Cornell-Kennon SA et al (2006) Validating cancer drug targets. Nature 441:451–456
7. Chin L, Hahn WC, Getz G et al (2011) Making sense of cancer genomic data. Genes Dev 25: 534–555
8. Sellers WR (2011) A blueprint for advancing genetics-based cancer therapy. Cell 147:26–31
9. Beroukhim R, Getz G, Nghiemphu L et al (2007) Assessing the significance of chromosomal aberrations in cancer: methodology and application to glioma. Proc Natl Acad Sci USA 104:20007–20012
10. Wiedemeyer R, Brennan C, Heffernan TP et al (2008) Feedback circuit among INK4 tumor suppressors constrains human glioblastoma development. Cancer Cell 13:355–364
11. Taylor BS, Barretina J, Socci ND et al (2008) Functional copy-number alterations in cancer. PLoS One 3:e3179
12. Tomlins SA, Rhodes DR, Perner S et al (2005) Recurrent fusion of TMPRSS2 and ETS transcription factor genes in prostate cancer. Science 310:644–648
13. Tibshirani R, Hastie T (2007) Outlier sums for differential gene expression analysis. Biostatistics 8:2–8
14. Kaelin WG Jr (2005) The concept of synthetic lethality in the context of anticancer therapy. Nat Rev Cancer 5:689–698
15. Urnov FD, Rebar EJ, Holmes MC et al (2010) Genome editing with engineered zinc finger nucleases. Nat Rev Genet 11:636–646
16. Waldman T, Lee C, Nishanian TG, et al (2003) Human somatic cell gene targeting. Curr Protoc Mol Biol Chapter 9, unit 9.15
17. Sharpless NE, DePinho RA (2006) The mighty mouse: genetically engineered mouse models in cancer drug development. Nat Rev Drug Discov 5:741–754
18. Haber D, Gray NS, Baselga J (2011) The evolving war on cancer. Cell 145:19–24
19. Caponigro G, Sellers WR (2011) Advances in the preclinical testing of cancer therapeutic hypotheses. Nat Rev Drug Discov 10:179–187
20. Barretina J, Caponigro G, Stransky N et al (2012) The cancer cell line encyclopedia: using preclinical models to predict anticancer drug sensitivity. Nature 483:603–607
21. Rubio-Viqueira B, Hidalgo M (2009) Direct in vivo xenograft tumor model for predicting chemotherapeutic drug response in cancer patients. Clin Pharmacol Ther 85:217–221
22. Knight ZA, Shokat KM (2007) Chemical genetics: where genetics and pharmacology meet. Cell 128:425–430
23. Ditchfield C, Johnson VL, Tighe A et al (2003) Aurora B couples chromosome alignment with anaphase by targeting BubR1, Mad2, and Cenp-E to kinetochores. J Cell Biol 161:267–280

24. Honda R, Körner R, Nigg EA (2003) Exploring the functional interactions between Aurora B, INCENP, and survivin in mitosis. Mol Biol Cell 14:3325–3341
25. Bishop AC, Shah K, Liu Y et al (1998) Design of allele-specific inhibitors to probe protein kinase signaling. Curr Biol 26:257–266
26. Huang SM, Mishina YM, Liu S et al (2009) Tankyrase inhibition stabilizes axin and antagonizes Wnt signalling. Nature 461: 614–620
27. Hopkins AL, Groom CR (2002) The druggable genome. Nat Rev Drug Discov 1: 727–730

Part II

DNA as a Tool to Modulate Drug Targets

Chapter 2

Identification of Aptamers as Specific Binders and Modulators of Cell-Surface Receptor Activity

Henning Ulrich and Carsten Wrenger

Abstract

In recent years, the SELEX (Systematic Evolution of Ligands by EXponential enrichment) technology has established itself as a powerful tool in basic research with promising applications in diagnostics and therapeutics. Oligonucleotides with high-affinities to their targets, denominated as aptamers, are obtained from partially random oligonucleotide pools by reiterative in vitro selection cycles and screening for binding activity. The original technique allowing the identification of aptamers binding to soluble targets, has recently been extended in order to produce aptamers binding to complex targets including receptors and ion channels embedded in the plasma membrane as well as whole cell surfaces or parasite organisms. In addition to discussing the most recent developments with focus on possible diagnostic and therapeutic application, we provide a simple protocol which has been successfully used to select for RNA aptamers as allosteric modulators of nicotinic receptor activity.

Key words: RNA aptamers, SELEX, Nicotinic acetylcholine receptors

1. Introduction

The SELEX (*S*ystematic *E*volution of *L*igands by *EX*ponential enrichment) was introduced in 1990 by Larry Gold and Jack Szostak (1, 2) which has been evolved into an extremely potent technology to isolate high-affinity oligonucleotide ligands and inhibitors from random oligonucleotide libraries of up to 10^{16} different molecules and possible secondary and tertiary structures. The isolated molecules were named aptamers (from aptus = latin to fit). In the beginning, RNA and DNA aptamers were identified which bind to proteins that naturally interact with nucleic acids or small molecules such as ATP. In the following years, the use of the SELEX technique was extended in order to isolate oligonucleotide ligands for a wide range of proteins of importance for therapy and diagnostics, such as growth factors (3), cell-surface antigens, entire

Jürgen Moll and Riccardo Colombo (eds.), *Target Identification and Validation in Drug Discovery: Methods and Protocols*, Methods in Molecular Biology, vol. 986, DOI 10.1007/978-1-62703-311-4_2,

cells, and even whole organisms (4–6). Since these RNA and DNA molecules bind their targets with similar affinities and specificities as antibodies, and are able to distinguish between isotypes of an enzyme, aptamers have been also called synthetic antibodies (7). Recently, the use of in vitro selection methods has been extended to complex targets, such as receptors that are only functional in their membrane-bound form (8), and living organisms such as bacteria (9), trypanosomes (10), and *Plasmodium falciparum*-infected erythrocytes (11). Moreover, since it became apparent that aptamers are capable to distinguish between little differences in cell-surface marker proteins expressed by, i.e., cancer cells, parasite-infected cells, or stem-cells in comparison to normal somatic cell (5), they have been developed into promising agents for diagnostic and therapeutic applications.

Aptamers are ideal candidates for studying protein interactions in vitro and in vivo and for developing novel molecules of pharmaceutical importance for the following reasons: aptamers isolated from combinatorial library have high dissociation constants, ranging from nanomolar to femtomolar, similar to the best affinity interactions between monoclonal antibodies and antigens. Most importantly, the aptamers can be chemically altered depending on their desired purpose. Chemical modifications that enhance aptamer stability for in vivo applications and possible therapy include substitutions of the 2′OH group of the ribose backbone of pyrimidines by amino, fluoride, or O-methylene functions. With such modifications, the half-life of RNA aptamers in biological fluids such as human serum or urine increases from approximately 8 s for unmodified RNA to 86 h for fluoro- or amino-modified RNA (12). Modifications at the 5′ position of pyrimidines expand the possibilities for chemically decorating oligonucleotide libraries including iodide or bromide substitutions or the introduction of SH-groups for chemical and photocross-linking of aptamers with their targets. Aptamers can be attached to radioactive or fluorescent molecules and can be used this way for in vitro and in vivo imaging of target proteins.

Aptamers can be selected against any proteins within a complex mixture of potential targets and then be used for ligand-mediated target purification. Larry Gold and coworkers developed aptamers against erythrocyte membranes that bind specifically to prior unidentified target proteins (4). The same strategy, also denominated as deconvolution SELEX, was used to stain rat brain tumor vessels and to identify the endothelial regulatory protein pigpen as aptamer-target protein (13). These studies indicate the possibility of purifying any protein target for which a high-affinity nucleic acid ligand has been identified. In order to generate covalent cross-links of selected aptamers with their targets, pyrimidines of the oligonucleotides are modified at the C5 position of the ribose. The aptamers can either be selected from oligonucleotide libraries that contain photo-cross linkable pyrimidines or already selected aptamers can be modified posterior to the SELEX process. These modified oligonucleotides attached to a reporter molecule generate reactive groups that, when irradiated by UV light, can form covalent linkage with another molecule in close proximity. For ligand-mediated target purification, aptamers are selected that recognize specifically an unknown target protein, i.e., in a cell membrane. Binding specificity is based on the fact that a given aptamer species (i.e., radiolabeled) can be eluted by a competitor (molar excess of unlabeled vs. labeled aptamer) from the cell membrane. In a second step, the 5′-biotin-labeled aptamer is immobilized on a solid surface, such as streptavidin-coated magnetic beads, and then used as an affinity matrix to purify and identify the bound target protein. The binding proteins retained on aptamer-coated magnetic beads are separated from the remaining protein mixture by applying a magnetic field. Alternatively, aptamers can be covalently photo-cross-linked to their targets and used as a tracer in order to separate the target protein from remaining proteins in a cell extract.

Aptamers are capable to identify individual epitopes on cell surfaces as it has been demonstrated for human membranes. Since aptamers differentiate in molecular signature of cell-surface antigens, this technology has been recognized as a powerful tool for biomarker discovery and targeting cancer cells. Aptamer probes were developed recognizing leukemia, prostate, or glioblastoma cancer cells (14–16). Molecular recognition is a key mechanism for the diagnosis of cancer which has been exploited by applying multiple anticancer cell aptamers conjugated on nanorods. Up-regulated ErbB2 expression in breast cancer cells was targeted by fluorescence-tagged RNA aptamers in surface plasmon resonance measurements (17). Aptamer-based therapeutic approaches are based on the specific delivery of si-RNAs or toxic loads to tumor cells (18, 19) or by inhibition of cancer cell-surface receptors, such as the EGF receptors involved in protection of tumor cells against apoptosis (20, 21).

Moreover, aptamers have been developed as specific ligands and/or inhibitors of neurotransmitter receptors, including GABA receptors, AMPA-glutamate receptors, as well as neuronal and nicotinic acetylcholine receptors (nAChRs) (22–25, 8). In addition to acting as inhibitors, such as anesthetics, anticonvulsives, or the abused drug cocaine, on nAChRs, aptamers can be developed as specific ligands of defined conformational sites of neurotransmitter receptors in order to reestablish dysfunctions in channel-open and -closing equilibrium. Such approaches aim at reestablishment of decreased GABA A-receptor activity in epilepsy involving mutation of this receptor (22), or at restoring of normal nicotinic receptor activity following inhibition by abused drugs such as cocaine and phencyclidine (23).

A simple protocol detailed in this chapter was established to develop RNA aptamers that bind to the electric organ nAChR and that are displaced by cocaine (8) (see Fig. 1 for a scheme). This protocol can be easily transferred to SELEX applications with other receptors or cell-surface epitopes, given that these are enriched in membrane preparations.

The aptamers obtained against the nAChR either are biologically active by themselves and inhibit the muscle-type nAChR activity as cocaine does or are by them biologically inactive and, therefore, protect the receptor against inhibition by cocaine (25).

Obtaining RNA aptamers by using the SELEX technology is a conceptually straightforward process. In the beginning, an oligonucleotide is synthesized by a DNA synthesizer that consists of a random sequence of typically 16–75 oligonucleotides flanked by two constant regions. A T7-promoter site is incorporated in one of the constant sequences by taking advantage of the introduced EcoRI and HindIII sites, which allows removal of the constant sites and to insert the random regions into the multi-cloning site of a bacterial vector. Random sequences are created by premixing all four nucleotides during DNA synthesis. The double-stranded DNA template is enzymatically generated and amplified by primer extension and error-prone PCR. The random nucleotide sequence of the double-stranded DNA template is verified by cloning and sequencing of around 40 individual RNA molecules. In the random sequence of these RNA molecules, sequence motifs (i.e., AA, AC, AG, AT) should be equally distributed.

For the in vitro selection process, the RNA pool containing 10^{13} different sequences and structural motifs is generated by an in vitro transcription reaction. Folding of the RNA molecules is induced by heat denaturation and renaturation at room temperature (26).

The RNA molecules are made nuclease resistant either by incorporating modified bases, i.e., 2′-fluoro- or by employing 2′-amino-modified pyrimidines during the in vitro transcription step (12), or by protection during the selection step using a dithiothreitol (DTT)-independent RNAse inhibitor (27). The RNA pool is exposed to its target in a 10–1,000 ratio of RNA:target-binding sites, and ligands are separated from nonfunctional molecules by filtration or by migration on polyacrylamide gel where target-bound RNA molecules migrate more slowly than unbound RNA which allows gel purification as indicated in Fig. 2 (28).

The recovered RNA is reverse transcribed to cDNA and amplified. The next-generation RNA pool, generated again by in vitro transcription, is already enriched in RNA molecules that bind to their target. The procedure is repeated with rising stringency until the random RNA pool is purified to a fraction of RNA

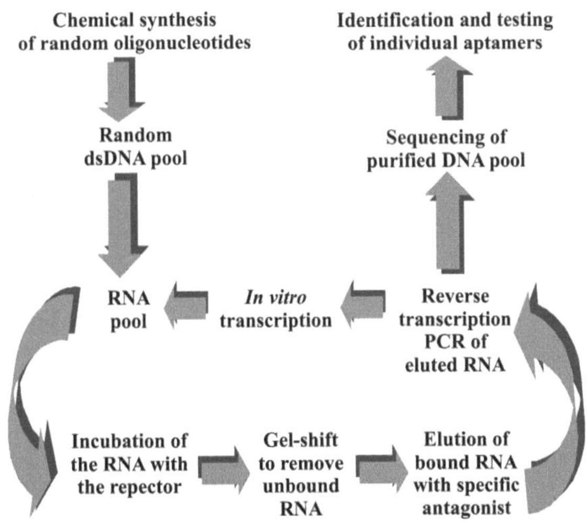

Fig. 1. Scheme for RNA aptamer identification targeting cocaine-binding sites on the nicotinic acetylcholine receptor (nAchR).

molecules with the desired binding properties and no further improvement of binding can be achieved. The final RNA pool is reverse transcribed into cDNA, amplified by PCR, and cloned into a bacterial vector. Clones containing the aptamer insert are picked and about 40 individual aptamers are identified by sequencing. The previously random regions are aligned and analyzed for consensus motifs, and the structures of RNA aptamers containing consensus motifs are predicted on the basis of free energy minimization (29). Aptamers containing consensus motifs are characterized regarding their binding affinity for their target and are screened for their biological activity by patch-clamping (30), using the whole-cell current-recording technique (31) in combination with rapid ligand delivery within 10 ms (called the cell-flow technique) (32, 33) (see Fig. 2.1 for a general scheme of RNA aptamer selection).

The protocol described here for the nAChR can be used with modification for the selection of aptamers that bind to other proteins expressed on a cell surface. In addition to the electrophysiological assay described here, screening for biological activity can include the quantification of ion flux into cells using fluorescent reporters (34, 35). For imaging purposes, fluorescence-tagged nucleotides can be enzymatically incorporated into the aptamer sequence (i.e., biotin-CTP or fluorescein-CTP). Alternatively, the aptamer can be modified at its 5′-end in order to attach a biotin moiety by substituting 5′-GTP for 5′-GDP-β-S and followed by covalent coupling of biotin molecule to this group and addition of a streptavidin-fluorophore (12, 36).

2. Materials

2.1. Target Preparation

1. Motor-driven blade homogenizer, low-speed refrigerated centrifuge, ultracentrifuge, spectrophotometer, and gamma counter.
2. Frozen electric organ from *Torpedo californica* (Aquatic Research Organisms, Inc., Hampton, NH): Maintain at −80°C.
3. Buffer H: 10 mM sodium phosphate, 400 mM sodium chloride, 5 mM Na_2EDTA, 10 mM sodium azide, at pH 7.4. Per liter: 2.68 g $Na_2HPO_4 \times 7H_2O$, 112.88 g NaCl, 1.86 g Na_2EDTA, 0.65 g NaN_3. Adjust to pH 7.4 with 1 M HCl.
4. Buffer A (buffer H without NaCl).
5. Phenylmethylsulfonylfluoride (PMSF) 0.1 M stock solution: 0.174 g PMSF in 10 ml isopropanol.
6. 60% (w/w) sucrose solution. Per 100 ml: 77.2 g D-glucose in buffer A.
7. 36% (w/w) sucrose solution. Per 100 ml: 60 ml 60% sucrose solution diluted to 100 ml with buffer A.
8. Bovine serum albumin and Lowry protein determination reagents.
9. Buffer B: 10 mM sodium phosphate, 50 mM NaCl, 0.1% (w/v) Triton X-100 at pH 7.5. Per liter: 2.68 g $Na_2HPO_4 \times 7H_2O$, 14.11 g NaCl, 1.0 ml Triton X-100. Adjust to pH 7.5 with 1 M HCl.
10. Unlabeled α-bungarotoxin (BGT) solution: Prepare a 10 mM BGT stock solution from lyophilized BGT (Sigma, St. Louis, MO).
11. ^{125}I-BGT solution: Prepare a 200 nM ^{125}I-BGT stock solution from lyophilized ^{125}I-BGT (Perkin Elmer/New England Nuclear, Boston, MA).
12. DE-81 filters (Whatman, Kent, UK).

2.2. SELEX Procedure

1. Equipment: PCR machine, scintillation counter, tabletop centrifuge, temperature-controlled water baths, equipment for horizontal and vertical electrophoresis, UV-illuminator, phosphor imager, automatic DNA sequencer, vacuum dot-blot manifold (Schleicher and Schuell). PCR 0.5 ml hot-start tubes, aerosol resistant pipette tips, autoclaved Eppendorf tubes (all from Fischer Scientific, Brightwaters, NY) and glassware, diethyl pyrocarbonate (DEPC, Sigma)-treated solutions.
2. Sterile disposable Petri dishes and sterile inoculating loops for bacterial culture.
3. BA-85 nitrocellulose and gel-blotting paper (GB002, both 102 × 133 mm, Schleicher and Schuell, Keene, NH).

4. GF/F glass fiber filters, 1.3 cm diameter (Whatman, Kent, UK).
5. Sterile syringe filters (0.45 μm; 25 mm diameter) (Whatman).
6. Sigmacote (Sigma).
7. Kodak X-Omat K Diagnostic Film for autoradiography or imaging plate for phospho imager.
8. Incubation buffer: 25 mM HEPES, 145 mM sodium chloride, 5.3 mM potassium chloride, 1.8 mM calcium chloride dihydrate, 1.7 mM magnesium chloride hexahydrate, pH 7.4. Per 1 l: 59.5 g HEPES, 8.47 g NaCl, 3.95 g KCl, 2.65 g $CaCl_2 \times 2H_2O$, 3.46 g $MgCl_2 \times 6H_2O$. Adjust to pH 7.4 with 1 N NaCl.
9. pGEM-3Z vector (cloning vector that allows highly efficient synthesis of RNA in vitro) (Promega, Madison, WI) and for transformation of *E. coli* competent cells (strains JM 29 or DH5α) (Promega).
10. pCR4-TOPO vector with One-shot cells (for fast cloning) (Invitrogen, Carlsbad, CA).
11. LB medium and agar: Per liter LB medium dissolve in distilled water (dd H_2O) and autoclave: 10 g tryptone, 5 g yeast extract, 5 g NaCl, 1 ml 1 N NaOH. Per liter LB agar dissolve in dd H_2O and autoclave: 10 g tryptone, 5 g yeast extract, 5 g NaCl, 1 ml 1 N NaOH, 15 g agar.
12. Partial random DNA template (108 bp) (synthesized by Biosource International, Foster City, CA).
    5′-ACC-GAG-TCC-AGA-AGC-TTG-TAG-TAC-TNN-NNN-NNN-NNN-NNN-NNN-NNN-NNN-NNN-NNN-NNN-NNN-NNN-NNG-CCT-AGA-TGG-CAG-TTG-AAT-TCT-CCC-TAT-AGT-GAG-TCG-TAT-TAC-3′ (N=A, C, G, or T are incorporated with equal probabilities). Restriction sites (HindIII and EcoRI) are underlined.
13. Primer region 1 (P-40, 40 bp, forward primer with T7 promoter site)
    5′-GTA-ATA-CGA-CTC-ACT-ATA-GGG-AGA-ATT-CAA-CTG-CCA-TCT-A-3′.
14. Primer region 2 (P-22, 22 bp, reverse primer).
    5′-ACC-GAG-TCC-AGA-AGC-TTG-TAG-T-3′
    Reverse primer for aptamer amplification in pGEM-3Z vector (P-22 pGEM, 22 bp)
    5′-GAA-TAC-TCA-AGC-TTG-TAG-TAC-T-3′ (forward primer is the same as above).
15. Kits and enzymes: Superscript reverse transcriptase (Invitrogen), Maxiscript and Megascript in vitro transcription kits (Ambion, Austin, TX). Taq-DNA polymerase, T4-polynucleotide kinase, EcoRI and HindIII restriction enzymes (Invitrogen).

16. Wizard Miniprep kit (Promega).
17. Big Dye sequencing kit (Applied Biosystems, Foster City, CA).
18. DTT-dependent and -independent RNAse inhibitors (Ambion), NTPs, dNTPs, $MgCl_2$, $MnCl_2$, yeast t-RNA, DNA, and RNA molecular weight standards.
19. Acrylamide/bisacrylamide, low-melting agarose (Sea Plaque GTG, Biocompare, Inc, South San Francisco, CA).
20. Ethanol, saturated phenol (pH 7 and 5.2), chloroform, isoamylalcohol (molecular biology grade, Sigma).
21. Tris–borate–EDTA (TBE) buffer (10× solution): 0.89 M TRIZMA base, 0.89 M H_3BO_3, 20 mM EDTA. Per liter: 108 g TRIZMA base, 55 g boric acid, 40 ml 0.5 M ethylenediaminetetraacetate (EDTA) solution.
22. Tris–EDTA (TE) buffer: 10 mM TRIZMA hydrochloride, 1 mM EDTA.
    10 mM Tris–HCl, 1 mM EDTA, pH 7.5. Per liter: 1.57 g Tris–HCl, 0.29 g EDTA. Adjust to pH 7.5.
23. Formamide-loading buffer for RNA gel: Per 10 ml add 0.2 ml 0.5 M EDTA, pH 8.0, 10 mg bromophenol blue, 10 mg xylene cyanol, 10 ml formamide (deionized).
24. Loading buffer for DNA gel: Per 10 ml: 1 ml 10× TBE, 10 mg bromophenol blue, 10 mg xylene cyanol, 2 g glycerol.
25. Urea, ammonium acetate, sodium acetate.
26. Spin columns S-30 (cut-off 30 nucleotides) (Sigma).
27. Cocaine hydrochloride (1 mg/ml in methanol) (Sigma): Evaporate the methanol and resuspend in incubation buffer at a concentration of 10 mM.
28. MK-801 (RBI Biochemicals, Natick, MA): Dissolve at a concentration of 10 mM in incubation buffer.
29. (α-^{32}P)-UTP (3,000 Ci/mmol), (γ-^{32}P)-ATP (5,000 Ci/mmol), both from Amersham Biosciences, Uppsala, Sweden.

2.3. Cell Culture and Electrophysiology

1. Equipment: CO_2-cell culture incubator, laminar flow cabinet, vibration–isolation table (TMC, Peabody, MA), Faraday cage (TMC), Axopatch 200B integrated patch-clamp amplifier with Digidata 1322A interface and pClamp program (Axon Instruments, Union City, CA), inverted microscope, one micro manipulator and two coarse manipulators (Narishige International, East Meadow, NY), computer-operated 12 V solenoid valves (Lee Company, Westbrook, CT), two peristaltic pumps for ligand and inhibitor delivery (Rainin, Oakland, CA), Kendall tubing diameter: 0.05, 0.16, and 0.42 cm (VWR International, Westchester, PA), U-shaped stainless steel capillary tubes (250 mM inner diameter) with a circular porthole of

150 mm at the base of the U (37) for ligand and inhibitor application made from Hamilton HPLC-tubing (Hamilton Corp., NV), patch-pipette puller (Sutter, Novato, CA), pipette polisher (Microforge, Narishige International, USA).

2. BC_3H1 cells (American Type Culture Collection, Manassas, VA).
3. DMEM high glucose, trypsin–EDTA solution, fetal bovine serum (FBS), heat inactivated (all from Invitrogen).
4. 25 cm^2 cell culture flasks (Costar, Corning, Acton, MA).
5. 15 ml screw capped vials (Greiner, Fischer Scientific).
6. 35 mm cell culture dishes (Corning, Acton, MA).
7. Incubation buffer (extracellular buffer): 25 mM HEPES, 145 mM NaCl, 5.3 mM KCl, 1.8 mM $CaCl_2 \times 2H_2O$, 1.7 mM $MgCl_2 \times 6H_2O$, pH 7.4.
8. Pipette solution (intracellular buffer): 25 mM HEPES, 140 mM potassium chloride, 10 mM sodium chloride, 2 mM magnesium chloride hexahydrate, 1 mM EGTA, pH 7.4. Per liter: 59.5 g HEPES, 10.44 g KCl, 0.58 g NaCl, 0.41 g $MgCl_2 \times 6H_2O$, 0.38 g EGTA. Adjust to pH 7.4 with 1 M KCl.
9. Carbamoylcholine (Sigma).
10. Cocaine hydrochloride (Sigma).
11. Borosilicate glass capillaries for patch pipettes (World Precision Instruments, Sarasota, FL).

3. Methods

3.1. Preparation and Evaluation of the nAChR-Enriched Plasma Membranes as a Target for the In Vitro Selection Process

1. Weigh 60–80 g of frozen electric organ and return it to the freezer. Measure a volume of buffer H equal in ml to the tissue weight in grams and add PMSF to a final concentration of 1 mM. Put the frozen pieces of organ in a sturdy cloth sack and pulverize with a hammer. Transfer the still frozen powder into a graduated cylinder, add the buffer containing PMSF, and leave the tissue/buffer mixture for about 10 min at room temperature to thaw. Then homogenize the tissue for 30 s four times using a motor-driven blade homogenizer at maximal speed, resting 10 s between each run. All subsequent steps should be performed on ice.
2. Transfer the homogenate into precooled centrifuge bottles and centrifuge for 10 min at 2,400 × *g*, 4°C. Collect the supernatant and discard the pellet. Transfer the supernatant to a precooled ultracentrifuge bottle and ultracentrifuge for 90 min at 34,000 × *g*, 4°C. Discard the supernatant and resuspend the

pellet in 60 ml of buffer A containing 1 mM PMSF by homogenizing the pellet three times for 10 s each using a motor-driven blade homogenizer at low speed. Wash the pellet again as above in buffer A and resuspend the pellet in 25 ml of buffer A containing 1 mM PMSF.

3. This step employs a discontinuous gradient ultracentrifugation method (38). Add 44 ml of 60% sucrose solution to the resuspended pellet, mix well, transfer to a 250 ml graduated cylinder, and add buffer A until the total volume is 108 ml to obtain a final sucrose concentration of 28%. Add 10 ml of a 36% sucrose solution to six swinging bucket rotor 30 ml ultracentrifuge tubes. Gently overlay the layers in each centrifuge tube with 18 ml of the resuspended pellet. Top off each tube with buffer A. Ultracentrifuge on a swinging bucket rotor for 4 h at 100,000 × *g*, 4°C. Aspirate and discard the floating pad at the buffer A/28% sucrose interface, together with most of the 28% layer. Collect the visible membrane layer containing nAChRs at the 28%/36% interface. Dilute the recovered 28%/36% interface with an equal volume of buffer A and collect the pellet by ultracentrifugation for 30 min at 38,000 × *g*, 4°C. Resuspend the pellet in 10 ml buffer A using a syringe equipped with a #18 needle. Aliquot and store the membranes frozen at −80°C.
4. The concentration of the membrane proteins is determined according to the method of Lowry et al. (39) using bovine serum albumin as the reference standard.

3.2. Determination of nAChR-Binding Sites in Torpedo Electric Organ Membrane Preparations

(^{125}I) BGT binding to the receptor is assayed using a filtration method (40).

1. Dilute 50 μl of the membrane suspension 1:100 in buffer B and then measure 50 μl aliquots of the diluted membranes into six Eppendorf vials.
2. To each of the three vials add 50 μl buffer B and to each of the other three vials add 50 μl unlabeled BGT solution.
3. Vortex and preincubate on a rotator at room temperature for 60 min. Then add 100 μl of ^{125}I-BGT to each vial.
4. Vortex and incubate on a rotator at room temperature for 60 min.
5. Load vacuum manifold ports with DE-81 filters (Whatman).
6. Prewash filters with 0.6 ml buffer B using vacuum.
7. Transfer 50 μl of each reaction mixture directly onto the filters (do this in triplicate) and wash each filter with 1 ml buffer B.
8. Allow buffer to filter through and air-dry the filters for 1 min using vacuum suction.

9. Transfer filters into counting vials and determine filter-bound radioactivity in a gamma counter. Subtract nonspecific binding (binding with unlabeled BGT present) to determine the concentration of specifically bound ^{125}I-BGT. The concentration of nAChR in the membrane suspension is one-half this value (×400) since there are two BGT-binding sites per nAChR molecule.

3.3. Preparation of the Random DNA Pool for In Vitro Transcription

1. Check purity by resuspending 35 nmol of the pre-purified single-stranded oligonucleotide in 100 μl TE buffer. End label 1 μl of the oligonucleotides with $(\gamma)^{32}$P ATP using T4-kinase and check for purity on 8% denaturing urea (6 M) PAGE (41). If there is not a single clean band, but shorter oligonucleotides are also present, the random oligonucleotides need to be purified.
2. Prepare a purification gel by adding an equal amount of formamide loading buffer to the resuspended DNA pool and a small amount of ^{32}P-labeled DNA from step 1.
3. Denature the mixture at 65°C for 10 min and load into nine lanes (each lane contains 500,000 cpm).
4. Expose gel using a phosphor imaging plate (15 min), and match the obtained image with the gel to localize the bands with the right size.
5. Cut out the bands and reimage the gel.
6. Crush the gel slices with a pipette tip and elute the DNA twice with 400 μl sodium acetate (0.3 M, pH 5.2) and 1 mM EDTA (for 12 h each).
7. Ethanol-precipitate the eluted purified DNA (see Note 1).

3.4. Second-Strand Synthesis and Amplification by Error-Prone PCR

Use hot-start tubes and assemble the bottom and top part of the reaction for second-strand synthesis and amplification of the DNA template by error-prone PCR. Hot-start PCR is the PCR technique of assembling the reaction mixture at a temperature that is greater than the annealing temperature. This procedure increases precision, yield, and specificity. The pre-adhered wax bead assures synchronous reaction start-up and eliminates the need for using mineral oil.

1. Assemble the reaction as reported in Table 1. This mix is enough for 30 PCR reactions of 120 μl volume each.
2. Second-strand synthesis: Execute the following program on a PCR machine:

   94°C 3 min, 42°C 2 min, and 72°C for 8 min. Then add 1.5 μl primer P-22 (100 pmol/μl) to each 120 μl PCR and start the following program: 94°C 2 min, 42°C 1 min, and 72°C 1 min for one cycle and 94°C 2 min, 42°C 1 min, and 72°C 1 min + 0.5 min increase each round for 11 cycles.

Table 1
Second-strand synthesis and amplification by error-prone PCR mix

Addition	Bottom (μl)	Top (μl)
H_2O	981	1,540
10× PCR buffer	180	180
dATP (100 mM)	12	
dGTP (100 mM)	12	
dCTP (100 mM)	30	
dTTP (100 mM)	30	
$MgCl_2$ (50 mM)	504	
P-40 (100 pM)	45	
ss DNA pool (60 pM)	6	
$MnCl_2$ (1 M)		1.8
Taq DNA polymerase		30

3. Analyze on native PAGE (8% in 1× TBE buffer).
4. Repeat the protocol that is described above three times (so that the total reaction volume is 10.8 ml).
5. Purify if necessary.
6. Ethanol precipitate and dissolve in 100 μl dd H_2O.

3.5. Evaluation of the Degree of Randomness of the Pool

PCR amplification of the single-stranded (ss) DNA pool will result in multiple copies of a double-stranded DNA pool. Chemical lesions occurring during the chemical synthesis of the ss DNA pool, and the possibility that some sequences are more amplified than others during PCR procedures, may result in a limited pool size and predominance of sequence motifs in the random sequence (see below). Therefore it is necessary to sequence and analyze a small number of individual sequences of the random pool (2).

1. Clone 1 μl of the purified PCR product into a pCR4-TOPO vector and transform to "One Shot" competent cells (Invitrogen) according to the manufacturer's recommendation.
2. Isolate and sequence 40 clones.
3. Use a word processing program to analyze random sequences for equal occurrence of AA, AC, AG, and AT motifs in the random regions (repeat for C, G, and T). If one or more of these motifs predominate or the base composition is not random in the random sequences, the pool may not contain enough different sequences for a successful SELEX experiment.

3.6. In Vitro Transcription and Purification of the SELEX RNA Pool

1. To perform in vitro transcription reaction: Save 20 µl of the total volume of 100 µl (for ^{32}P-UTP transcription for binding studies with nAChRs in *Torpedo* electric organ membranes, and in case you have to amplify the DNA again) and use 80 µl for transcription.
2. Assemble the following reaction: DNA solution (80 µl); H_2O (DEPC-treated) (74 µl); 10× synthesis buffer (25 µl); ATP, CTP, GTP, UDP 10 mM solutions (10 µl) each; DTT (100 mM) (6 µl); and T7 RNA polymerase (50 U/µl) (25 µl).
3. Incubate the reaction at 37°C overnight.
4. Add 25 µl DNAse (1 U/µl) and incubate for 20 min.
5. Analyze 10 µl of the reaction product on a denaturing PAGE (8%, 6 M urea) to verify the right size of the RNA (90 nt) and to determine if smaller products are present.
6. If smaller reaction products are present, purify the full-length RNA from the gel to avoid the selection and amplification of shorter sequences. Alternatively, for generation of nuclease-resistant transcripts substitute CTP and GTP by 2′-fluoro- or 2′-amino-pyrimidines (TriLink BioTechnologies, San Diego, CA) at final concentrations of 1.2 mM.
7. For radiolabeling of the RNA pool, add 50 µCi (α-^{32}P)-ATP (10 µCi/µl).
8. To purify RNA, extract the reaction product with 275 µl of a saturated solution of phenol in 0.1 M NaOAc, pH 5.2.
9. The supernatant is then extracted with an equal volume of chloroform in order to remove remaining phenol and then ethanol-precipitated.

3.7. Selection of Aptamers that Bind to the nAChR and Displace Cocaine

The first two rounds are performed under low stringency conditions to enhance RNA–protein binding and to avoid early depletion of sequences present in the SELEX RNA pool. For SELEX cycles 1–3 a nitrocellulose-filter binding assay is used to separate receptor-bound from free aptamers. Beginning from SELEX cycle 4, the nitrocellulose-filter binding and a gel-shift selection step are employed as two consecutive selection processes (see Note 2).

1. The respective RNA pool is heated up for 10 min to 85°C prior to selection and allowed to cool down to room temperature in order to allow for proper formation of secondary and tertiary structures.
2. To perform SELEX experiment using nitrocellulose-filter binding, prepare filtration unit by preincubating a nitrocellulose sheet in incubation buffer.
3. Mount nitrocellulose sheet in autoclaved filtration unit. Assemble the following reaction mixtures for the SELEX process: nAChR-enriched plasma membranes (800 µg/ml

protein; 1.6 μM receptor) (40 μl); incubation buffer (280 μl); anti-RNAse (40 U/μl) (40 μl); RNA pool (50 μM) (40 μl).

4. Incubate for 40 min at room temperature.
5. Place 4 × 100 μl of the selection mixture into each of the four wells of the filtration unit.
6. Wash each well twice with 200 μl incubation buffer.
7. Cut out the pieces of the nitrocellulose filters that cover the used wells of the filtration unit.
8. For the elution of receptor-bound, cocaine-displaceable RNA aptamers, incubate each of these filter pieces in 100 μl of 1 mM cocaine (or 100 μl of 1 mM MK-801 that binds to the same site as cocaine on the nAChR) in incubation buffer for 20 min at room temperature.
9. For the recovery and purification of the eluted RNA, add 5 μg t-RNA as a carrier, and phenol- and chloroform-extract the eluate, precipitate with ethanol, and resuspend the pellet in 12 μl DEPC-treated H_2O.
10. Then perform reverse transcription (RT-PCR) assembling four separate reactions as follows: 2.5 μl of the recovered RNA, 2 μl P-22 (100 μM), and 6.5 μl DEPC-treated H_2O.
11. Incubate at 70°C for 10 min, and then place on ice for 1 min.
12. Add to the following RNA/P-22 solution: 14 μl RNA/P-22 solution, 2 μl 10× synthesis buffer, 2 μl dNTP mix (10 mM), 2 μl DTT (0.1 M), 1 μl $MgCl_2$ (50 mM), and 1 μl superscript RT (200 U/μl).
13. Incubate at room temperature for 10 min, then at 46°C for 50 min, and then at 70°C for 10 min.
14. Place the vial on ice for 5 min, add 1 μl RNAse H, and incubate at 37°C for 15–20 min (volume is now 21 μl).
15. Transfer each reaction into a hot-start tube and prepare a PCR mix as reported in Table 2 and perform the following program: cycle 1 (94°C, 5 min; 60°C, 5 min; 72°C, 1 min), 5–10 cycles (94°C, 1 min; 60°C, 1 min; 72°C, 1 min), and final cycle (94°C, 1 min; 60°C, 1 min; 72°C, 10 min).
16. Analyze 10 μl aliquot of the PCR product on a native PAGE (see Note 3).
17. Gel-purify PCR products if necessary.
18. Phenol- and chloroform-extract, ethanol-precipitate, and resuspend in 100 μl DEPC-treated H_2O.
19. In vitro transcribe and purify RNA as already described.
20. To perform gel-shift selection experiments, prepare a native gel, containing 3% acrylamide (stock: 38.5% acrylamide and 1.5% bis-acrylamide) and 1× TBE, adjusted to pH 7.4.

Table 2
PCR mix

Addition	Bottom (μl)	Top (μl)
RT-reaction product	21	
dd-H_2O	21	44
10× buffer	3	5
P-40 primer (50 μM)	2	
$MgCl_2$ (50 mM)	3	
Taq polymerase (5 U/μl)		1

21. The RNA pool unlabeled or radiolabeled - as detailed later - is diluted in incubation buffer prior to the SELEX step, denatured and renatured as detailed in Subheading 3.7.
22. Assemble the following reaction: 20 μl nAChR-enriched plasma membranes (800 μg/ml protein; 1.6 μM receptor), 9 μl incubation buffer, 1 μl anti-RNAse (40 U/μl), and 10 μl RNA (320 μM).
23. The reaction is carried out for 40 min at room temperature in incubation buffer.
24. Add one part of 4× loading buffer (20% glycerol, 0.2% bromophenol blue in 1× TBE) to three parts of reaction mixture.
25. Load the samples on the gel and electrophorese the gel at 10 V/cm for 3 h in 1× TBE, adjusted to pH 7.4.
26. Stain the gel with 0.5 μg/ml ethidium bromide.
27. Excise the band containing the RNA–protein complex.
28. Elute the bound RNA with 0.5 M NaOAc pH 5.2 containing 1 U/μl anti-RNAse.
29. Phenol- and chloroform-extract and ethanol precipitate the recovered RNA (see Subheading 3.7). The RNA is reverse transcribed, purified, and precipitated as detailed in Subheadings 3.6 (25) and 3.7, and then RNA is used for the nitrocellulose filter selection step and cocaine displacement of nAChR-bound RNA molecules (Fig. 2).

3.8. Synthesis and Purification of ^{32}P-Labeled RNA

Radiolabeled RNA can be generated either by incorporation of a (α^{32}P) nucleoside triphosphate during an in vitro transcription reaction or by the transfer of (γ^{32}P)-ATP to the 5′ terminus of a dephosphorylated RNA molecule (41). The authors prefer the first mentioned method, as it needs only a single enzymatic reaction.

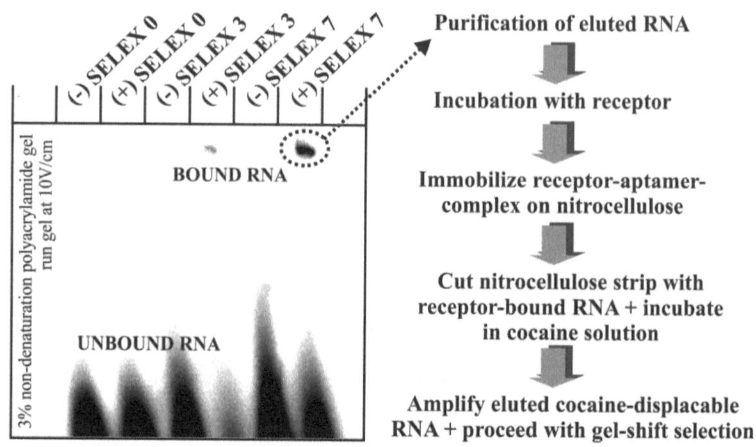

Fig. 2. Alternation of gel-shift and filter-binding selection steps: Target-bound and unbound radiolabeled RNA aptamers are separated by polyacrylamide gel electrophoresis, visualized by autoradiography, purified from the gel, and used for the subsequent nitrocellulose-filter binding selection step. The experiments are carried out in the presence (+) and absence (–) of target protein using the SELEX cycles 0 (control), 3, and 7. The figure illustrates the increase of binding affinity of selected RNA pools, seen as augmented quantity of RNA retained together with the receptor protein at the top of the gel (modified from ref. (8)).

1. Gel-purify DNA pools of various SELEX cycles.
2. Ethanol-precipitate the DNAs and resuspend at a concentration of 200 ng/μl for in vitro transcription reactions.
3. Make labelling cocktail: (Volume needs to be multiplied by the number of reactions to be set up). Add 6 μl DEPC-treated H_2O, 2 μl 10× transcription buffer, 1 μl ATP (0.5 mM), 1 μl CTP (10 mM), 1 μl GTP (10 mM), and 1 μl UTP (10 mM).
4. Add to the 12 μl mix: 2 μl DNA (SELEX pools), 5 μl (α-^{32}P)-ATP (10 μCi/μl), and 1 μl T7 RNA polymerase (50 U/μl) for a total volume of 20 μl.
5. Incubate at 37°C for at least 2 h.
6. Add 2 μl DNAse I (1 U/μl) and incubate for 15 min at 37°C.
7. Add 3 μl 0.5-M EDTA and purify the ^{32}P-RNAs from unincorporated ^{32}P-nucleotides using a spin column 30 (Sigma).
8. Run an 8% denaturing gel (6 M urea) to check the size and the purity of the (^{32}P)-RNA (see Note 4). (For nuclease-resistant (^{32}P)-RNA aptamers set up transcription reactions in the presence of 1.5 mM 2′-fluoro- or 2′-amino-modified pyrimidines instead of CTP and UTP).
9. Gel-purify the (^{32}P)-RNA if necessary.
10. For gel purification of ^{32}P-labeled RNA, pour denaturing gel (8% acrylamide, 6 M urea) (see Note 3).

11. Load left and right outer lanes with loading dye, and pre-run the gel at 250 V until the dye has migrated down approximately 75% of the gel.
12. Mix (^{32}P)-RNA reaction mixture with 15 µl formamide loading buffer, and incubate at 85°C for 15 min.
13. Load samples and run the gel at 250 V until the rapidly migrating dye (bromophenol blue) has almost moved out of the gel.
14. When finished, visualize the location of the (^{32}P)-RNA on the gel by autoradiography or phospho imaging, and cut out the gel slices containing the full-length (^{32}P) RNA using a fresh, sterile razor blade for each sample.
15. Transfer the gel slabs into a 2 ml Eppendorf tube, crush with a pipette tip, and extract overnight with 1 ml 0.3 M NaOAc, pH 5.2, 10 µl DTT, and 5 µl RNAsin.
16. Pass the suspension through a syringe filter (0.45 µm sterile filter) to remove the gel, add an equal volume of a saturated solution of phenol in 0.3 M NaOAc pH 5.2, and vortex hard.
17. Centrifuge for 3 min at 12,000×*g* in a tabletop centrifuge.
18. Collect the supernatant and concentrate it by ethanol precipitation (see Note 1).

3.9. Determination of Specific Binding to the nAChR

1. Soak GF/F filters (diameter 1.3 cm, Whatman) and a sheet of filter paper (GB002, 102×133 mm, Schleicher and Schuell) for 1 h in incubation buffer containing 1% Sigmacote to reduce unspecific binding of the RNA molecules to the filters.
2. The (^{32}P) RNAs are diluted in incubation buffer containing 10 mg/ml to a specific activity of 5×10^5 cpm/µl and denatured and renatured before the experiment.
3. The specific binding of the selected aptamers to their target is determined as the difference between the total binding and the unspecific binding in the presence of an excess of unlabeled competitor. Assemble the following reactions for duplicate estimations in a total volume of 100 µl (for determination of total and unspecific binding each). The specific binding is the difference between total and unspecific binding. Binding of the selected RNA molecules to the nAChR reaction: 89 µl incubation buffer, 10 µl nAChR-enriched plasma membranes (800 µg/ml protein; 1.6 µM receptor), and 1 µl ^{32}P-RNA dilution containing 10 mg/ml t-RNA anti-RNAse (40 U/µl).

   Determination of unspecific binding reaction by competition of the RNA aptamers with cocaine: 79 µl incubation buffer, 10 µl nAChR-enriched plasma membranes (800 µg/ml protein; 1.6 µM receptor), 1 µl ^{32}P-RNA dilution containing 10 mg/ml t-RNA anti-RNAse (40 U/µl), and 10 µl cocaine (10 mM). The percentage of binding of the (^{32}P) RNA aptamers to the filters (background binding) is determined in the absence of receptor protein.

4. Incubate reactions for 40 min at room temperature.
5. In the meantime, mount the glass fiber filters on top of the filter sheet in the dot-blot filtration unit (Schleicher and Schuell). For duplicate determinations, two samples each of 45 μl solution are spotted on separate filters.
6. Gentle suction is applied to the filtration unit. The filters are washed with 200 μl incubation buffer.
7. The filters are removed and transferred into 5 ml scintillation fluid.
8. The filter-bound radioactivity is measured by scintillation counting.

3.10. Determination of the Binding Affinities Using IC_{50} Determinations

The IC_{50} value is the concentration at which a competitor (cocaine) displaces 50% of the radioactive ligand (^{32}P-RNA) from the nAChR. On the assumption that cocaine binds with an apparent dissociation constant (K_d) of 50 μM to the membrane-bound nAChR (42), the K_i of the RNA:nAChR complex is calculated from this data, using the equation of Cheng and Prusoff (43) where $K_i = IC_{50} / (1 + [^{32}P - RNA] / K_d)$. (^{32}P-RNA) is the radioligand concentration used in the experiment.

1. For IC_{50} determinations, incubate 5×10^5 cpm (^{32}P)-ATP-labeled RNA in the presence of 20 nM unlabeled RNA with constant concentrations of nAChR protein.
2. Increase cocaine concentration from 0 to 10 μM and separate the reaction mixtures as described in Subheading 3.9.

3.11. Identification and Characterization of Individual Aptamers

1. For cloning and sequencing of individual RNA aptamers, nine SELEX cycles are necessary to obtain high-affinity RNA ligands for the nAChR.
2. Reverse transcribe the final RNA pool, and amplify and gel-purify the resulting cDNA (see Subheading 3.7).
3. Cut the constant regions using the restriction enzymes EcoRI and HindIII.
4. Purify the 69-bp reaction product on a gel made of low-melting agarose (Sea Plaque GTG).
5. Cut pGEM-3Z vector (Promega) using EcoRI and HindIII and gel-purify the linearized vector.
6. Ligate the purified DNA into the linearized pGEM-3Z vector.
7. Transform *E. coli* JM109 or DH5α cells with the vector according to the instructions of Sambrook and Russell (40).
8. After streaking the transformed cells on ampicillin-selective LB-agar plates, grow the cells overnight.

9. Isolate individual white colonies, each containing one aptamer sequence.
10. Grow the isolated cells to the desired optical density (1.8–2.0 O.D. at 600 nm).
11. Purify the plasmids using the Wizard DNA Miniprep kit (Promega), and sequence the inserts according to the protocol supplied with the Big Dye sequencing kit (Applied Biosystems).
12. Then identify the consensus sequences and predict the aptamer secondary structure. The previous random sequences of the cloned RNA molecules are aligned and compared for consensus motifs found in almost all RNA molecules by simply aligning the sequences in a word processing program and visualizing sequence similarities by eye or the "find" function of the program, or by using a sequence alignment computer program. A thorough analysis of identifying aptamer consensus motifs was published (44). The corresponding aptamer secondary structures are predicted using the multifold computer program (41).
13. Amplify plasmid inserts coding for individual RNA aptamers by PCR using the primers P-40 and P-22pGEM and purify the PCR products on a low-melting agarose gel.
14. The purified DNAs are used as templates for in vitro transcription reactions. Forward primer (P-40, same sequence as primer used during SELEX process) and reverse primer (p-22 pGEM).

3.12. Screening for Binding Affinity and Biological Activity

1. Screening for binding affinity: Sequences containing consensus regions are in vitro transcribed in the presence of (^{32}P) UTP and tested for their affinity towards membrane preparations from *Torpedo* organs containing nAChRs (see Subheading 3.9).
2. Screening for biological activity: The biological activity of the selected RNA aptamers are determined in vitro, as to whether they inhibit the nAChR function as cocaine does or whether they compete with cocaine but do not have any biological activity by themselves and, therefore, protect the receptor against inhibition by cocaine (25).
3. For the electrophysiological assay, BC_3H1 cells that express the muscle-type nAChR are plated in cell culture dishes.
4. A patch-clamp (recording) pipette is attached to an individual cell and a gigaohm seal is formed between the cell and the patch-clamp pipette by applying suction through the pipette.
5. The whole-cell configuration is achieved by breaking the cell membrane; the buffer in the recording pipette is now in free exchange with the cytosol of the cells.

6. The flow of ions across the plasma membrane is recorded as current. An extensive description of the patch-clamp technique is given in ref. (30).
7. Currents induced by carbamoylcholine in the absence or the presence of cocaine and/or RNA aptamers are recorded by using the whole-cell recording technique (31) and the cell-flow method for rapid solution exchange and correction of the current for desensitization that occurs during the rising phase of the current (32). The recordings are made with a cell bathed in incubation buffer at a holding potential of −60 mV, pH 7.4, and 22°C. The recorded currents are amplified by using a patch-clamp amplifier and low-pass Bessel filtered at 1–2 kHz. The filtered signal is digitized on a 1322A-Digidata interface controlled by the Axon pCLAMP software. The recorded currents are corrected for receptor desensitization as has been described (32).
8. In vitro transcribe the large amounts of aptamers needed for electrophysiological experiments using the Megascript kit (Ambion).
9. Prepare BC_3H1 cells by plating them on 35-mm cell culture dishes and culture them for 14 days in subconfluent conditions in DMEM medium (high glucose) containing 1% FBS at 37°C and 7.5% CO_2.
10. Use filtered incubation buffer as the extracellular buffer and pipette solution (intracellular buffer), to fill the recording pipette.
11. Pull and fire polish patch pipettes when filled with buffer solution show a resistance of 3–5 MΩ and a series resistance of 5–6 MΩ.
12. Use the cell-flow technique (32) to determine the inhibition of the receptor. In these experiments, cells containing the nAChR are used. The maximum current amplitude, a measure of the concentration of open receptor-channels, is determined. The experiments are done in the presence of 100 μM carbamoylcholine (a stable analog of acetylcholine) and in the absence and presence of 150 μM cocaine.
13. Repeat the experiment with 100 μM carbamoylcholine in the presence of increasing concentrations of the RNA aptamer to be tested, in order to determine if the RNA aptamer inhibits receptor function. If an RNA aptamer does not inhibit receptor function, experiments are performed in the presence of a constant concentration of carbamoylcholine and cocaine with increasing concentrations of the RNA aptamer to determine whether the aptamer alleviates inhibition (25).

4. Notes

1. For precipitation of DNA and RNA, add 10% of the volume sodium acetate, pH 7 or 5.2, respectively. Add four parts of ice-cold 98% ethanol and mix well. Add 1 μl of linear acrylamide (3%). Incubate for 2 h at −20°C and centrifuge at 12,000 × *g* for 30 min in a tabletop centrifuge. Carefully remove the supernatant—the DNA or RNA should be visible against the light as one or several small shiny specks, usually attached to the tube, but they may be loose—and wash the pellet with 200 μl 80% ethanol. Remove the supernatant, and dry the DNA or RNA (3 min in a speed vacuum concentrator) until all visible liquid has gone, but avoid overdrying of the pellet because the DNA or RNA will then be difficult to redissolve. Dissolve the DNA in 100 μl dd H_2O.
2. In order to avoid the selection and amplification of RNA molecules that do not bind to the target site, two assays for in vitro selection are both employed after SELEX cycle 3. In addition a negative preselection step can be used at which unspecific binders to nitrocellulose (used for separation of receptor-bound from unbound RNA molecules) are discarded (see Subheading 3.7).
3. Too many PCR cycles will result in the occurrence of multimers of denatured DNA molecules that will migrate in the high molecular weight range in an analyzing PAGE. These sequences cannot be renatured to double-stranded DNA molecules and are lost, as they cannot be in vitro transcribed.
4. For denaturing PAGE for RNA: Prepare a 40% stock solution weigh out 38.5 g acrylamide ultrapure (pay attention is carcinogenic) and 25 g *N,N'* methylenebisacrylamide ultrapure. Dissolve in 100 ml DEPC-treated H_2O, microwave the solution for 40s, filter it and warrant protection from light. Prepare 10× TBE and 10 M urea solution in DEPC-treated H_2O (8 ml 40% acrylamide solution, 4 ml 10× TBE buffer, 28 ml dd H_2O). Fill between the glass plates and start polymerization. Add 30 μl *N,N,N',N'*-tetramethyl-ethylenediamine (TEMED) and 300 μl ammonium persulfate (APS) (10% solution).

Acknowledgments

The authors would like to thank FAPESP (Fundação de Amparo à Pesquisa do Estado de São Paulo), Brazil, for financial support via the respective grants (Project No. 06/61285-9 to H.U.; Project No. 2009/54325-2 to C.W.). H.U. also acknowledges financial support from CNPq (Conselho Nacional de Desenvolvimento Científico e Tecnológico), Brazil.

References

1. Tuerk C, Gold L (1990) Systematic evolution of ligands by exponential enrichment: RNA ligands to bacteriophage T4 DNA polymerase. Science 249:505–510
2. Ellington AD, Szostak JW (1990) *In vitro* selection of RNA molecules that bind specific ligands. Nature 346:818–822
3. Ng EW, Shima DT, Calias P et al (2006) Pegaptanib, a targeted anti-VEGF aptamer for ocular vascular disease. Nat Rev Drug Discov 5:123–132
4. Morris KN, Jensen KB, Julin CM et al (1998) High affinity ligands from in vitro selection: complex targets. Proc Natl Acad Sci USA 95:2902–2907
5. Ulrich H, Wrenger C (2009) Disease-specific biomarker discovery by aptamers. Cytometry A 75:727–733
6. Homann M, Goringer HU (1999) Combinatorial selection of high affinity RNA ligands to live African trypanosomes. Nucleic Acids Res 27:2006–2014
7. Jayasena SD (1999) Aptamers: an emerging class of molecules that rival antibodies in diagnostics. Clin Chem 45:1628–1650
8. Ulrich H, Ippolito JE, Pagan OR et al (1998) *In vitro* selection of RNA molecules that displace cocaine from the membrane-bound nicotinic acetylcholine receptor. Proc Natl Acad Sci USA 95:14051–14056
9. Li H, Ding X, Peng Z et al (2011) Aptamer selection for the detection of *Escherichia coli* K88. Can J Microbiol 57:453–459
10. Ulrich H, Magdesian MH, Alves MJ et al (2002) *In vitro* selection of RNA aptamers that bind to cell adhesion receptors of *Trypanosoma cruzi* and inhibit cell invasion. J Biol Chem 277:20756–20762
11. Barfod A, Persson T, Lindh J (2009) In vitro selection of RNA aptamers against a conserved region of the *Plasmodium falciparum* erythrocyte membrane protein. Parasitol Res 105: 1557–1566
12. Ulrich H, Martins AH, Pesquero JB (2004) RNA and DNA aptamers in cytomics analysis. Cytometry A 59:220–231
13. Blank M, Weinschenk T, Priemer M et al (2001) Systematic evolution of a DNA aptamer binding to rat brain tumor microvessels. Selective targeting of endothelial regulatory protein pigpen. J Biol Chem 276: 16464–16468
14. Shangguan D, Li Y, Tang Z et al (2006) Aptamers evolved from live cells as effective molecular probes for cancer study. Proc Natl Acad Sci USA 103:11838–11843
15. Lupold SE, Hicke BJ, Lin Y et al (2002) Identification and characterization of nuclease-stabilized RNA molecules that bind human prostate cancer cells via the prostate-specific membrane antigen. Cancer Res 62: 4029–4033
16. Bayrac AT, Sefah K, Parekh P et al (2011) *In vitro* selection of DNA aptamers to glioblastoma multiforme. ACS Chem Neurosci 2: 175–181
17. Kim MY, Jeong S (2011) *In vitro* selection of RNA aptamer and specific targeting of ErbB2 in breast cancer cells. Nucleic Acid Ther 21: 173–178
18. Dassie JP, Liu XY, Thomas GS et al (2009) Systemic administration of optimized aptamer-siRNA chimeras promotes regression of PSMA-expressing tumors. Nat Biotechnol 27: 839–849
19. Yu C, Hu Y, Duan J et al (2011) Novel aptamer-nanoparticle bioconjugates enhances delivery of anticancer drug to MUC1-positive cancer cells *in vitro*. PLoS One 6:e24077
20. Li N, Nguyen HH, Byrom M et al (2011) Inhibition of cell proliferation by an anti-EGFR aptamer. PLoS One 6:e20299
21. Cerchia L, de Franciscis V (2011) Nucleic acid aptamers against protein kinases. Curr Med Chem 18:4152–4158
22. Cui Y, Rajasethupathy P, Hess GP (2004) Selection of stable RNA molecules that can regulate the channel-opening equilibrium of the membrane-bound gamma-aminobutyric acid receptor. Biochemistry 43:16442–16449
23. Du M, Ulrich H, Zhao X, Aronowski J et al (2007) Water soluble RNA based antagonist of AMPA receptors. Neuropharmacology 53: 242–251
24. Huang Z, Han Y, Wang C et al (2010) Potent and selective inhibition of the open-channel conformation of AMPA receptors by an RNA aptamer. Biochemistry 49:790–798
25. Hess GP, Ulrich H, Breitinger HG et al (2000) Mechanism-based discovery of ligands that prevent inhibition of the nicotinic acetylcholine receptor. Proc Natl Acad Sci USA 97: 13895–13900
26. Marshall KA, Ellington AD (2000) *In vitro* selection of RNA aptamers. Methods Enzymol 318:193–214
27. Kusser W (2000) Chemically modified nucleic acid aptamers for in vitro selections: evolving evolution. J Biotechnol 74:27–38
28. Blackwell TK (1995) Selection of protein binding sites from random nucleic acid sequences. Methods Enzymol 254:604–618

29. Zuker M, Mathews DH, Turner DH (1999) Algorithms and thermodynamics for RNA secondary structure prediction: a practical guide. In: Barciszewski J, Clark BFC (eds) RNA biochemistry and biotechnology. NATO ASI series. Kluwer Academic Publishers, Dordrecht, pp 11–43
30. Sakmann B, Neher H (1995) Single channel recording, 2nd edn. Plenum, New York
31. Hamill OP, Marty A, Neher E et al (1981) Improved patch clamp techniques for high-resolution current recording from cells and cell-free membrane patches. Pflugers Arch 391:85–100
32. Udgaonkar JB, Hess GP (1987) Chemical kinetic measurements of a mammalian acetylcholine receptor using a fast reaction technique. Proc Natl Acad Sci USA 84:8758–8762
33. Hess GP (2002) Rapid chemical reaction techniques developed for use in investigations of membrane-bound proteins (neurotransmitter receptors). Biophys Chem 100:493–506
34. Moreton RB (1997) Optical methods for imaging ionic activities. Scanning Microsc Suppl 8:371–390
35. Sabri S, Richelme F, Pierres A et al (1997) Interest of image processing in cell biology and immunology. J Immunol Methods 208:1–27
36. Davis KA, Lin Y, Abrams B et al (1998) Staining of a cell surface human CD4 with 2′F-pyrimidine-containing RNA aptamers for flow cytometry. Nucleic Acids Res 26: 3915–3924
37. Milburn T, Matsubara N, Billington AP et al (1989) Synthesis, photochemistry, and biological activity of a caged photolabile acetylcholine receptor ligand. Biochemistry 28:49–55
38. Sczczawinska K, Ferchmin PA, Hann RM et al (1992) Electric organ polyamines and their effect on the acetylcholine receptor. Cell Mol Neurobiol 12:95–106
39. Lowry OH, Roseborough NJ, Farr AL et al (1951) Protein measurement with the Folin phenol reagent. J Biol Chem 193:265–275
40. Schmidt J, Raftery MA (1973) A simple assay for the study of solubilized acetylcholine receptors. Anal Biochem 52:349–354
41. Sambrook J, Russell DW (2001) Molecular cloning: a laboratory manual, 3rd edn. Cold Spring Harbor Laboratory, Cold Spring Harbor, New York
42. Niu L, Abood LG, Hess GP (1995) Cocaine: mechanism of inhibition of a muscle acetylcholine receptor studied by a laser-pulse photolysis technique. Proc Natl Acad Sci USA 92: 12008–12012
43. Cheng CY, Prusoff WH (1973) Relationship between the inhibition constant (K1) and the concentration of inhibitor which causes 50 per cent inhibition (IC50) of an enzymatic reaction. Biochem Pharmacol 22:3099–3108
44. Bailey TL, Boden M, Buske FA et al (2009) MEME SUITE: tools for motif discovery and searching. Nucleic Acids Res 7:202–208

Chapter 3

The Design and Structure–Functional Properties of DNA-Based Immunomodulatory Sequences

Nikolai V. Kuznetsov

Abstract

DNA-based immunomodulatory sequences (DIMS) are promising compounds for the treatment of different diseases, including inflammation and cancer. They act through the interaction with TLR9, a member of the Toll-like receptor family whose essential role in innate immunity was recently recognised by being awarded the Nobel Prize 2011. Combining the data obtained from in vitro and in vivo models with circular dichroism spectroscopy approach, we could show that formation of certain tertiary structures by DIMS can be connected to their specific physiologic effects such as activation of immune cells, induction of interferons and delay of the disease progression. Moreover the ability of selected DIMS compounds to form certain tertiary structures must be regarded as important for biological activities as is the presence of functional primary structure motifs such as unmethylated deoxyribodinucleotide CpG. These findings are useful when considering the design of DNA-based immunomodulators.

Key words: Immunomodulatory sequence, DIMS, Oligodeoxyribonucleotide, CpG, G-rich, DNA tertiary structure, Toll-like receptor (TLR)

1. Introduction

DNA-based immunomodulatory sequences (DIMS) can be used for the treatment of different diseases within a variety of therapeutic areas, including inflammation and cancer. Chemically, DIMS are short modified DNA oligonucleotides (oligodeoxyribonucleotides, ODNs) that can be easily synthesized and highly purified in significant quantities. A variety of immunostimulatory ODNs containing unmethylated deoxyribodinucleotide CpG motifs (CpG ODNs) that mimic prokaryotic DNA have been described. It was shown that recognition of CpG ODNs requires Toll-like receptor 9 (TLR9). Cells that express TLR9, which include plasmacytoid dendritic cells (PDC) and B-cells, produce proinflammatory cytokines,

Jürgen Moll and Riccardo Colombo (eds.), *Target Identification and Validation in Drug Discovery: Methods and Protocols*, Methods in Molecular Biology, vol. 986, DOI 10.1007/978-1-62703-311-4_3,

interferons (IFN), and chemokines inducing Th1-like immune response (1, 2).

The importance of TLR9-based immunity became obvious during the last decade that brought new possibilities in DNA-based therapeutic compound design. The Nobel Prize 2011 in Physiology or Medicine (31) has highly supported the significance of Toll-like receptor (TLR) mechanisms in innate immunity and landmarked their wide application opportunities within the area of anti-inflammatory drug development. Beyond this perspective it is important to remember that the ability of DNA compounds to form certain secondary and tertiary structures needed to be accounted at least as important as the presence of functional primary structure motifs such as CpG. Manipulating DNA sequences as a nucleotide-based text is an attractive way because of its simplicity; however, additional two- and three-dimensional approaches can help to understand how these types of compounds elicit different biological functions. Despite their short length (10–30 nt) many DIMS are able to form high-order structures, which are crucial for their functions. In this chapter we describe the principles of DIMS design and the impact for its biological effects using several DIMS examples and considering primary, secondary, and tertiary DNA structure design aspects.

2. Materials

2.1. DNA-Based Immunomodulatory Sequences

DIMS ODNs synthesized by Biomers.net GmbH as follows (5′–3′):

1. DIMS9051	G*G*G*GCGTCTGCC*G*G*G
2. DIMS9052	G*G*G*GTCGTCTG*C*G*G
3. DIMS9054	G*G*G*GTCGTCTGC*G*G*G
4. DIMS9058	G*A*T*CGTCCGTCGG*G*G*G
5. DIMS9059	G*A*T*CGTCCG*G*G*G
6. DIMS9011	T*C*A*CGACCGTCAAAC*T*C*C
7. DIMS0400	TGGGGT

(* indicates the sites of phosphorothioate modification of sugar-phosphate DNA backbone)

2.2. Reagents

1. PBMCs medium: RPMI1640 containing 10% FCS, 2 mM L-glutamine, 100 U/ml penicillin, 100 μg/ml streptomycin, 10 mM HEPES, and 5 μg/ml gentamicin.
2. 1× PBS buffer: 140 mM NaCl, 27 mM KCl, 10 mM phosphate buffer, pH 7.4.
3. Human IFNα ELISA kit (PBL, Biomedical Laboratories, NJ, USA).

4. Human IFNβ ELISA kit (Fujirebio INC., Tokyo, Japan).
5. Human IFNγ Cytometric Bead Array (CBA) Flex set (Becton Dickinson, CA, USA).
6. Myelin Oligodendrocyte Glycoprotein (MOG) Peptide Fragment Rat (Sigma-Aldrich, MO, USA).

3. Methods

3.1. Principles of DIMS Design at Primary Structure Level

3.1.1. Definition of Primary Structure

The sequence of nucleotides in a nucleic acid is defined as the primary structure of nucleic acid.

3.1.2. DNA Sequence Motifs

Certain short DNA sequences with lengths of 3–12 nucleotides can be defined as more or less stable sequence motifs. Many DNA motifs play a known functional role, e.g., promoter-binding sites for various transcription factors. The functions of other DNA motifs might be yet unknown. The variety of motif sequences can be found in the scientific literature and databases or can be determined empirically (Table 1).

3.1.3. Combination of DNA Motifs

The primary structure of DNA is a one-dimensional system similar to four-letter text and can be subjected to the simplest combinatory rules. The particular motifs can be combined with one or several other motifs in a way similar to using building blocks. For instance, G-rich motif can be added to one or both ODN flanks. A certain sequence, e.g., a sequence containing unmethylated deoxyribodinucleotide CpG motifs that mimic prokaryotic DNA (1), can be placed between similar or different motifs, like GC-rich palindrome and/or G-rich motifs (Fig. 1). Various motif combinations will yield a number of putative DNA sequence variants that can be used for further tests and selection of perspective ODN compounds (see Notes 1–4).

3.2. Principles of DIMS Design at Secondary Structure Level

3.2.1. Definition of Secondary Structure

The secondary structure of a nucleic acid molecule refers to the base-pairing interactions within a single molecule or a set of interacting molecules (3).

3.2.2. DNA Bending Rigidity

It is known that double-stranded DNA (dsDNA) is a relatively rigid structure in standard conditions in physiologically relevant solutions like 1× PBS buffer. For example, typical dsDNA has a

Table 1
Examples of motifs used in composition of biologically active oligonucleotides

#	Sequence 5′–3′	Description	ODN group/action	Company	Ref.
1	GTCGTT	Hexanucleotide human CpG motif	Immunostimulatory oligodeoxynucleotides of C-class inducing high levels of IFNα and stimulating B-cells	Coley	(28, 29)
2	GACGTT	Hexanucleotide murine CpG motif			
3	CGGCGCGCGCCG	GC-rich palindromic sequences			
4	CGGCGGCCGCCG				
5	CCGGCGCGCCGG				
6	GAC[a]GTT	N3-Me-2′-deoxycytidine-modified CpG motif	Immunomodulatory oligodeoxynucleotides recognized by mouse immune cells	Idera	(30)
7	GACG[a]TT	N1-Me-2′-deoxyguanosine-modified CpG motif			
8	GTC[a]GTT	N3-Me-2′-deoxycytidine-modified CpG motif	Immunomodulatory oligodeoxynucleotides recognized by human immune cells		
9	GTCG[a]TT	N1-Me-2′-deoxyguanosine-modified CpG motif			
10	CGTC	Tetranucleotide human CpG motif	DNA-based immunomodulatory sequences (DIMS)	InDex	unpublished
11	TCGTC	Pentanucleotide human CpG motif	Modulating response of immune cells in various human and rodent biological systems		
12	GGG	G-rich flanking sequences			
13	GGGG				
14	GGGGG				
15	GCCGGC	GC-rich palindromic sequences			
16	GCCGCGGC				
17	GCCGCGCGCGGC				

[a]2′-Deoxy-*N*-methyl modification; G-rich motifs have *black* background; sequences containing unmethylated CpG motifs are shown with *grey* background (corresponding to drawing in Fig. 1)

persistence length (l_p) in the range of 50–100 nm (4–6) that corresponds to the length of 150–300 bp. This means that dsDNA with a length shorter than 150 bp can be considered as a stiff macromolecular rod. However as soon as dsDNA melts to single strands the rigidity of the structure will be lost. Basically single-stranded DNA (ssDNA) is 50–100 times more flexible having an l_p in a range of 1–3 nm (7–9) that corresponds to the length of 3–9 nucleotide residues. How the ssDNA ODN is bent in an aqueous solution in standard conditions will be dependent on many factors including ODN primary structure that will determine base-pairing interactions within a single ODN or between two or more interacting molecules.

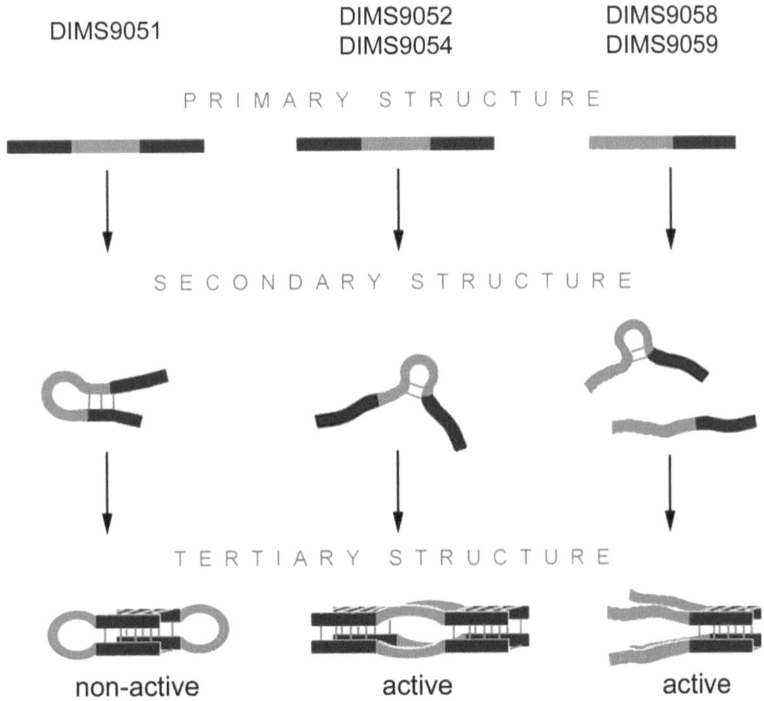

Fig. 1. The three structural levels of DNA-based immunomodulatory sequence (DIMS). G-rich motifs are shown in *black*, sequences containing unmethylated CpG motifs shown in *grey* (corresponding to the color of motif background in Table 1); thin *grey lines* indicate hydrogen bonds formed by Watson–Crick nucleotide pairs (DNA secondary structure level) or by Hoogsteen hydrogen bonding (DNA tertiary structure level).

3.2.3. Thermodynamic Aspect of DNA Folding

In the same conditions different secondary structures will have different probabilities of formation that could be evaluated with the help of the second law of thermodynamics. The spontaneity of these processes could be roughly estimated by comparing values of change in Gibbs free energy of the formation (ΔG) of these structures. The reaction in an isolated system (in case of ODN this corresponds to the formation of particular secondary structures) will be spontaneous if the process is characterized by negative ΔG value. Furthermore if an ODN molecule is able to fold in several alternative secondary structures the one characterized by lowest ΔG value will be formed preferentially. This feature could allow predictions on the probability that a particular secondary structure will be formed. Examples of predicted secondary structures for DIMS with G-rich motifs and/or CpG motifs and corresponding ΔG values are shown in Fig. 2 (see Notes 5–8).

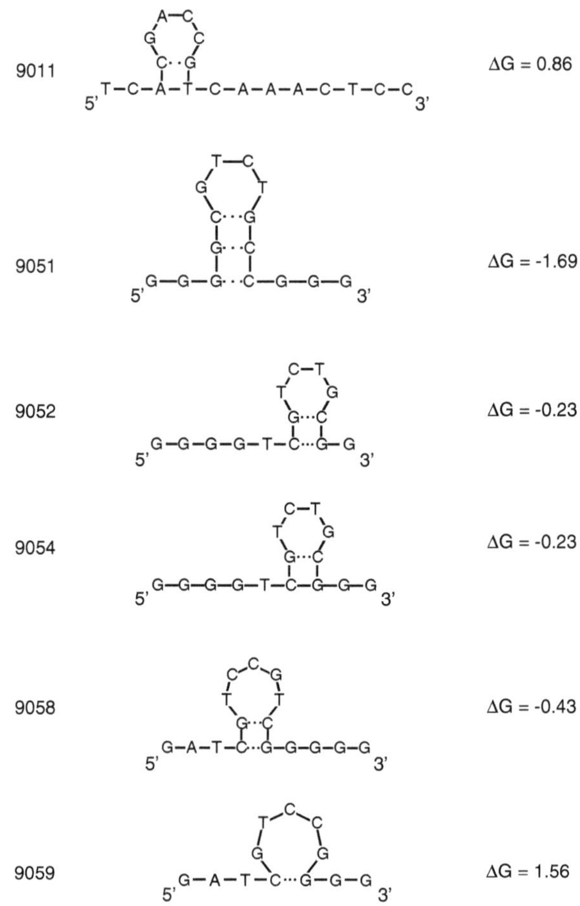

Fig. 2. The secondary structure prediction of DIMS with flanking G-rich motifs and DIMS9011 (negative control). Gibbs free energy values (ΔG) corresponding to secondary structures are indicated; software used: GeneWalker; Quikfold.

3.3. The Tertiary Structure Level: Principles of Design and Analysis

3.3.1. Definition of Tertiary Structure

The tertiary structure of a macromolecule is defined by IUPAC as the spatial organization (including conformation) of an entire macromolecule consisting of a single chain (10).

It is worth to mention that this definition is not always consistent with the nature of the macromolecular assembly when applying to nucleic acids. For example, from all different types of quadruplex nucleic acids only quadruplex monomers are covered by IUPAC definition of tertiary structure being a single chain of DNA or RNA. However, also the quadruplexes with higher molecularity of the formed structures (dimers, tetramers) belong to this important tertiary structure family.

3.3.2. Divergency of DNA Tertiary Structures

The tertiary structure of DNA is the structural level that is most relevant to 3-D reality. Traditionally, ODNs in a physiologically relevant aqueous solution are considered to be in a random-coiled ssDNA state or in the form of dsDNA helix in the presence of a complementary DNA, including the case of self-complementarity. The double helix is the dominant tertiary structure for biological DNA that can be in one of the three DNA conformations found in nature, A-DNA, B-DNA, and Z-DNA. The B-conformation described by Watson and Crick (11) is believed to predominate in cells (12). However other types of nucleic acid tertiary structures different from random or classical double-stranded helix forms can also be observed. Among them are triplexes, quadruplexes, and several other nucleic acid structures (13, 14).

3.3.3. G-quadruplex DNA

G-rich DNA sequences are capable of forming stable four-stranded structures known as G-quadruplexes (G-quartets, G-tetrades, G4 DNA) (15, 16). G-quadruplexes arise from the association of four G-bases into a planar assembly that is stabilized by von Hoogsteen hydrogen bonds and by base stacking interactions. G-quadruplexes have been shown to be relevant in a number of biological processes (17): they are an important component of human telomeres (18), and play a role in regulation of transcription (19, 20) as well as in translation (21). A G-quadruplex ODN aptamer (AS-1411, Antisoma plc.) is the first ODN aptamer which reached clinical trials for the potential treatment of cancers, including acute myelogenous leukemia (AML) (22, 23). G-rich ODNs can form a variety of possible G-quadruplex structures, depending on their thermodynamic and kinetic characteristics. The structures formed can be influenced by ODN primary structure and concentration, as well as the conditions (temperature, buffer content, etc.) used for annealing. G-quadruplexes can be formed by one, two, or four molecules of oligonucleotide, which are referred to as G4-monomers, G4-dimers, and G4-tetramers, respectively (24).

3.3.4. Tertiary Structure Affected by Multiple Factors

Keeping in mind all three DNA structure levels, primary, secondary, and tertiary, it is essential to understand that the lower level will mediate but not fully determine the higher structural level. In other words, the secondary as well as tertiary DNA structures of ODN in solution will be affected by many physical and chemical parameters, such as temperature, pH, salt content, compound concentration, etc. When evaluating complex biochemical systems, additional factors have to be taken into consideration: possible interactions of ODN with a variety of other molecules and macromolecules in solution, local concentration effects and compartmentalization, biological half-life, etc. Hence when designing a DIMS ODN compound, its 3-D structure will not be fully predictable.

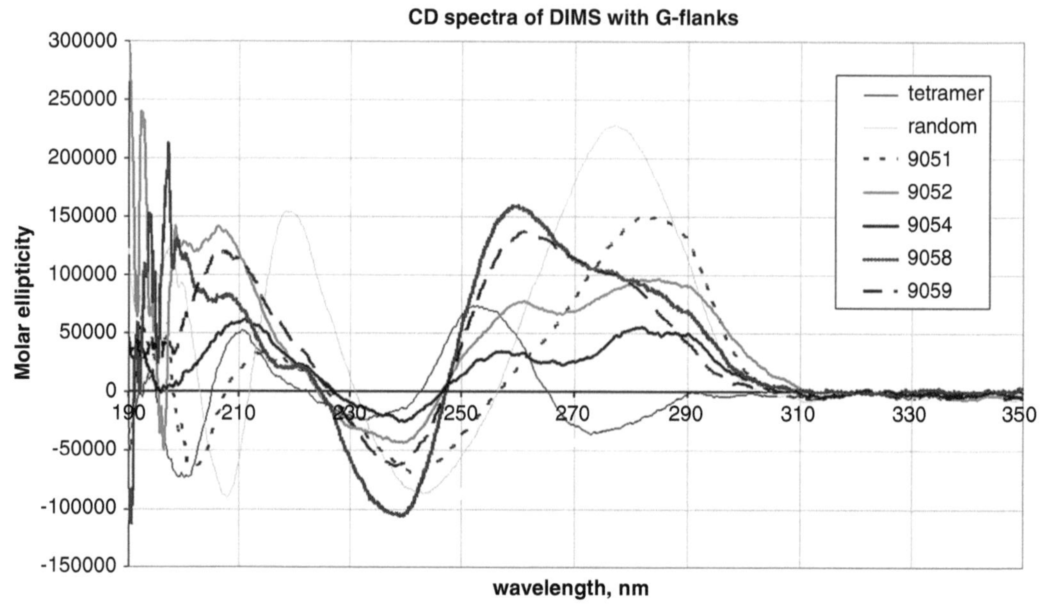

Fig. 3. CD spectra of DIMS with flanking G-rich motifs. A spectrum with characteristic positive peaks in the range of 252–262 nm is typical of a parallel G-4 tetramer structure and is observed for DIMS9052, DIMS9054, DIMS9058, and DIMS9059 but not for DIMS9051. Two reference CD spectra are shown by thin lines: parallel G4-tetramer formed by DIMS0400—thin *black line*; random-coiled DIMS9011—thin *grey line* and indicated in the legend as "tetramer" and "random," respectively.

3.3.5. Tertiary Structure Analysis

Several direct methods are available to analyze the tertiary structure of ODNs like nuclear magnetic resonance (NMR) and X-ray crystallographic (XRC) techniques, which needs a sophisticated setup and infrastructure. An alternative but indirect method to study the structure and conformations of nucleic acids is circular dichroism spectroscopy (CD spectroscopy) (25, 26), where circular dichroism refers to the differential absorption of left and right circularly polarized light (27).

3.4. Circular Dichroism Spectroscopy

Using CD spectroscopy and published data it is possible to predict the tertiary structure of ODN compounds similar to a reference ODN (e.g., DIMS0400 which forms a type of structure called parallel tetramer of G-quadruplex (24) (Fig. 3)), whose structural characteristics had been determined previously by NMR and/or XRC techniques (see Notes 9–15).

1. Prior to CD spectroscopy measurements, the oligonucleotide samples were heated for 1 min at 90°C and then slowly cooled to room temperature (20°C) over a period of 8 h.
2. CD spectra of 20 μM DIMS samples in 1× PBS buffer were measured on a Jasco J-720 spectropolarimeter (Jasco Corp., Tokyo, Japan).
3. Seven spectra measurements were taken for each sample in the range from 190 to 350 nm at 25°C (±0.1°C) at the rate of

100 nm/min with the final spectrum being the average of seven measurements.

4. Data collection and evaluation were carried out by supplied Jasco software. A custom-built computer program developed using MatLab (Mathworks Corp.) was used for quantitative analysis of CD spectra to calculate the relative composition of the spectra.

3.5. In Vivo EAE Rat Treatment with DIMS

Multiple sclerosis (MS) is the most frequent inflammatory demyelinating disease of the central nervous system that affects worldwide about 2.5 million people with no cure. Myelin oligodendrocyte glycoprotein-induced experimental autoimmune encephalomyelitis (MOG-induced EAE) in DA rats is an appropriate model for therapeutic testing, sharing many features with human multiple sclerosis.

1. Adult rats were injected with MOG, and then monitored daily to register the parameters, including EAE severity, incidence, cumulative score, and mortality. The study involved groups of 12 animals each.
2. Three doses of 150 μg of selected DIMS compounds were injected subcutaneously on days 9, 15, and 21.
3. MOG-induced EAE DA rats were treated with DIMS9052, DIMS9054, or PBS. During disease course there was no difference observed regarding all studied parameters between the control group receiving PBS and the group receiving DIMS9052. However, the group treated with DIMS9054 developed milder disease with respect to all studied parameters (Fig. 4) indicating that DIMS9054 has a therapeutic effect in the EAE rat model (see Notes 3, 13 for comments on results).

3.6. Human PBMC Isolation and Stimulation with DIMS

Peripheral blood mononuclear cells (PBMCs) that include mainly monocytes, T-cells, and B-cells, and smaller portions of NK cells and dendritic cells of myeloid and plasmacytoid origin, are a critical component of immune system. PBMC-based models are widely used in research of cell-mediated immunity and for screening of compounds for immunomodulatory effects.

1. PBMCs were isolated from buffy coats of healthy blood donors by density centrifugation according to the standard procedures and cultured in complete cell medium.
2. PBMCs were cultured in 48-well culture tissue plates at a concentration of 5×10^6 cells/ml with 10 μM DIMS or with medium alone as a negative control in a total volume of 400 μl/well.
3. Cells were cultured for 48 h at 37°C, 5% CO_2.
4. The cell supernatants were collected and frozen at −20°C for later analysis.

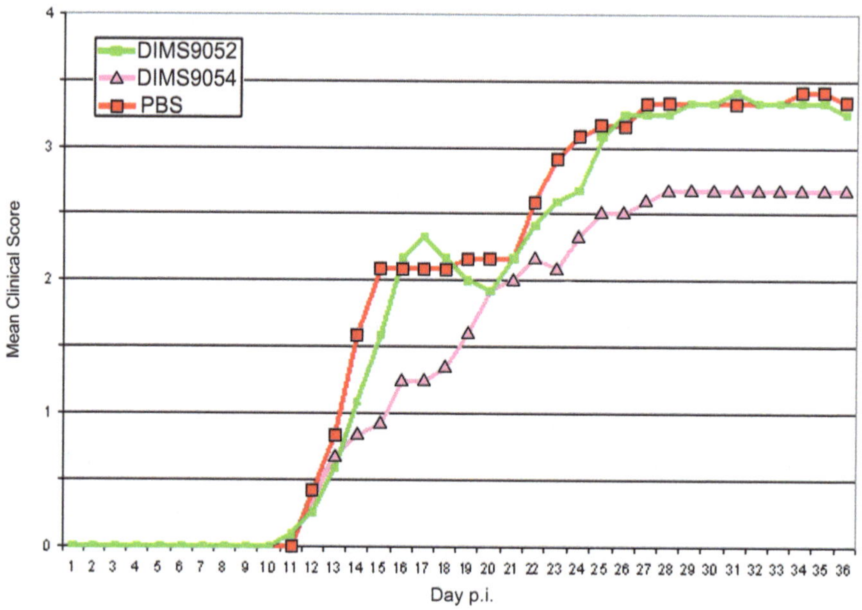

Fig. 4. The time course of in vivo EAE rat model treated with DIMS9052, DIMS9054, or PBS.

5. The cell culture supernatants from PBMCs stimulated with DIMS were analyzed by ELISA for the presence of human interferon α and β (IFNα and IFNβ).
6. The absorbance was measured on a microplate reader (Tecan, Switzerland).
7. Data were analyzed using Magellan software (Tecan, Switzerland).
8. The presence of IFNγ in cell culture supernatants from PBMCs stimulated with DIMS compounds was detected using CBA flex kit (Becton Dickinson) on a FACSArray flow cytometer (Becton Dickinson). The data were analyzed using FCAP Array software (Becton Dickinson).
9. IFN production of PBMCs cultured in the presence of selected DIMS is shown from six healthy individuals (ELISA mean data of IFNα and IFNβ) and CBA mean data (IFNγ) from four healthy individuals (Table 2). Different DIMS are capable of inducing IFN with different efficiency in human PBMCs (see Notes 1, 2, 5, 7, 8, 10–15 for comments on results).
10. In another experimental setting in vitro data on the cell surface antigen expression were obtained for PBMCs isolated from chronic lymphocytic leukemia (CLL) patient blood samples and cultured in the presence of DIMS (according to standard procedures). Cultured cells were collected and stained with the corresponding antibodies for detection of the expression of cell surface antigens. Subsequently they were analyzed by flow cytometry.

Table 2
The three levels of DIMS structures and their correlation with the induction of IFNs in human PBMCs in vitro

DIMS #	Primary structure 5′–3′	Major secondary structure	Major tertiary structure (%)	IFN induction		
				IFNα	IFNβ	IFNγ
DIMS9011	T*C*A*CGACCG TCAAAC*T*C*C	Random	Random coil (100%)	–	–	–
DIMS9051	**G*G*G*G**CGTCT GCC***G*G*G**	Stable hairpin	"Chair" dimer of G-quadruplex (>50%)	–	–	–
DIMS9052	**G*G*G*G**TCGTC T**G***C***G*G**	Labile hairpin	Parallel tetramer of G-quadruplex (35%)	++	–	–
DIMS9054	**G*G*G*G**TCGTC TGC***G*G*G**	Labile hairpin	Parallel tetramer of G-quadruplex (>45%)	+++	+++	+
DIMS9058	**G***A*T*CGTCCGT CG**G*G*G*G**	Labile hairpin	Parallel tetramer of G-quadruplex (>45%)	+++	+++	+
DIMS9059	**G***A*T*CGTCCG* **G*G*G**	Random	Parallel tetramer of G-quadruplex (>45%)	+++	+++	+

Flanking G-nucleotides are bold
CpG-dinucleotide underlined
The score for the mean concentration was:
– no induction (less than 125 pg/ml of IFNα, IFNβ; less than 65 pg/ml of IFNγ)
+ weak induction (125–250 pg/ml of IFNα, IFNβ; 65–125 pg/ml of IFNγ)
+ + moderate induction (250–750 pg/ml of IFNα, IFNβ; 125–500 pg/ml of IFNγ)
+ + + strong induction (more than 750 pg/ml of IFNα, IFNβ; more than 500 pg/ml of IFNγ)
Interferon production of PBMCs cultured in the presence of selected DIMS determined by ELISA for IFNα, IFNβ (mean data from six healthy individuals) and by Cytometric Bead Array (Becton Dickinson) for IFNγ (mean data from four healthy individuals)
*Phosphorothioate modification

11. PBMCs isolated from CLL patients were cultured in the presence of DIMS9052 or DIMS9058. The first compound induces only weak expression of surface antigens CD23 and CD69, whereas DIMS9058 was able to induce high levels of CD20, CD23, CD25, CD80, and CD69 antigens indicating general immune cell activation (Table 3) (see Notes 4, 14 for comments on results).

4. Notes

1. Testing a number of DIMS in several biological systems showed differential immunostimulatory properties or other physiological effects. However, it is not always easy to explain such a variation of DIMS stimulatory capacity looking just to the primary structure. For example, changing only two nucleotides in

Table 3
DIMS induce cell surface antigen expression in PBMCs derived from CLL patients

DIMS	Relative cell surface antigen expression				
	CD20	CD23	CD25	CD80	CD69
DIMS9052	0	1	–	0	3
DIMS9058	5	5	3	5	5

The protein cell surface antigen expression mean value in PBMC of 10 CLL patients is reported as relative units on scale 0–5

CpG context in DIMS9051, but otherwise maintaining the same set and pattern of motifs as in DIMS9054, will totally abolish the DIMS ability to induce IFNs in cultured healthy human PBMCs (Table 2).

2. In the second example, DIMS9011 and DIMS9059 both have a set of two CpG motifs but strongly differ in their IFN stimulatory capacity: DIMS9059 is one of the best IFN inducers, whereas DIMS9011 does not induce IFNs at all (Table 2). The difference in primary structures of these compounds is that DIMS9059 possesses a 3′-flanking G-rich motif that DIMS9011 does not have, providing a possible explanation for functional divergences of these DIMS (see Notes 8, 15).
3. In the third example, DIMS9054 but not DIMS9052 was able to reduce the rate of disease progression in the EAE in vivo rat model (Fig. 4) despite the fact that both DIMS have similar primary structures (see Notes 7, 13).
4. In another experimental setting, in vitro data were obtained for PBMCs isolated from CLL patients and incubated with DIMS9052 or DIMS9058 that share similar sets of DNA motifs: G-rich motifs and CpG motifs. However the first compound induces only weak expression of surface antigens CD23 and CD69, whereas DIMS9058 was able to induce high levels of CD20, CD23, CD25, CD80, and CD69 antigens indicating general immune cell activation (Table 3) (see Note 14).
5. Part of a possible explanation of such a difference is that the secondary structure of DIMS affects their biological properties. For example, the secondary structure of DIMS9051 has lowest negative ΔG value of formation among five DIMS molecules with G-flanks. Furthermore, the secondary structure prediction for DIMS9051 reveals a comparatively stable hairpin having the stem consisting of three Watson–Crick GC pairs (Fig. 2). The formation of a stable secondary structure by DIMS9051

corresponds to its decreased ability to induce IFNs in comparison to other four DIMS with G-flank motifs (Table 2). The folding into such a stable secondary structure probably will lead to the formation of a specific tertiary structure of DIMS9051 that will block its immunostimulatory properties (see Note 12).

6. The secondary structure of DIMS9054 has Δ*G* formation value greater than one for DIMS9051 and is represented by a relatively labile hairpin that melts easily when it comes to intermolecular interactions between four molecules of DIMS9054 (Figs. 1 and 2) (see Notes 10–12).
7. The secondary structures of three active DIMS with G-flank motifs, DIMS9052, DIMS9054, and DIMS9058 are characterized by negative Δ*G* formation value (but higher than Δ*G* value for DIMS9051) and contain relatively labile hairpins with stems of two Watson–Crick GC pairs only (Fig. 2). Regarding their immunostimulatory activity, DIMS9052 induced intermediate levels of IFNα but not IFNβ or IFNγ whilst DIMS 9054 and 9058 perform best in the induction of IFNs (Table 2). However the difference in IFN stimulatory capacity of DIMS9052 and DIMS9054 cannot be clearly connected to secondary structures formed as they look quite similar at this structure level. This can be explained considering their tertiary structures (see Note 13).
8. The predicted secondary structures for DIMS9059 (an effective IFN inducer) and for DIMS9011 (functionally inactive negative control), both characterized by positive Δ*G* value of formation, suggest that in standard conditions these DIMS most probably will have a random secondary structure. The possible explanation of functional difference for these two compounds could be found on the tertiary structure level as well (see Note 15).
9. Our experimental data showed that the ability of DIMS to form certain tertiary structures correlated with their specific physiologic effects can be summarized as follows: the tertiary structure of CpG-containing oligonucleotides might play a role regarding recognition by the TLR9 receptor. Moreover, the folding of ODN into certain tertiary structure is considered to be at least as important as the presence of functional motifs, e.g., an unmethylated deoxyribodinucleotide CpG motif.
10. DIMS capability to induce IFNs (α, β, γ) in cultured healthy human PBMCs in vitro well correlates with its ability to form parallel tetramer of G-quadruplex (parallel G4-tetramer) (Table 2; Fig. 1). CD spectra demonstrated appearance of the characteristic peaks in the range of 252–264 nm, typical of a parallel G-4 tetramer structure, for DIMS9052, DIMS9054, DIMS9058, and DIMS9059 but not for DIMS9051 (Fig. 3).

11. The quantitative CD spectra analysis showed highest values of parallel G4-tetramer structure for compounds DIMS9054, DIMS9058, and DIMS9059 that perform best in the induction of IFNs, whereas compound DIMS9052 which shows a low parallel G4-tetramer value is able to induce intermediate levels of IFNα but not IFNβ or IFNγ (Table 2).
12. CD spectra quantitative analysis shows that compound DIMS9051 can form another type of G-quadruplex called a dimer "chair" (24). In cultured healthy human PBMCs in vitro DIMS9051 shows no IFN-inducing potential similarly to the random-coiled negative control DIMS9011 (Table 2). The formation of a G-quadruplex dimer "chair" probably is facilitated by locking of DIMS9051 in a comparatively stable hairpin secondary structure that could serve as an intermediate conformation (Figs. 1 and 2).
13. DIMS9052 and DIMS9054 have similar primary and secondary structures, but nevertheless differ in their ability to form specific tertiary structures of parallel G4-tetramers (Fig. 3). This can be correlated to the difference in their IFN stimulatory capacity (Table 2) as well as in their ability to reduce the rate of disease progression in the EAE in vivo rat model (Fig. 4).
14. DIMS9052 and DIMS9058 share similar sets of DNA motifs and have similar secondary structures but nevertheless differ in their ability to form a specific tertiary structure of parallel G4-tetramers (Fig. 3) that can be associated to the difference in their IFN stimulatory capacity (Table 2) as well as to their ability to activate immune cells in cultured PBMC samples of CLL patients (Table 3).
15. Compounds DIMS9011 and DIMS9059 have a set of two CpG motifs and probably, random secondary structures. Nevertheless, they strongly differ in their IFN stimulatory capacity: DIMS9059 is one of the best IFN inducers whereas DIMS9011 does not induce IFN at all (Table 2). The major difference in primary structures of these compounds is that DIMS9059 possesses 3′-flanking G-rich motifs that DIMS9011 does not have. Such G-stretches allow DIMS9059 to efficiently form the tertiary structure of parallel G4-tetramer, which can be important for its biological activity and that the functionally negative random-coiled DIMS9011 is not able to form (Fig. 3).

Acknowledgments

I would like to thank all colleagues at InDex Pharmaceuticals who made this publication possible. Special thanks to Dr. Lisa Bandholts and Dr. Natalia Markova for their input and to Dr. Oliver von Stein for his useful comments.

References

1. Krieg AM (2002) CpG motifs in bacterial DNA and their immune effects. Annu Rev Immunol. 20:709–760
2. Krieg AM (2006) Therapeutic potential of Toll-like receptor 9 activation. Nat Rev Drug Discov 5(6):471–484
3. Dirks RM, Lin M, Winfree E et al (2004) Paradigms for computational nucleic acid design. Nucleic Acids Res 32:1392–1403
4. Bustamante C, Marko JF, Siggia ED et al (1994) Entropic elasticity of λ-phage DNA. Science 265:1599–1600
5. Leger JF, Robert J, Bourdieu L et al (1998) RecA binding to a single double-stranded DNA molecule: a possible role of DNA conformational fluctuations. Proc Natl Acad Sci USA 95:12295–12299
6. Diamant H, Andelman D (2011) General criterion for controllable conformational transitions of single and double stranded DNA, Cornell University Library. http://www.arxiv.org/ftp/cond-mat/papers/0109/0109253.pdf Accessed 30 Oct 2011.
7. Smith SB, Cui YJ, Bustamante C (1996) Overstretching B-DNA: the elastic response of individual double-stranded and single-stranded molecules. Science 271:795–799
8. Tinland B, Pluen A, Sturm J et al (1997) Persistence length of single stranded DNA. Macromolecules 30:5763–5765
9. Mills JB, Vacano E, Hagerman PJ (1999) Flexibility of single-stranded DNA: use of gapped duplex helices to determine the persistence lengths of poly(dT) and poly(dA). J Mol Biol 285:245–257
10. International Union of Pure and Applied Chemistry (IUPAC) (2011) Compendium of Chemical Terminology Internet edition. Online corrected version: (2006) "tertiary structure". http://goldbook.iupac.org/T06282.html Accessed 30 Oct 2011.
11. Watson JD, Crick FHC (1953) A structure for deoxyribose nucleic acid. Nature 171:737–738
12. Richmond TJ, Davey CA (2003) The structure of DNA in the nucleosome core. Nature 423:145–150
13. Soyfer VN, Potaman VN (1995) Triple-helical nucleic acids. Springer, New York
14. Neidle S, Balasubramanian S (2006) Quadruplex nucleic acids. RSC Publishing, Cambridge, UK
15. Burge S, Parkinson GN, Hazel P et al (2006) Quadruplex DNA: sequence, topology and structure. Nucleic Acids Res 34:5402–5415
16. Huppert JL (2008) Four-stranded nucleic acids: structure, function and targeting of G-quadruplexes. Chem Soc Rev 37:1375–1384. Epub 6 May 2008.
17. Patel DJ, Phan AT, Kuryavyi V (2007) Human telomere, oncogenic promoter and 5′-UTR G-quadruplexes: diverse higher order DNA and RNA targets for cancer therapeutics. Nucleic Acids Res. 35:7429–7455. Epub 2 Oct 2007.
18. Oganesian L, Bryan TM (2007) Physiological relevance of telomeric G-quadruplex formation: a potential drug target. Bioessays 29:155–165
19. Qin Y, Hurley LH (2008) Structures, folding patterns, and functions of intramolecular DNA G-quadruplexes found in eukaryotic promoter regions. Biochimie 90:1149–1171. Epub 29 Feb 2008.
20. Siddiqui-Jain A, Grand CL, Bearss DJ et al (2002) Direct evidence for a G-quadruplex in a promoter region and its targeting with a small molecule to repress c-MYC transcription. Proc Natl Acad Sci USA 99:11593–11598. Epub 23 Aug 2002.
21. Kumari S, Bugaut A, Huppert JL et al (2007) An RNA G-quadruplex in the 5′ UTR of the NRAS proto-oncogene modulates translation. Nat Chem Biol 4:218–221. Epub 25 Feb 2007.
22. Ireson CR, Kelland LR (2006) Discovery and development of anticancer aptamers. Mol Cancer Ther 12:2957–2962
23. Mongelard F, Bouvet P (2010) AS-1411, a guanosine-rich oligonucleotide aptamer targeting nucleolin for the potential treatment of cancer, including acute myeloid leukemia. Curr Opin Mol Ther 1:107–114
24. Dapić V, Abdomerović V, Marrington R et al (2003) Biophysical and biological properties of quadruplex oligodeoxyribonucleotides. Nucleic Acids Res 31:2097–2107
25. Baase WA, Johnson WC Jr (1979) Circular dichroism and DNA secondary structure. Nucleic Acids Res 2:797–814
26. Paramasivan S, Rujan I, Bolton PH (2007) Circular dichroism of quadruplex DNAs: Applications to structure, catione effects and ligand binding. Methods 43:324–331
27. Atkins P, de Paula J (2005) Elements of physical chemistry. Oxford University Press, UK
28. Hartmann G, Krieg AM (2000) Mechanism and function of a newly identified CpG DNA motif in human primary B cells. J Immunol 164:944–953

29. Vollmer J, Weeratna R, Payette P et al (2004) Characterization of three CpG oligodeoxynucleotide classes with distinct immunostimulatory activities. Eur J Immunol 34:251–262
30. Putta MR, Zhu F, Li Y et al (2006) Novel oligodeoxynucleotide agonists of TLR9 containing N3-Me-dC or N1-Me-dG modifications. Nucleic Acids Res 34:3231–3238
31. "The 2011 Nobel Prize in Physiology or Medicine—Press Release." Nobelprize.org. http://www.nobelprize.org/nobel_prizes/medicine/laureates/2011/press.html Accessed 23 Nov 2011

Part III

RNA Interference: From Design to Data Analysis

Chapter 4

siRNA Design Principles and Off-Target Effects

Sebastian Petri and Gunter Meister

Abstract

Short interfering RNAs (siRNAs) are a major research tool that allows for knock-down of target genes via selective mRNA destruction in almost all eukaryotic organisms. siRNAs typically consist of a synthetic ~21 nucleotide (nt) RNA-duplex where one strand is designed with perfect complementarity to the target mRNA. Although siRNAs were initially thought to be very target-specific because of their design, it turned out during the last years that all siRNAs have a more or less pronounced intrinsic off-target activity which can make the interpretation of data from siRNA experiments difficult. Here we describe essential rules for siRNA design that should be taken into account in order to obtain potent siRNAs with minimal off-target activity. In addition, we describe how to control for off-target activity in siRNA experiments.

Key words: siRNA, Argonaute proteins, siRNA off-target activity, siRNA on-target activity, siRNA design, siRNA modifications

1. Introduction

The phenomenon known as RNA interference or short RNAi was first discovered in worms where it was found that double-stranded (ds) RNA inhibits the expression of complementary mRNAs (1). However, long dsRNA induces an interferon response and is toxic to mammalian cells. Further work in *C. elegans* and Drosophila demonstrated that long dsRNA is processed to shorter, about 21 nt long RNAs, the so-called short interfering RNAs (siRNAs), which are the actual active RNA species in RNAi (2, 3). A major breakthrough for the application of RNAi as potential drug was achieved when Tuschl and colleagues found that short synthetic siRNAs with one strand complementary to a target mRNA do not induce an interferon response in mammalian tissue culture cells and are extremely powerful tools to inactivate target gene expression (4). This discovery paved the way for a broad application of RNAi as research tool and even more complex applications such as

Jürgen Moll and Riccardo Colombo (eds.), *Target Identification and Validation in Drug Discovery: Methods and Protocols*, Methods in Molecular Biology, vol. 986, DOI 10.1007/978-1-62703-311-4_4,

genome-wide RNAi-based functional screening approaches (5). In addition, tremendous effort was made to further improve siRNA-based knockdown strategies in order to use them for therapeutic purposes (6).

Today, the RNAi machinery is reasonably well understood. Long dsRNA is processed to generate siRNAs by the RNase III enzyme Dicer resulting in 21 nt long siRNAs. In many organisms, Dicer enzymes act in concert with dsRNA-binding proteins. In flies, for example, R2D2 is important for the selection of the correct siRNA strand. Similar functions have been proposed for the mammalian TAR RNA-binding protein (TRBP). One strand of the siRNAs is selected and gives rise to the guide strand. The other strand, often referred to as the passenger strand, is degraded. In all organisms, the guide strand directly interacts with a member of the Argonaute protein family, which form the core of the RNA-induced silencing complex (RISC). The siRNA guides RISC to fully complementary sequences on mRNAs and RISC cleaves the target RNA sequence-specifically, which is subsequently fully degraded (for reviews see (3, 7–9)).

It turned out that the effect of siRNAs in mammalian cells is mediated by an endogenous small RNA pathway that regulates gene expression. This pathway is based on microRNAs (miRNAs). miRNAs are processed from endogenous stem-loop-structured precursors to generate 21 nt long mature miRNAs. Similar to siRNAs, miRNAs are also processed by Dicer and directly interact with Argonaute proteins. The Argonaute–miRNA complex, however, is often referred to as microribonucleoprotein particle (miRNP) or miRISC. In contrast to siRNAs, miRNAs recognize target sites on mRNAs with only partial complementarity. These sites are preferentially located in the 3′ untranslated regions (UTRs) of mRNAs and miRISC binding leads to translation repression or deadenylation-dependent mRNA decay.

In humans, eight Argonaute genes exist. The Argonaute family can be subdivided into the Ago subfamily (Ago1-4) and the Piwi subfamily (HILI, HIWI1-3). Piwi proteins are restricted to the male germline where they are involved in silencing of mobile genetic elements. Ago proteins are ubiquitously expressed and all four members interact with endogenous miRNAs or siRNAs in knockdown experiments (for reviews on Argonaute proteins see (10–15)). Structural studies have shown that Argonaute proteins are similar to RNase H and indeed some Ago proteins are endonucleases. In mammals, however, only Ago2 has endonuclease activity and is often referred to as "slicer" (16, 17). Although Ago1, Ago3, and Ago4 are catalytically inactive and therefore not important for mRNA cleavage, they can contribute to knockdown phenotypes (18). In addition, siRNAs can function like miRNAs and interact with partially complementary sequences (19) and these effects can be mediated by all four Ago proteins. The minimal complementarity that is needed for efficient inhibition of gene silencing by a miRNA is composed of nt 2–8 of the miRNA. This sequence is known as the seed sequence of the miRNA. Nucleotides 2–8 of an siRNA may also pair with unrelated mRNAs and lead to repression of gene expression. Consequently, siRNAs often lead to unpredictable off-target effects. During the past decade, it became clear that siRNAs can induce a broad spectrum of off-target activity, which is considered as one major challenge in siRNA-based drug development (20).

Global transcription profiling revealed that a high number of nontarget mRNAs are down- and up-regulated in RNAi experiments (21, 22). More detailed analysis of such off-target mRNAs showed that mRNAs containing a motive complementary to the seed region of the siRNA are highly enriched, indicative of unspecific "miRNA-like" behavior of the siRNAs (23–26). In certain cases this siRNA off-target effect can reduce protein levels nearly as efficient as the intended siRNA on-target effect (27). A rough calculation of the theoretical number of seed matches in the expressed 3′ UTRs of a human cell reveals that approximately 300 3′ UTRs in every cell contain an siRNA seed match motif (seed frequency: $4^7 = 16{,}384$; assumptions for expressed 3′ UTRs: average 3′ UTR length 1,000 nt, 5,000 genes expressed at the same time in a cell). Even if only 10% of those 3′ UTRs would indeed bind to an siRNA at a significant level, this would already affect expression of about 30 genes. Another source of siRNA-mediated off-target is based on the activity of Ago proteins. Upon transfection of cells, siRNAs and miRNAs compete for the cellular Ago protein pool and siRNAs can displace miRNAs from Ago proteins leading to miRNA-loss of function phenotypes (28, 29). In addition, certain siRNA sequence motifs have been shown to induce cytoxicity by a so far unknown mechanism (30). Finally, another source of siRNA off-target effects is the potential of siRNAs to trigger immune responses via toll-like receptors. Here, certain sequence motifs within siRNAs are especially efficient in triggering an immune response (31, 32).

Taken together, siRNAs can induce pronounced off-target effects and therefore it is important to design siRNAs and perform siRNA-based experiments in a way that reduces off- and increases on-target activities.

2. Methods

2.1. RNA Sequence Features Affecting siRNA Design

For conclusive RNAi experiments, the design of potent and also highly specific siRNAs is essential. Therefore, several mechanistic aspects of siRNAs should be considered when siRNAs are designed. In addition, siRNAs can be improved by the insertion of chemically modified nucleotides. So far, numerous siRNA design

Table 1
Summary of relevant topics and respective strategies for siRNA design

siRNA feature	Strategy	
Strand selection	Asymmetry rule	Strand with less stable paired 5′ end is preferentially selected as guide strand
Strand selection and general siRNA functionality	Nucleotide specificity	Strands with U, alternatively A, at the 5′ end have higher affinity for Ago proteins. Guide strands should start with U or A, passenger strands with G or C
General siRNA functionality	Overall GC content 5′ terminal GC content	GC content is ideally between 30 and 65%. GC stretches of 9 or more nucleotides should be avoided The 5′ terminal third of the antisense strand should not contain more than two GC base pairs
Immune stimulation	Avoid immune stimulatory motifs	The motifs GUCCUUCAA and UGUGU should not be included in siRNAs
Cytotoxicity	Avoid cytotoxic motif	The motif UGGC should be avoided in siRNAs
Off-target mRNAs with homology to target-mRNA	Sequence comparison	Comparison of siRNA sequences to EST databases (Is the EST expressed in the cell-line of interest?)
Cross species functionality	siRNA target conserved sequence	Compare mRNA sequences of, e.g., mouse and human and target 100% conserved motifs
Rescue experiments	Target noncoding sequence	3′ untranslated regions (UTRs), alternatively 5′ UTRs, can be targeted by siRNA; rescue can be performed by ectopically expressed genes lacking UTRs

algorithms have been created (33). However, such programs can only predict siRNA on- and potential off-target effects with limited reliability and therefore it is absolutely essential to validate siRNA knockdown efficiency experimentally.

The design of a siRNA based on sequence properties has been implemented in most publicly available design algorithms (34) and is the method of choice when starting RNAi experiments on so far uncharacterized genes for which no well-characterized/published siRNAs are available. It is likely that siRNAs with IC_{50}-values in the range of 50–500 pM will be obtained and with state-of-the-art transfection reagents more than 90% reduction of the targeted mRNA can be achieved with siRNA concentrations ranging from 1 to 10 nM. The main sequence features that could affect siRNA activity are here described and summarized in Table 1 (see Note 1).

2.2. siRNA Overall Sequence Properties

2.2.1. GC Content

How strong the overall GC content of an siRNA influences its activity remains controversial in the literature. While some studies claim that the optimal GC content of an siRNA is 30–50%, others have found that also GC-rich siRNAs with GC contents of about 60% are highly efficient (35, 36). We therefore recommend using siRNAs with an overall GC content between 30 and 65% of base pairing nucleotides. It has been shown that GC stretches of 9 or more nucleotides anywhere in the base pairing sequence of an siRNA reduce its efficiency. In addition, siRNAs with low GC content (not more than two GC base pairs) in the 5′ terminal third of the guide (antisense) strand are likely to be potent siRNAs (36). Therefore these two parameters should be taken into account when designing siRNAs.

2.2.2. Immune Stimulatory and Cytotoxic Sequence Motifs

Two sequence motifs in siRNAs are able to induce a potent stimulation of an immune response via binding to toll-like receptors. These siRNA motifs are "GUCCUUCAA" and "UGUGU" (31, 32). In addition, the sequence "UGGC" is strongly enriched in the guide strand of siRNAs with high cytotoxic potential (30). Therefore, the abovementioned sequence motifs should be avoided in order to reduce cytotoxicity.

2.2.3. Sequence Redundancy

Target sites of siRNAs should be carefully checked for similar sequences in the genome. Sequence complementarity of nontarget mRNAs with both siRNA strands should be minimal to avoid off-target activity of the used siRNAs. In order to check for sequence redundancy, candidate siRNA sequences can be compared with EST databases via, e.g., BLAST. Notably, comparing any 19 nt random sequence against human EST databases usually gives a number of hits with 15 or more matching nucleotides. However, not every mRNA is expressed in the cell-line targeted in a particular RNAi experiment. Therefore, several siRNAs with completely independent sequences against a target gene should be used Controlling siRNA Off-Target Effects and should result in similar phenotypes (see also off-target section 2.6 for a more detailed discussion).

2.3. siRNA Strand Selection

siRNAs are duplex RNAs and in principle both strands have the potential to be loaded onto Ago proteins as guide strands. Only the guide (also referred to as the antisense strand), which is complementary to the target mRNA, has activity towards this RNA. The opposing strand, the so-called passenger strand (often referred to as the sense strand), can only cause off-target activity (Fig. 1). Therefore it is absolutely essential to design siRNAs in a way that allows for a maximum asymmetric loading of the siRNA guide strand into the Ago protein. Two siRNA sequence parameters are known so far, which determine the guide strand selection.

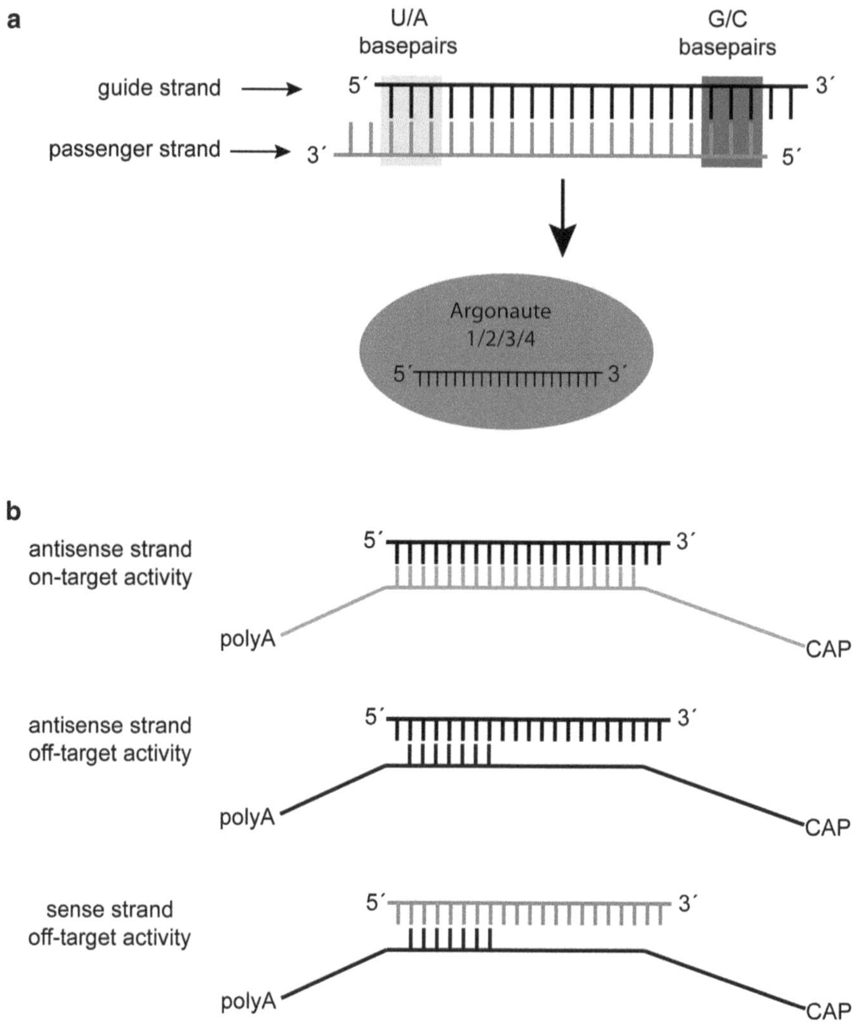

Fig. 1. Sequence requirements for siRNA strand selection and guide (antisense) and passenger (sense) strand-mediated on- and off-target effects. (**a**) The strand that is antisense to the target RNA (*black*) should be predominantly selected as the guide strand. This strand becomes stably incorporated into one of the Ago proteins, while the sense strand (*light brown*) is discarded as passenger strand. According to the asymmetry rule, the strand with a less stably paired 5′ end is selected (*light grey box*). The strand with the more stably paired 5′ end becomes the passenger strand (*dark grey box*). In addition, the 5′ end of the guide strand should start with a U or A and the 5′ end of the passenger strand with a G or C, according to Ago 5′ nucleotide specificity. (**b**) The guide strand (*black*) can mediate on-target activity by perfect base pairing with the respective target mRNA (*light grey*). The guide strand can also mediate off-target activity by base pairing to any nontarget mRNA with sequence complementarity to the so-called seed region (nt 2–8 of the guide strand). When selected, the passenger strand can only mediate off-target activity by base pairing to any nontarget mRNA with sequence complementarity to the seed region (nt 2–8).

2.3.1. The Asymmetry Rule

Both endogenous and synthetic siRNAs are generated from double-stranded molecules and selection of the correct strand has to be determined. Generally, strand selection follows the so-called asymmetry rule. The asymmetry rule predicts that the siRNA strand with the less stably paired 5′ end will be selected and incorporated as guide strand into Ago protein complexes (37, 38). The strand with the more stably paired 5′ end will be mainly discarded as the passenger strand (Fig. 1a). Consequently, when designing siRNAs the desired guide (antisense) strand should always have the less stably paired 5′ end.

2.3.2. 5′ Nucleotide Specificity

Structural studies on Ago proteins revealed that the so-called MID domain binds the 5′ end of the guide strand (39, 40). Because of its central localization in the Ago protein, this domain has been named MID domain. Crystal co-structures and K_D measurements of the MID domain in combination with all four nucleotides at the 5′ end revealed that uridine (U) binds with the highest affinity to the MID domain, adenosine (A) with a slightly reduced affinity, and cytosine (C) and guanosine (G) with more than tenfold less affinity (41). Therefore, siRNA guide (antisense) strands should ideally contain a U or A at the 5′ end. C and G should be avoided. For the passenger (sense) strand 5′ end, C and G should be selected in order to minimize strand incorporation (Fig. 1a). Based on our own unpublished data the nucleotide specificity is not only a tool to manipulate strand selection but siRNA strands with U or A at the 5′ end show also a higher absolute affinity for Ago proteins and therefore are more likely to be potent siRNAs (see Notes 2 and 3).

2.4. Chemical Modifications Influencing siRNA Activity and Selectivity

A number of modified nucleotides have been tested and described in siRNA design. These are mostly modifications of the 2′ OH group of the ribose. By the incorporation of chemically modified nucleotides into siRNAs, the on-target efficiency of the siRNAs can be increased (42–44). On the other hand, different types of siRNA off-target effects can be reduced by the use of chemically modified nucleotides: immunostimulatory effects (reviewed in (20)) as well as sequence-dependent miRNA-like off-target effects (26, 44, 45) and passenger strand incorporation (46, 47). In addition, chemical modifications can be used to improve the cellular delivery of siRNAs in living animals and are important tools to enhance the serum stability of siRNAs (48).

For more detailed information about chemical modifications of siRNAs, we would like to refer to the following review articles: (20, 49) (see Note 4).

2.5. Avoiding siRNA Off-Target Activity by Experimental Design

Because of the intrinsic property of siRNAs to induce off-target effects in several ways as described in the introduction, we would like to underline the importance of minimizing siRNA off-target activity. We will focus mainly on siRNA activity towards modulation of off-target transcript levels. For an excellent review about siRNA-mediated off-target effects in general and for unspecific immune responses in particular, the reader is referred to reference (20) and Table 2.

Table 2
Summary of relevant topics and respective strategies to minimize and control for siRNA off-target activity

Feature	Strategy	
Minimizing off-target activity in general	Use minimal siRNA conc	Titrate siRNAs, measure efficiency, and use them at the lowest possible concentration
Controlling general off-target effects	Control-siRNAs	Use control-siRNAs against unrelated targets or with "random" sequences. Design control siRNAs in accordance with rules for target siRNAs. Compare effects to untransfected cells and non-efficient on-target siRNAs. At least two (ideally three) different siRNA sequences should lead to the same phenotype
Sequence-dependent off-target effects	Comparison of siRNA effects	Comparison of knockdown efficiencies and resulting phenotypes of sequence-independent siRNAs against the same target
Rescue experiments	Target noncoding sequence	3′ UTRs, alternatively 5′ UTRs, can be targeted by siRNA; rescue can be performed by vector-encoded target genes lacking UTRs

2.5.1. siRNA Concentration

The most simple, efficient, and low-priced strategy to minimize siRNA off-target activity is to use as low siRNA concentrations as possible. miRNA-like off-target effects as well as siRNA off-target effects caused by competition of siRNA- and miRNA-loading onto Ago proteins have been shown to be concentration dependent (26, 29, 50). An essential prerequisite to use siRNAs at low concentrations is the highest possible efficiency in siRNA delivery and activity. The design criteria described before may lead to siRNAs with IC_{50}-values in the range of 25–500 pM for unmodified siRNAs in cell-lines which are easy to transfect like HeLa, Cos-7, HEK293-T, etc. (51, 52). In cell culture-based experiments using state-of-the-art transfection reagents (examples: RNAiMAX-Invitrogen, Hyperfect-Qiagen) those siRNAs should reach maximum on-target activity at concentrations below 10 nM. In order to estimate IC_{50}-values and the minimal concentration needed for maximum siRNA activity, we do recommend to roughly titrate siRNA concentrations (use, e.g., 50; 10; 2; 0.4; 0.08 nM final siRNA concentration and compare to an unrelated siRNA and determine mRNA levels, e.g., by real-time PCR). We would recommend not to use siRNAs for follow-up experiments that show IC_{50}-values above 500 pM in the case where cell-lines are used which can be transfected with high efficiency.

2.5.2. Control siRNAs

In each RNAi experiment unrelated siRNAs are essential controls that have to be included. These can be siRNAs against a completely unrelated target-mRNA, which do not show any obvious phenotype compared to untreated cells in the experimental context. Otherwise randomly chosen sequences or siRNAs against mRNAs that are not expressed in the biological system chosen can be used (for example siRNAs against Photinus luciferase or green fluorescent protein in cell-lines derived from human sources). As discussed under the topic "sequence redundancy," also randomly chosen siRNA sequences are likely to have relatively high sequence complementarity (>15 nt) when comparing them to, e.g., human EST databases. Therefore, ideally two or more siRNAs should be used as unrelated control siRNAs. As mentioned before, control siRNA-treated biological systems should ideally behave similar as untransfected or transfection reagent-treated systems. Especially at higher siRNA concentrations (>10–20 nM), control siRNAs will induce off-target effects as well. For example, when control siRNAs are used at higher concentrations, this leads frequently to reduced cell proliferation, compared to untransfected cells. In addition, control siRNAs should cause similar effects as very inefficient on-target siRNAs (e.g., <20–30% knockdown efficiency). Generally, control siRNAs have to be designed according to the same principles as on-target siRNAs. This is particularly important with respect to off-target effects caused by siRNA/miRNA competition for Ago proteins (28, 29).

2.5.3. siRNA Pools

It has been suggested that pooling, e.g., four independent siRNAs against a certain target mRNA might be a valid strategy in order to reduce off-target effects of individual siRNAs but keep the on-target efficiency high. This might not be useful in standard experiments with a limited number of samples. In our opinion, it is more reliable to test the individual siRNAs independently, applying the concentrations at which each individual siRNA would be used in the siRNA-pool. An siRNA pool might be used as an additional sample. The following example might argue against a pooling strategy: a typical distribution of IC_{50}-values of four siRNAs used as a pool in a knockdown experiment could be #1: 0.05 nM; #2: 0.2 nM; #3: 0.6 nM; and #4: 2 nM. In such a case, almost the complete knockdown could be mediated by oligo #1, nevertheless oligos #2, #3, and #4 are likely to contribute to off-target effects. An experimental setup using siRNA pools might be a valid strategy in order to keep sample numbers and costs at a reasonable limit such as for genome-wide siRNA screens. However, also in these screens, pools might cause more off-target effects than beneficial effects on on-target activity. In any case, individual "hits" should be validated by using several individual siRNA-oligos (see Note 5).

2.6. Controlling siRNA Off-Target Effects

Since the described strategies to reduce off-target effects provide no guarantee that siRNA off-target activity can be neglected, it is essential to control all effects observed in RNAi experiments for specificity. The most simple way to do so is to use several siRNAs (at least three) with completely independent sequences against the same target mRNA. In such a setup, it is important that siRNAs with similar knockdown efficiencies result in similar phenotypes. Knockdown efficiencies can be controlled most simply by real-time PCR-based quantification of mRNA levels. If an antibody for the protein corresponding to the knocked down mRNA is available, testing the level of the protein is the method of choice (see Note 6).

The "gold-standard" experiment to validate phenotypes observed in RNAi experiments is to perform rescue experiments, in which the recombinant equivalent of the knocked down gene is expressed from a vector that is co-transfected with the respective siRNA or transfected at a later time point. An essential point of this strategy is to design either the siRNA or the vector in a way that does not allow for targeting of the recombinant open reading frame by the siRNA. Therefore, the ectopic open-reading-frame sequence targeted by the siRNA can be mutated using translationally silent mutations. Usually, the more convenient way is to design siRNAs against the 5′ or 3′ UTR of the target mRNA. Consequently, endogenous mRNAs are targeted, but the transfected version lacking the corresponding UTRs is not affected (see Note 7).

3. Notes

1. "Unmodified" siRNA in this context is restricted to the 19 base pairing nucleotides of siRNA duplexes. Most siRNAs are synthesized with two nucleotide 3′ overhangs. The last or the last two nucleotides are often desoxy-thymidines in order to increase the resistance of the siRNAs against exonucleases and because of technical reasons in RNA synthesis. siRNAs modified with 3′ desoxy-thymidines are classified as "unmodified" in the context of this chapter.
2. Our unpublished results indicate that asymmetry rule and nucleotide specificity can be similarly strong parameters for guide strand selection. Each parameter can overrule the other one, always depending on the individual siRNA. Therefore both parameters should be ideally combined in order to design highly asymmetric siRNAs.
3. Depending on the experimental context, it might be advantageous to choose siRNAs against mRNA target sequences that are conserved across species like human and mouse or rat. Such siRNAs allow for usage of the same siRNA for example in human tissue culture experiments as well as later on in living animals such as mice or rats.
4. Chemical modifications of siRNAs have been extensively studied. Nevertheless, chemical modifications that enhance siRNA

on-target activity or reduce activity towards off-target mRNAs might give expected results for a number of individual siRNAs, but might fail to work in other siRNA sequence contexts. Therefore, each chemically modified siRNA should be validated individually. This is highly recommended for complex siRNA experiments, e.g., in living animals, for experiments that are often repeated, or when large amounts of siRNAs have to be used. For standard use of siRNAs in cell culture, especially of uncharacterized siRNAs, it is often cheaper and of less effort to start with unmodified siRNA sequences. Highly efficient siRNAs can later on be further improved by chemical modifications if needed.

5. siRNA efficiency will strongly vary among different cell-lines as it is mainly determined by two parameters: (1) the transfection efficiency, which might be low, e.g., in many primary cells, and (2) the Ago2 expression level, which is the most relevant Ago protein for on-target effects (53). Therefore, if one or both parameters are low, siRNA IC_{50}-values will be significantly higher and also IC_{50}-values of efficient siRNAs might be above 1 nM.
6. Determining knockdown efficiencies using detection of protein levels by western-blotting should be the method of choice, because there can be strong discrepancies of protein reduction compared to mRNA reduction. Proteins might have a half-life of days and also if the mRNA level is rapidly reduced after transfection of siRNAs, the protein levels might remain high for days.
7. Under specific conditions, rescue experiments might fail: first, the recombinant protein might be not fully functional. Second, the expression levels might be far away from the physiological expression level. If a rescue experiment works, it provides very specific evidence that a phenotype caused by the knockdown of a siRNA target mRNA is specific. In addition, rescue experiments with mutant variants of the respective gene provide an elegant way to analyze protein properties in relation to the observed phenotypes, for example, if the catalytic activity or other domains of the protein are needed for a particular function.

Acknowledgments

Our research is supported in part by the BMBF (NGFN+, FKZ PIM-01GS0804-5 to G.M.), the Bavarian Genome Research Network (BayGene to G.M.), the Deutsche Forschungsgemeinschaft (DFG), and Roche Kulmbach GmbH. S.P. received a fellowship from the Roche Postdoc Fellowship Program.

References

1. Fire A, Xu S, Montgomery M et al (1998) Potent and specific genetic interference by double-stranded RNA in *Caenorhabditis elegans*. Nature 391:806–811
2. Tuschl T (2001) RNA interference and small interfering RNAs. ChemBioChem 2:239–245
3. Meister G, Tuschl T (2004) Mechanisms of gene silencing by double-stranded RNA. Nature 431:343–349
4. Elbashir SM, Harborth J, Lendeckel W et al (2001) Duplexes of 21-nucleotide RNAs mediate RNA interference in mammalian cell culture. Nature 411:494–498
5. Dorsett Y, Tuschl T (2004) siRNAs: applications in functional genomics and potential as therapeutics. Nat Rev Drug Discov 3:318–329
6. Kim DH, Rossi JJ (2007) Strategies for silencing human disease using RNA interference. Nat Rev Genet 8:173–184
7. Kim VN, Han J, Siomi MC (2009) Biogenesis of small RNAs in animals. Nat Rev Mol Cell Biol 10:126–139
8. Siomi H, Siomi MC (2009) On the road to reading the RNA-interference code. Nature 457:396–404
9. Filipowicz W, Jaskiewicz L, Kolb FA et al (2005) Post-transcriptional gene silencing by siRNAs and miRNAs. Curr Opin Struct Biol 15:331–341
10. Hutvagner G, Simard MJ (2008) Argonaute proteins: key players in RNA silencing. Nat Rev Mol Cell Biol 9:22–32
11. Ender C, Meister G (2010) Argonaute proteins at a glance. J Cell Sci 123:1819–1823
12. Peters L, Meister G (2007) Argonaute proteins: mediators of RNA silencing. Mol Cell 26:611–623
13. Tolia NH, Joshua-Tor L (2007) Slicer and the argonautes. Nat Chem Biol 3:36–43
14. Jinek M, Doudna JA (2009) A three-dimensional view of the molecular machinery of RNA interference. Nature 457:405–412
15. Carmell MA, Xuan Z, Zhang MQ et al (2002) The argonaute family: tentacles that reach into RNAi, developmental control, stem cell maintenance, and tumorigenesis. Genes Dev 16:2733–2742
16. Meister G, Landthaler M, Patkaniowska A et al (2004) Human Argonaute2 mediates RNA cleavage targeted by miRNAs and siRNAs. Mol Cell 15:185–197
17. Liu J, Carmell MA, Rivas FV et al (2004) Argonaute2 is the catalytic engine of mammalian RNAi. Science 305:1437–1441
18. Wu L, Fan J, Belasco JG (2008) Importance of translation and nonnucleolytic ago proteins for on-target RNA interference. Curr Biol 18:1327–1332
19. Doench JG, Petersen CP, Sharp PA (2003) siRNAs can function as miRNAs. Genes Dev 17:438–442
20. Jackson AL, Linsley PS (2010) Recognizing and avoiding siRNA off-target effects for target identification and therapeutic application. Nat Rev Drug Discov 9:57–67
21. Semizarov D, Frost L, Sarthy A et al (2003) Specificity of short interfering RNA determined through gene expression signatures. Proc Natl Acad Sci USA 100:6347–6352
22. Persengiev SP, Zhu X, Green MR (2004) Nonspecific, concentration-dependent stimulation and repression of mammalian gene expression by small interfering RNAs (siRNAs). RNA 10:12–18
23. Jackson AL, Bartz SR, Schelter J et al (2003) Expression profiling reveals off-target gene regulation by RNAi. Nat Biotechnol 21: 635–637
24. Scacheri PC, Rozenblatt-Rosen O, Caplen NJ et al (2004) Short interfering RNAs can induce unexpected and divergent changes in the levels of untargeted proteins in mammalian cells. Proc Natl Acad Sci USA 101:1892–1897
25. Birmingham A, Anderson EM, Reynolds A et al (2006) 3′ UTR seed matches, but not overall identity, are associated with RNAi off-targets. Nat Methods 3:199–204
26. Jackson AL, Burchard J, Schelter J et al (2006) Widespread siRNA "off-target" transcript silencing mediated by seed region sequence complementarity. RNA 12:1179–1187
27. Lin X, Ruan X, Anderson MG et al (2005) siRNA-mediated off-target gene silencing triggered by a 7 nt complementation. Nucleic Acids Res 33:4527–4535
28. Grimm D, Streetz KL, Jopling CL et al (2006) Fatality in mice due to oversaturation of cellular microRNA/short hairpin RNA pathways. Nature 441:537–541
29. Khan AA, Betel D, Miller ML et al (2009) Transfection of small RNAs globally perturbs gene regulation by endogenous microRNAs. Nat Biotechnol 27:549–555
30. Fedorov Y, Anderson EM, Birmingham A et al (2006) Off-target effects by siRNA can induce toxic phenotype. RNA 12:1188–1196
31. Hornung V, Guenthner-Biller M, Bourquin C et al (2005) Sequence-specific potent induction of IFN-alpha by short interfering RNA in

plasmacytoid dendritic cells through TLR7. Nat Med 11:263–270
32. Judge AD, Sood V, Shaw JR et al (2005) Sequence-dependent stimulation of the mammalian innate immune response by synthetic siRNA. Nat Biotechnol 23:457–462
33. Chaudhary A, Srivastava S, Garg S (2011) Development of a software tool and criteria evaluation for efficient design of small interfering RNA. Biochem Biophys Res Commun 404:313–320
34. Wang X, Varma RK, Beauchamp L et al (2009) Selection of hyperfunctional siRNAs with improved potency and specificity. Nucleic Acids Res 37:e152
35. Tafer H, Ameres SL, Obernosterer G et al (2008) The impact of target site accessibility on the design of effective siRNAs. Nat Biotechnol 26:578–583
36. Ui-Tei K, Naito Y, Takahashi F et al (2004) Guidelines for the selection of highly effective siRNA sequences for mammalian and chick RNA interference. Nucleic Acids Res 32: 936–948
37. Khvorova A, Reynolds A, Jayasena SD (2003) Functional siRNAs and miRNAs exhibit strand bias. Cell 115:209–216
38. Schwarz DS, Hutvágner G, Du T, Xu Z et al (2003) Asymmetry in the assembly of the RNAi enzyme complex. Cell 115:199–208
39. Parker JS, Roe SM, Barford D (2005) Structural insights into mRNA recognition from a PIWI domain-siRNA guide complex. Nature 434:663–666
40. Ma JB, Yuan YR, Meister G et al (2005) Structural basis for 5'-end-specific recognition of guide RNA by the A. fulgidus Piwi protein. Nature 434:666–670
41. Frank F, Sonenberg N, Nagar B (2010) Structural basis for 5'-nucleotide base-specific recognition of guide RNA by human AGO2. Nature 465:818–822
42. Allerson CR, Sioufi N, Jarres R et al (2005) Fully 2'-modified oligonucleotide duplexes with improved in vitro potency and stability compared to unmodified small interfering RNA. J Med Chem 48:901–904
43. Addepalli H, Meena, Peng CG et al (2010) Modulation of thermal stability can enhance the potency of siRNA. Nucleic Acids Res 38:7320–7331
44. Petri S, Dueck A, Lehmann G et al (2011) Increased siRNA duplex stability correlates with reduced off-target and elevated on-target effects. RNA 17:737–749
45. Bramsen JB, Pakula MM, Hansen TB et al (2010) A screen of chemical modifications identifies position-specific modification by UNA to most potently reduce siRNA off-target effects. Nucleic Acids Res 38:5761–5773
46. Bramsen JB, Laursen MB, Damgaard CK et al (2007) Improved silencing properties using small internally segmented interfering RNAs. Nucleic Acids Res 35:5886–5897
47. Chen PY, Weinmann L, Gaidatzis D et al (2008) Strand-specific 5'-O-methylation of siRNA duplexes controls guide strand selection and targeting specificity. RNA 14:263–274
48. Soutschek J, Akinc A, Bramlage B et al (2004) Therapeutic silencing of an endogenous gene by systemic administration of modified siRNAs. Nature 432:173–178
49. Bramsen JB, Kjems J (2011) Chemical modification of small interfering RNA. Methods Mol Biol 721:77–103
50. Caffrey DR, Zhao J, Song Z et al (2011) siRNA off-target effects can be reduced at concentrations that match their individual potency. PLoS One 6:e21503
51. Grunweller A, Wyszko E, Bieber B et al (2003) Comparison of different antisense strategies in mammalian cells using locked nucleic acids, 2'-O-methyl RNA, phosphorothioates and small interfering RNA. Nucleic Acids Res 31:3185–3193
52. Vickers TA, Lima WF, Nichols JG et al (2007) Reduced levels of Ago2 expression result in increased siRNA competition in mammalian cells. Nucleic Acids Res 35:6598–6610
53. Diederichs S, Jung S, Rothenberg SM et al (2008) Coexpression of Argonaute-2 enhances RNA interference toward perfect match binding sites. Proc Natl Acad Sci USA 105: 9284–9289

Chapter 5

Western Blot Evaluation of siRNA Delivery by pH-Responsive Peptides

Wanling Liang, A. James Mason, and Jenny K.W. Lam

Abstract

Gene silencing, via RNA interference (RNAi) technologies using small interfering RNA (siRNA), has been developed as an important tool for target identification and validation in drug discovery and has huge therapeutic potential. However, effective delivery into cells presents a major challenge to the use of siRNA. pH-responsive cell-penetrating peptides have attracted considerable attention in recent years as delivery vectors due to their ability to transport their cargos across the biological membrane and/or to promote endosomal escape and prevent lysosomal degradation. To evaluate the in vitro transfection efficiency of the pH-responsive peptide-based siRNA delivery system, the western blotting technique is commonly employed. This method offers a simple, efficient and economical way to study the gene silencing effect of the siRNA by analysing the protein of interest in a sample with minimum equipment requirement. This chapter provides a description of siRNA delivery and analysis by western blotting protocols for qualitatively and quantitatively assessing gene silencing efficiency and selectivity.

Key words: Transfection, siRNA, GAPDH, Western blotting, Densitometry, pH-responsive peptide, Chemiluminescence

1. Introduction

Substantial progress has been made in RNA interference (RNAi) technology in the last decade and small interfering RNA (siRNA) has become an important agent for target identification and validation in drug discovery (1). Tremendous efforts have also been invested in recent years to develop siRNA-based therapeutics (2). Effective and nontoxic delivery of siRNA into cells presents one of the major obstacles that hinder the use of siRNA in the drug discovery process and clinical applications (3). Being a negatively charged hydrophilic macromolecule that is highly susceptible to nuclease degradation, siRNA is incapable of crossing the biological membrane on its own to exert the gene silencing effect in the cytosol.

Jürgen Moll and Riccardo Colombo (eds.), *Target Identification and Validation in Drug Discovery: Methods and Protocols*, Methods in Molecular Biology, vol. 986, DOI 10.1007/978-1-62703-311-4_5,

A delivery vector is therefore required to protect the siRNA from enzymatic degradation, facilitate cellular uptake to the target cells, and release the siRNA at the site of action. Although viruses are efficient nucleic acid delivery vectors in nature, safety concerns regarding their immunogenicity and toxicity have led to the development of safer nonviral vector alternatives. In recent years, cationic cell-penetrating peptides (CPPs) have attracted a lot of attention as siRNA delivery agents due to their ability to interact and cross cell membranes while binding to the negatively charged siRNA. CPPs either are covalently conjugated to siRNA through disulfide bond formation or bind to siRNA through electrostatic interaction to form complexes in a non-covalent manner. The former approach may have a negative effect on cellular delivery as the biological activity of the peptides may be altered during the modification (4), whereas the latter approach offers a simpler way to achieve siRNA delivery and the peptide/siRNA complexes can be prepared easily. In addition, CPPs with pH-responsive components can further enhance the intracellular delivery efficiency by promoting endosomal escape, thereby preventing the siRNA from degradation by enzymes in the late endosomes/lysosomes (5).

A good siRNA delivery system is crucial during the investigation of the gene silencing effect of siRNA whether it is used for target identification and validation or RNAi therapeutics. To evaluate siRNA transfection efficiency of a delivery system and the level of gene silencing in vitro, the western blotting technique is commonly employed as it offers a relatively simple, efficient, and economical way to measure the expression of proteins of interest. Western blotting is a powerful tool to analyze protein samples from cells or tissues. Protein samples undergo electrophoresis and are transferred onto a nitrocellulose membrane. The transferred protein is probed with a specific primary antibody followed by an enzyme-conjugated secondary antibody. The activity of an enzyme such as horseradish peroxidase (HRP) is used for signal generation. The membrane is finally incubated with appropriate enzyme substrate to produce a recordable signal (Fig. 1). Enhanced chemiluminescence (ECL) is one of the commonly employed methods that provide precise detection of proteins from western blots. In this method, HRP catalyzes the oxidation of luminol into a reagent that emits light when it decays. Since HRP is complexed with the protein of interest on the membrane through antibodies, the location as well as the amount of light emitted is directly correlated to the location and amount of protein on the membrane. The light can be detected by photographic film or captured as digital image by CCD camera. The density of protein bands on the western blot image can be measured using appropriate gel image analysis software, and the gene silencing effect can be evaluated by comparing the relative protein expression of the sample with the untreated or negative control.

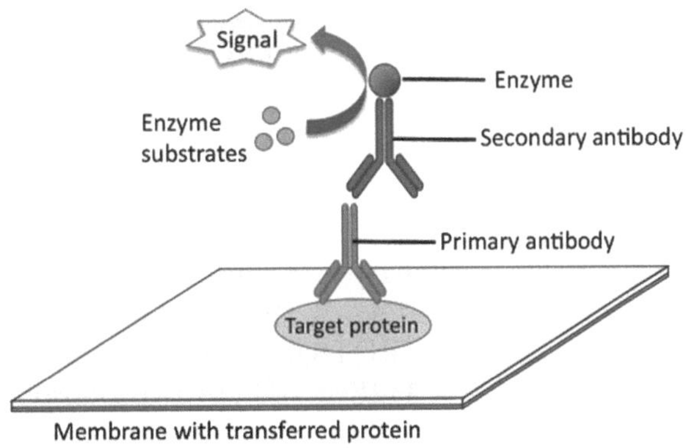

Fig. 1. Schematic diagram to illustrate the principle of western blotting in generating signals from target proteins through antibody probing.

In this chapter, a pH-responsive histidine-rich amphipathic peptide, LAH4-L1 (6–8), is employed as siRNA delivery vector to demonstrate the use of the western blotting technique to evaluate the in vitro siRNA-mediated gene silencing efficiency of a delivery system. We describe the entire procedure from the preparation of peptide/siRNA complexes, through the transfection of the cells, the techniques of western blotting, densitometry analysis of the blot to the calculation of gene silencing efficiency. siRNA targeting glyceraldehyde 3-phosphate dehydrogenase (GAPDH), a housekeeping gene that is constitutively expressed at high levels in most cells and tissues, is used as positive control here. Non-targeting siRNA (a negative control) is also used to examine if the delivery agent has caused any effect to the cells. (Control siRNAs are often selected on the basis that they produce limited off-target effects) (9).

2. Materials

2.1. Peptide and siRNA

1. pH-responsive peptide: LAH4-L1 (available from, e.g., EZBiolab Inc., USA; Pepceuticals Ltd, UK; ChinaPeptides Co., China). Resuspend the peptide in filtered ultrapure water to give a stock solution with concentration of 2 mg/ml. Store in aliquots at −20°C.
2. siRNAs for western blotting experiment: Silencer® Select GAPDH siRNA for human, mouse, and rat; Silencer® Select negative control siRNA for human, mouse, and rat (Ambion). Resuspend the siRNA in nuclease-free water to give a stock solution with concentration of 40 μM. Store in aliquots at −20°C.

2.2. Cell Culture

1. A549 cells (Human lung adenocarcinoma epithelial cell line). Subculture once a week, maintain in 5% CO_2 at 37°C.

2. Dulbecco's modified Eagle's medium (DMEM) supplemented with 10% FBS, 100 units/ml penicillin, 100 μg/ml streptomycin, and 0.25 μg/ml amphotericin B.
3. Transfection medium: Opti-MEM® I Reduced Serum Medium (Invitrogen).

2.3. Sample Preparation

1. Cell lysis buffer stock solution (10×): 1.21 g Tris base (pH 7.4), 4.39 g NaCl, 0.19 g ethylenediaminetetraacetic acid (EDTA), 0.19 g ethylene glycol tetraacetic acid (EGTA), 0.28 g sodium pyrophosphate, 0.11 g β-glycerolphosphate, 0.09 g sodium orthovanadate, and 5 ml Triton X-100. Make up to 50 ml with distilled water. Store at 4°C.
2. Cell lysis buffer (1×): Dilute 100 μl of cell lysis buffer stock solution (10×) to 1 ml with distilled water. Add 1 μl leupeptin (5 mg/ml in distilled water), 1 μl aprotinin (1 mg/ml in distilled water), 1 μl pepstatin A (1 mg/ml in EtOH), 10 μl dithiothreitol (15.4 mg/ml in distilled water), and 10 μl PMSF (17.4 mg/ml in isopropyl alcohol).
3. Sodium dodecyl sulfate (SDS) sample buffer stock solution: Dissolve 1.51 g Tris base to 15 ml of distilled water, and then add 25 ml of 50% glycerol into the mixture. Stir to dissolve and then adjust pH to 6.8 with HCl. Dissolve 5 g of SDS in this buffer and make up to 50 ml with distilled water. Dissolve 0.01 g of bromophenol blue and stir overnight.
4. Bradford protein assay reagent.
5. Bovine serum albumin.

2.4. SDS Polyacrylamide Gel Electrophoresis

1. 1.30% Acrylamide:bis-acrylamide solution 37.5:1.
2. Stacking gel buffer: 0.5 M Tris–HCl, 0.4% SDS. Weigh 6.06 g Tris base, 0.4 g SDS, make up to 100 ml with distilled water, and adjust pH to 6.8 with HCl. Filter and store at 4°C.
3. Resolving gel buffer: 1.5 M Tris–HCl, 0.4% SDS. Weigh 18.16 g Tris base, 0.4 g SDS, make up to 100 ml with distilled water, and adjust pH to 8.8 with HCl. Filter and store at 4°C.
4. Water-saturated butanol.
5. SDS-PAGE running buffer: Weigh 30.3 g Tris base, 144.0 g glycine, 10.0 g SDS, make up to 1,000 ml with distilled water to give a 10× running buffer stock solution. Dilute the running buffer stock solution using distilled water to 1× before use.
6. 10% Ammonium persulfate (APS).
7. *N, N, N, N'*-tetramethyl-ethylenediamine (TEMED). Store at 4°C.
8. Pageruler™ prestained protein ladder.
9. Electrophoretic blotting system.
10. Electrophoresis power supply (EPS).

2.5. Electrophoretic Transfer

1. Transfer membrane, nitrocellulose, 30 cm × 3 m roll, pore size 0.2 μm (Pall Corporation).
2. SDS-PAGE transfer buffer stock solution (10×): Weigh 30.30 g Tris base, 144.0 g glycine. Make up to 1,000 ml with distilled water.
3. SDS-PAGE transfer buffer (1×): Dilute 100 ml of transfer buffer stock solution (10×) to 800 ml with distilled water and then add 200 ml of methanol.
4. Tris-buffered saline (TBS): Weigh 24.2 g Tris base (0.2 M), 87.8 g NaCl (1.5 M), adjust pH to 7.4, and make up to 1,000 ml with distilled water to give a 10× TBS stock solution. Dilute the stock solution to 1× with distilled water before use.
5. TBST: TBS containing 0.05% Tween-20.
6. Blocking solution: 5% nonfat dry milk in TBS.

2.6. Antibodies and Conjugates

1. Primary antibody: Monoclonal anti-GAPDH antibody and monoclonal anti-β-actin antibody from mouse (see Note 1).
2. Secondary antibody: Amersham ECL™ anti-Mouse IgG, HRP-linked whole antibody.
3. Amersham ECL™ western blotting detection reagents.

2.7. Densitometry Analysis

1. Image J 1.45 software (NIH) (see Note 2).

3. Methods

3.1. Preparation of Peptide/siRNA Complexes for Transfection

1. Dilute the stock solution of Silencer®Select GAPDH siRNA or Silencer®Select negative control siRNA to a concentration of 100 pmol/500 μl in Opti-MEM® I in a microcentrifuge tube.
2. Dilute the peptide stock solution in Opti-MEM® I Reduced Serum Medium to 500 μl in a microcentrifuge tube so that the weight ratio of peptide to siRNA (prepared in step 1) is 10 to 1 (see Notes 3 and 4).
3. Add the peptide solution (500 μl) to the siRNA solution (500 μl) to give a final volume of 1 ml.
4. Mix the suspension by gently inversing the microcentrifuge tube several times. The mixture is incubated at room temperature for 30 min to allow the formation of stable peptide/siRNA complexes.

3.2. siRNA Transfection and Protein Preparation for Western Blotting

1. Seed A549 cells in 6-well plates with a seeding density of 1×10^5 cells per well (see Note 5).
2. After 24 h, the cells are washed twice with warm PBS. Add the peptide/siRNA complexes prepared in Opti-MEM® I Reduced

Serum Medium (as prepared in Subheading 3.1) to the cells with a final volume of 1 ml per well (each well contains 100 pmol siRNA) (see Note 6).

3. The cells are incubated for 4 h at 37°C, and are washed with warm PBS twice. Fresh DMEM supplemented with 10% FBS are added to the cells.
4. After 72 h of incubation, the cells are washed with cold PBS twice (see Note 7).
5. Scrape cells in 1 ml PBS (on ice). Transfer the cell suspension into a microcentrifuge tube which is centrifuged at 10,000 × *g* for 3 min at 4°C. Remove all the supernatant and add 100 μl cell lysis buffer to the cell pellet.
6. Briefly sonicate the sample and centrifuge at 10,000 × *g* for 3 min at 4°C to remove the water-insoluble cell debris.
7. Protein concentration of the sample is quantitated by the Bradford protein assay with bovine serum albumin used as standard.
8. 20 μg of protein from the cell lysate is transferred to a microcentrifuge tube and make up to 20 μl with distilled water.
9. Add 5 μl of 2-mercaptoethanol to 95 μl of SDS sample buffer stock solution to give a sample buffer mixture. (SDS may precipitate at room temperature. Warm up the SDS sample buffer stock solution before use) (see Note 8).
10. Add the protein sample to the sample buffer mixture at 4:1 volume ratio. Heat the samples at 95°C for 10 min. The samples are now ready for SDS-PAGE.

3.3. SDS Polyacrylamide Gel Electrophoresis

1. Set up the gel apparatus. Clean the glass plates with 70% ethanol and assemble them in a casting stand.
2. Prepare resolving gel mixture and stacking gel mixture as showed in Table 1.
3. Pour the resolving gel mixture in between the glass plates and add sufficient amount of water-saturated butanol on top to level off the gel. Wait for approximately 45 min.
4. After the resolving gel is ready, rinse with distilled water to remove butanol.
5. Add the stacking gel mixture on top and insert the gel comb. Wait for another 30 min. After the stacking gel is set, remove the comb and put the gel with the stand into the running tank.
6. Fill the running tank with cold SDS-PAGE running buffer. Carefully load the samples and the protein ladder into the wells (see Note 9).
7. Run the samples at 120 V, 500 mA for approximately 50 min. When the dye front reaches the end of the gel, switch off the

Table 1
The components of the resolving gel and stacking gel for separation of proteins with different molecular weights

	Resolving gel			Stacking gel
	7.5% Proteins >100 kDa	10% Proteins 30–100 kDa	12.5% Proteins 10–20 kDa	
30% Acrylamide/bis (29:1)	2.5 ml	3.4 ml	4.0 ml	1.3 ml
Resolving/stacking gel buffer	2.5 ml	2.5 ml	2.5 ml	2.5 ml
Distilled H_2O	5.0 ml	4.1 ml	3.5 ml	6.0 ml
10% APS (fresh)	50 μl	50 μl	50 μl	30 μl
TEMED	5 μl	5 μl	5 μl	10 μl

power. Separate the gel plates by a spatula and remove the stacking gel.

8. Wash the gel (still supported by the bottom glass plate) carefully with distilled water.

3.4. Electrophoretic Transfer of Protein from Polyacrylamide Gels to Nitrocellulose Membrane

1. Cut a piece of nitrocellulose membrane to the size of gel.
2. Place nitrocellulose membrane, filter paper, and fibrous pad in SDS-PAGE transfer buffer for at least 1 h before use. All fibrous pad and filter papers should be totally wet before use.
3. Place the gel on a filter paper which has been soaked in transfer buffer.
4. Assemble the blotting sandwich cassette as follows (Fig. 2):

   Support grid → fibrous pad → filter paper → gel → nitrocellulose membrane → filter paper → fibrous pad → support grid (see Note 10).
5. Close the clamp and place the cassette in gel blotting apparatus such that the membrane is close to the anode and the gel is close to the cathode.
6. Fill the tank 3/4 full with SDS-PAGE transfer buffer, put a stir bar into it, and install the frozen ice pad. Top up with buffer, turn on magnetic stirrer, and close the lid.
7. Run the blot at maximum voltage with a constant current of 300 mA for 90 min. High-molecular-weight protein may need longer transfer time.
8. After transfer is complete, shut off power and remove cassette holder.

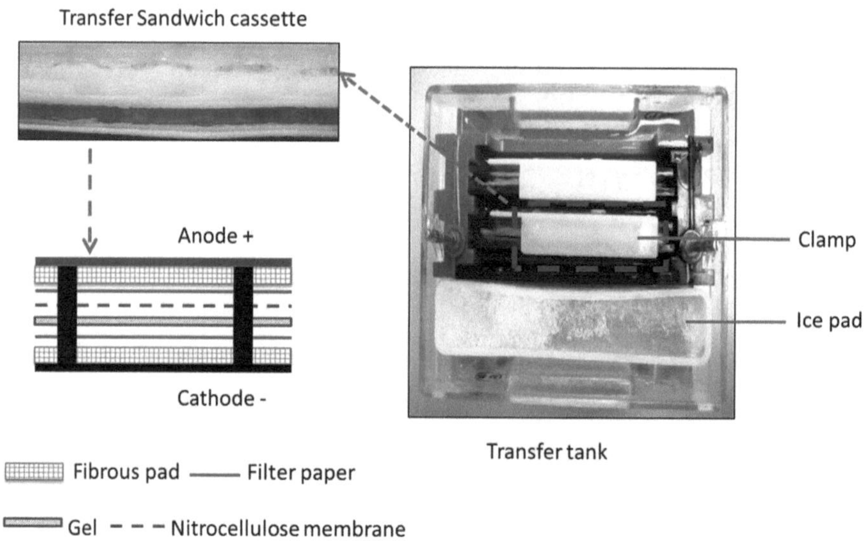

Fig. 2. Illustration of the setup of the sandwich cassette in the transfer tank observed from the top for the western blotting experiment.

9. Mark the membrane with a ballpoint pen for sample identification purpose. Do not let the membrane dry.
10. Minimize the background signal by blocking the membrane in 5% nonfat dry milk blocking solution for 2 h with shaking at room temperature (see Note 11).
11. Remove the blocking solution and wash the membrane thrice with TBST.

3.5. Antibody Probing (Anti-GAPDH)

1. The membrane is incubated with primary antibody (anti-GAPDH) overnight at 4°C on shaking.
2. Quickly rinse the membrane with TBST followed by washing thrice with TBST for 10 min each.
3. Incubate the membrane with HRP-conjugated secondary antibody for 2 h at room temperature on shaking.
4. Rinse the membrane two to three times followed by 10-min wash twice with TBST and 10-min wash twice with TBS (large volume). Remove buffer.

3.6. ECL Detection

1. Add ECL reagent mixture on the membrane for exactly 1 min.
2. Blot the membrane dry with soft tissue papers. Wrap the membrane in plastic wrap.
3. Expose the wrapped membrane to the film in ECL cassette and develop the film in darkroom: 1 min in developer, 1 min in water, 1 min in fixer, and 1 min in water. Repeat the ECL exposure if necessary (see Notes 12 and 13).

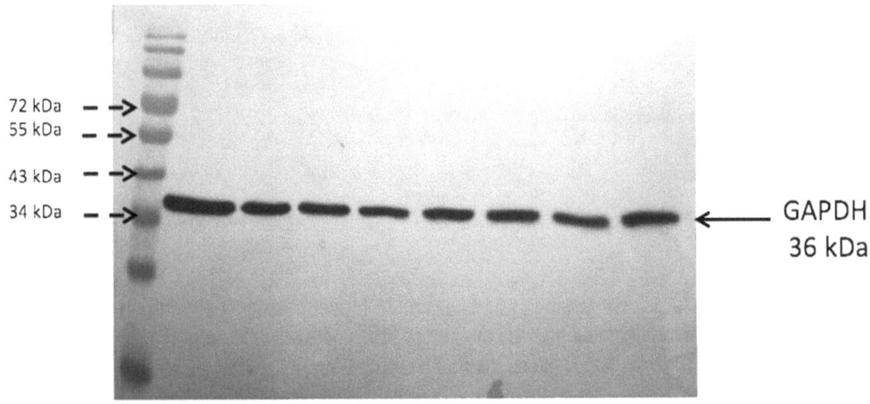

Fig. 3. Identification of the target protein (GAPDH) by comparing the molecular weight of the band on the film with the protein ladder with known molecular weight on the nitrocellulose membrane.

4. After developing the film, overlay the film and nitrocellulose membrane to confirm the molecular weight of the band with the aid of a protein ladder (Fig. 3). Identify the target band according to the molecular weight of the protein.
5. If a gel documentation imaging system with CCD camera installed is available (which can be expensive), the light output of the membrane can be captured automatically without the need of developing the film in darkroom (see Note 14).

3.7. Probing with Another Primary Antibody (Anti-β-Actin)

1. To detect another protein expression, the membrane is rinsed twice with TBST and incubated with another primary antibody (anti-β-actin) for 1 h at room temperature on shaking.
2. Quickly rinse the membrane with TBST followed by 10-min wash thrice with TBST.
3. Incubate the membrane with HRP-conjugated secondary antibody for 1 h at room temperature on shaking.
4. Rinse the membrane twice followed by 10-min wash twice with TBST and another 10-min wash twice with TBS (large volume). Remove buffer.
5. Repeat Subheading 3.6 to detect the expression of β-actin.
6. Since the molecular weight of GAPDH is around 36 kDa and β-actin is about 42 kDa, their bands appear in distinct areas of the membrane and are unlikely to interfere with each other. However if the two or more proteins to be detected on the same membrane have a similar molecular weight, stripping and re-probing are necessary (see Note 15).

3.8. Western Blot Image

1. From the western blot image (Fig. 4), the GAPDH protein band of the untreated sample and the negative control sample (in which the Silencer® Select negative control siRNA is used)

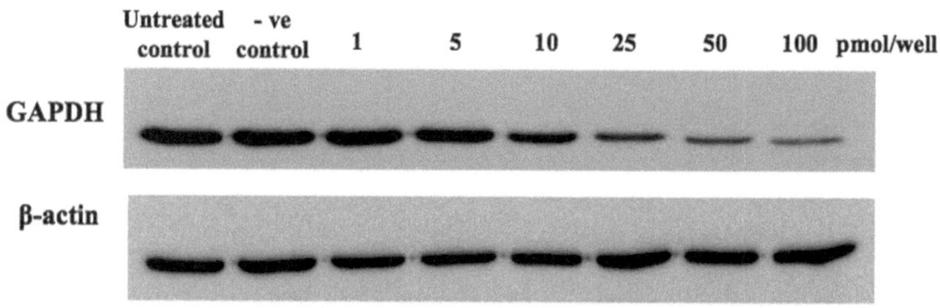

Fig. 4. Western blot image of GAPDH expression in A549 cells at 72 h post transfection with LAH4-L1/siRNA complexes. The intensity of GAPDH protein band decreases as the amount of siRNA per well increases. For the negative control (–ve), 100 pmol siRNA/well was used. β-Actin was used as internal control.

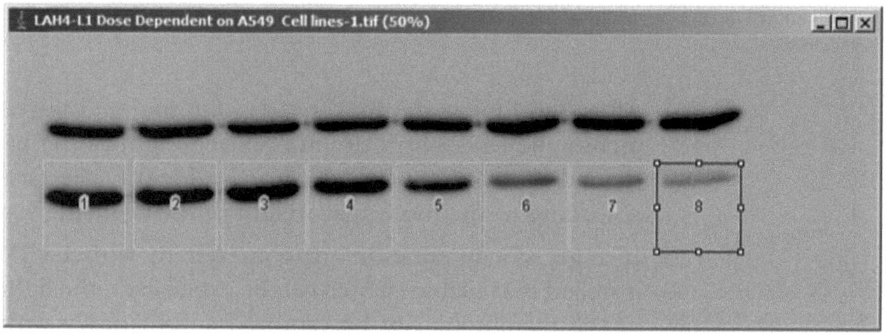

Fig. 5. Rectangles are drawn around all the individual bands to be analyzed. All the *rectangles* encompass the entire band patterns and all *rectangles* are of the same dimension.

should give the highest intensity, as the GAPDH protein is not silenced. For LAH4-L1/siRNA complex-treated samples (Silencer® Select GAPDH siRNA is used), the intensity of the GAPDH protein band should decrease as the amount of siRNA per well increases, indicating the increase of gene silencing efficiency.

2. β-actin protein (not targeted by siRNA) is used as the internal control to ensure that the amount of total protein added per well is the same; therefore the band intensity of β-actin should be about the same among all samples. To calculate the gene silencing efficiency (as shown in Subheading 3.9), the GAPDH expression for each sample is normalized with the β-actin expression of the corresponding sample.

3.9. Densitometry and Protein Band Analysis

1. Open the Image J software and the western blot image (scanned from the film or captured by a camera) (see Notes 16 and 17).
2. Draw a rectangle around each single band. The size of rectangle should be large enough to encompass the entire band. The

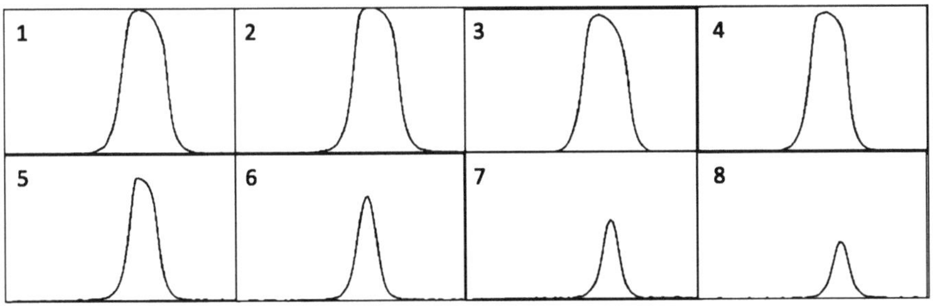

Fig. 6. The profile plots of the eight samples. The peak of each plot represents the relative density of the band in each rectangle. The larger the area under the curve (AUC), the higher the intensity of the band.

dimension of the rectangle should be kept the same for all other bands to be analyzed in the same experiment (Fig. 5).

3. Create a profile plot which represents the relative density of the contents of the rectangle over each band (Fig. 6). The peak corresponds to the dark bands in the rectangle. The area under the curve (AUC) is proportional to the intensity of the band.
4. To calculate the gene silencing efficiency of the peptide/siRNA complexes, the expression of the protein of interest (in this case: GAPDH) has to be first normalized with the expression of β-actin (protein for internal control) of the same sample.

The normalized intensity of GAPDH of sample treated with siRNA targeting GAPDH is compared with the normalized intensity of GAPDH of sample treated with negative control siRNA, Eq. 1. The percentage is the remaining expression of GAPDH of the sample after transfection with peptide/siRNA complexes. The data is represented in a bar chart (Fig. 7). Experiments are repeated independently at least three times for detection of significant differences in silencing activity.

Equation 1. Formula to calculate gene silencing efficiency of the delivery system. The percentage represents the remaining GAPDH protein expression after transfection:

$$\frac{\dfrac{\text{Intensity of GAPDH}_{(\text{sample treated with GAPDH siRNA})}}{\text{Intensity of }\beta\text{-actin}_{(\text{sample treated with GAPDH siRNA})}}}{\dfrac{\text{Intensity of GAPDH}_{(\text{sample treated with negative control siRNA})}}{\text{Intensity of }\beta\text{-actin}_{(\text{sample treated with negative control siRNA})}}} \times 100\%. \quad (1)$$

4. Notes

1. Antibody dilution is established empirically. The dilution ratio is usually ranged from 1:500 to 1:10,000. The signal intensity

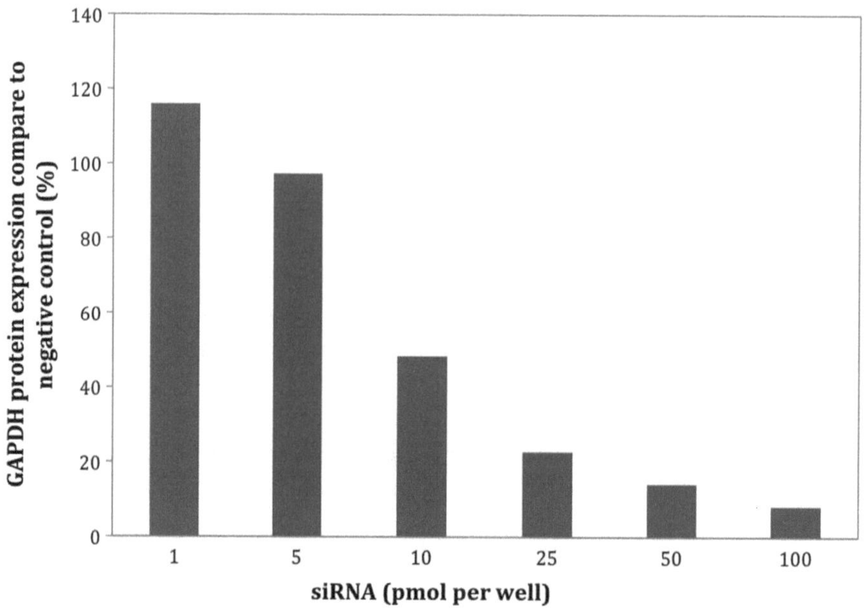

Fig. 7. Gene silencing efficiency in A549 cells analyzed by western blotting. Cells were transfected with LAH4-L1/siRNA complexes containing GAPDH siRNA of increasing amounts (from 1 to 100 pmol per well). Protein analysis was carried out at 72 h post transfection. The level of GAPDH protein expression of each sample was compared with the cells transfected with negative control siRNA.

should increase with increasing antibody concentration. Protein expression levels vary in different cell lines; therefore the dilution ratio of antibody should be optimized in accordance with each experiment/cell line to get satisfactory results. The final dilution also depends on the detection system used. Initially a lower dilution ratio (e.g., 1:500) can be tried. Also, monoclonal antibodies are recommended over polyclonal antibodies as the former are more specific and generally produce a cleaner background for protein analysis. Ensure that the type of antibodies can target the species of the cells used for transfection.

2. Image J is a public programme designed by National Institute of Health (NIH), USA, for image processing and analysis. It can be downloaded for free from NIH Web site (http://www.rsb.info.nih.gov/ij/). There are many other software that can be used to analyze and quantitate protein expression on western blot including GelQuant, ImageQuant™, and GelEval™, some of which are supplied with gel documentation imaging system.

3. The transfection efficiency and stability of some peptide/siRNA complexes may be affected by the presence of serum. It is suggested to prepare the complexes in the absence of serum initially. Serum-supplemented culture medium can be used with complexes that show good transfection efficiency in the absence of serum.

4. The optimal weight ratio of LAH4-L1 to siRNA for transfection was determined previously as 10:1. A range of different ratios should be tried for other peptides if the optimal ratio has not been determined.
5. Depending on the cell division rate and the adhesive property, different cell types may have different optimal cell seeding density for siRNA transfection. It is suggested to plate cells at densities ranging from 1×10^4 to 1×10^6 cells per well in a 6-well plate to find the optimal condition initially. If the cell density is too low, there may not be enough protein collected to perform western blotting analysis. Also, microplates with more wells than a 6-well plate are not recommended as there may not be enough protein collected for proper analysis.
6. To optimize the amount of siRNA for in vitro transfection, prepare the peptide/siRNA complexes with various concentrations of siRNA, ranging from 1 pmol to 100 pmol per ml. The optimal amount of siRNA required for efficient gene silencing effects varies between the cell types used.
7. Depending on the cell division rate, optimal gene silencing effects may be observed at different time points post transfection. It is suggested to perform western blotting at 48–96 h post transfection to determine the optimal time point.
8. 2-Mercaptoethanol is added to reduce the disulfide bond to disrupt the tertiary and quaternary structure of some proteins for analysis. It has a strong pungent smell. Handle sample containing 2-mercaptoethanol in a fume hood.
9. Different proteins have dissimilar expression in differing cell types. To get a good signal for analysis, the total amount of protein loaded into each well should be optimized for different cell types.
10. It is important to avoid air bubbles when assembling the blotting sandwich cassette in order to allow proteins to be transferred properly to the nitrocellulose membrane.
11. The purpose of blocking the membrane is to minimize any nonspecific binding which may give rise to background, affecting the protein analysis at later stage. Apart from 5% nonfat dry milk in TBS, 3–5% bovine serum albumin in TBS is an alternative blocking solution that is commonly used.
12. Since different proteins have distinct expression levels in each cell line, the exposure time of protein should be adjusted accordingly. If the expression of a particular protein in a particular cell line is very weak, the exposure time of this protein can be increased to improve the intensity of signal. However the background signal will increase with extending of exposure time. If exposure time has been prolonged and the signal is still

too weak to be detected, then increasing total amount of protein loaded into the well should be considered.

13. Besides normal ECL detection reagent, ECL prime (Amersham) or other similar products from other vendors (e.g., Millipore) are now available new-generation ECL products for chemiluminescent detection which provide higher sensitivity over the normal ECL reagent. If the concentration of target protein is too low and the primary antibody is limited, these products may be considered as an alternative.
14. If a gel documentation imaging system with CCD camera installed is available, chemiluminescence can be captured by such a system after the nitrocellulose membrane is exposed to ECL. Definite time intervals can be set and the camera is able to record the signal continuously to accumulate the signal automatically. The image which has optimal signal intensity can be selected and there is no need to adjust the exposure time. This is the key advantage of such an imaging system.
15. Stripping and re-probing membranes: Wash membrane with TBS. Soak membrane in pH2 TBST for 5 min. Rinse the membrane and soak it in pH2 TBST for a further 5 min. Wash membrane with TBST twice. Block membrane and repeat antibody probing (Subheading 3.5).
16. Image J supports various formats of image, including TIFF, GIF, JPEG PNG, BMP, and FITS. Since JPEG files use compression that will lead to loss of original data, it is recommended to use JPEG files only for illustration purpose since they may not be suitable for quantitative analysis. This method described here is modified from the gel analysis procedure from Image J manual. There are many ways to perform densitometric analysis. Similar results should be achieved with different methods.
17. Images of the blots sometimes have background noise. This can be compensated for by subtracting the background signal from the signal of the band.

Acknowledgments

This work was supported by Seed Funding for Basic Research (201011159060), The University of Hong Kong, and by the Medical Research Council (NIRG G0801072/87482 to AJM). The authors would like to thank the Vascular & Metabolic Research Unit, The University of Hong Kong, for their assistance with the western blotting experiment.

References

1. Seyhan AA, Rya TE (2010) RNAi screening for the discovery of novel modulators of human disease. Curr Pharm Biotechnol 11:735–756
2. McManus MT, Sharp PA (2002) Gene silencing in mammals by small interfering RNAs. Nat Rev Genet 3:737–747
3. Singh SK, Hajeri PB (2009) siRNAs: their potential as therapeutic agents—Part II. Methods of delivery. Drug Discov Today 14:859–865
4. Meade BR, Dowdy SF (2007) Exogenous siRNA delivery using peptide transduction domains/cell penetrating peptides. Adv Drug Deliv Rev 59:134–140
5. Varkouhi AK, Scholte M, Storm G, Haisma HJ (2011) Endosomal escape pathways for delivery of biologicals. J Control Release 151: 220–228
6. Lam JKW, Liang W, Lan Y et al (2012) Effective endogenous gene silencing mediated by pH responsive amphipathic peptides proceeds via multiple pathways. J Control Release 158: 293–303
7. Kichler A, Mason AJ, Bechinger B (2006) Cationic amphipathic histidine-rich peptides for gene delivery. Biochim Biophys Acta 1758: 301–307
8. Langlet-Bertin B, Leborgne C, Scherman D et al (2010) Design and evaluation of histidine-rich amphipathic peptides for siRNA delivery. Pharm Res 27:1426–1436
9. Samuel-Abraham S, Leonard JN (2010) Staying on message: design principles for controlling nonspecific responses to siRNA. FEBS J 277:4828–4836

Chapter 6

High-Throughput RNAi Screening for the Identification of Novel Targets

Meredith C. Henderson and David O. Azorsa

Abstract

Gene silencing through RNA interference has provided researchers with an effective way to study gene function. High-throughput RNA interference (HT-RNAi) screening has further permitted researchers to identify functionally relevant mediators of cellular response on a large scale. These screens have greatly expedited the discovery of novel targets and pathway mediators. Here, we describe the methodology for performing HT-RNAi screening of HeLa cells transfected with short interfering RNA (siRNA) libraries in 384-well microplate format. Using this plate format, the HT-RNAi assay can be easily adapted to semi-automated or fully automated platforms. The library siRNA are introduced into the cells through reverse transfection using cationic lipids. HT-RNAi screening for modulators of cell proliferation can be accomplished using a single read out reagent. This type of RNAi screening can be used with most plate-based cellular assays and can be optimized for most cultured cells lines, thus becoming a powerful tool to identify specific gene modulators and targets for drug discovery.

Key words: High-throughput, RNAi, siRNA, Library, Assay, Screening

1. Introduction

RNA interference (RNAi) technology provides a powerful mechanism for systematic single-gene inhibition. High-throughput RNAi (HT-RNAi) makes use of this technique on a much larger scale, such that hundreds or thousands of genes can be silenced in a single experiment. This type of functional genomic screen allows researchers to rapidly identify genes that affect cellular functions such as cell survival (1–7), signal transduction (8–12), differentiation (13), migration (14), and drug response (15–18).

HT-RNAi screens of mammalian cells primarily use either short interfering RNA (siRNA) or small-hairpin RNA (shRNA) libraries. Although there are similarities between the two platforms, screening using siRNA libraries has the advantages of having minimal

Jürgen Moll and Riccardo Colombo (eds.), *Target Identification and Validation in Drug Discovery: Methods and Protocols*, Methods in Molecular Biology, vol. 986, DOI 10.1007/978-1-62703-311-4_6,

library maintenance and showing more consistent quality of the RNAi duplexes. HT-RNAi screening is facilitated by using reverse transfection of nucleic acids as originally described by Ziauddin and Sabatini (19) and applied to screening of siRNA libraries (12). We have found that using commercially available transfection reagents, many cell types can be effectively transfected with siRNA and thus used for HT-RNAi assays.

We have utilized HT-RNAi screens to identify novel therapeutic targets in several different disease cell models (2, 20, 21) and to identify potentiating targets of chemotherapeutic response (6, 17). Our standard assay involves using siRNA libraries in 384-well format and using cell number as a readout, which can be determined in a single addition step using homogenous viability assays. A typical viability HT-RNAi screen consists of using printed siRNA library assay plates, which can be prepared beforehand, to which is added transfection reagent for siRNA:cationic lipid complexing followed by cell addition. After approximately 72–96 h, the assay is terminated by the addition of the homogenous viability substrate and read for signal. A diagram of our assay is shown in Fig. 1. This assay can be easily adapted to most cell-based assays and various types of readouts.

2. Materials

2.1. Printing Library siRNA

1. Library siRNA in 384-well format (10 μM stock) (see Note 1).
2. AllStars Negative and Positive Controls (Qiagen, Valencia, CA).
3. siRNA buffer: 100 mM KOAc, 30 mM HEPES, 2 mM MgOAc, adjust pH to 7.6 with dilute acetic acid or KOH. Filter sterilize by passing through a 0.22 μm filter.
4. 384-well clear microplates (see Note 2).
5. White 384-well tissue culture treated plates (Greiner Bio-One, Monroe, NC).
6. Aluminum 384-well plate seals.

2.2. siRNA Library Screen

1. HeLa cell line (ATCC catalog # CCL-2, ATCC Manassas, VA).
2. Dulbecco's modified Eagle's medium (DMEM) containing 10% (v/v) fetal bovine serum and 2 mM L-glutamine.
3. 384-well daughter plates prepared with diluted library and control siRNA.
4. Blank solid white-bottom and black clear-bottom 384-well plates (two of each).
5. Lipofectamine™ RNAiMAX Transfection reagent (Life Technologies, Carlsbad, CA).

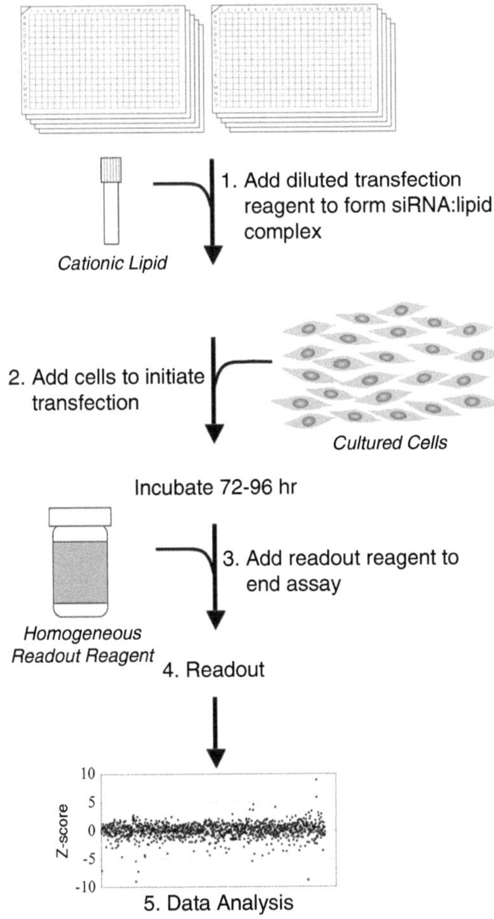

Fig. 1. High-throughput RNAi screening using siRNA libraries. Assay plates containing library siRNA are treated with (1) diluted transfection reagent to complex the siRNA and lipid. (2) Cells are added to initiate the transfections. Assay plates are incubated for 72–96 h then (3) readout reagent is added. Assay plates are read and the data is analyzed (4–5).

6. Opti-MEM® reduced serum media (Life Technologies) (see Note 3).
7. CellTiter-Glo® (Promega, Madison, WI).

3. Methods

3.1. Printing Library siRNA

1. Transfer 3 μL of stock library siRNA to 384-well clear microplate (daughter plate) and insert control siRNA in designated wells as shown in Fig. 2 (see Note 4).
2. Add 87 μL of siRNA dilution buffer to each well.

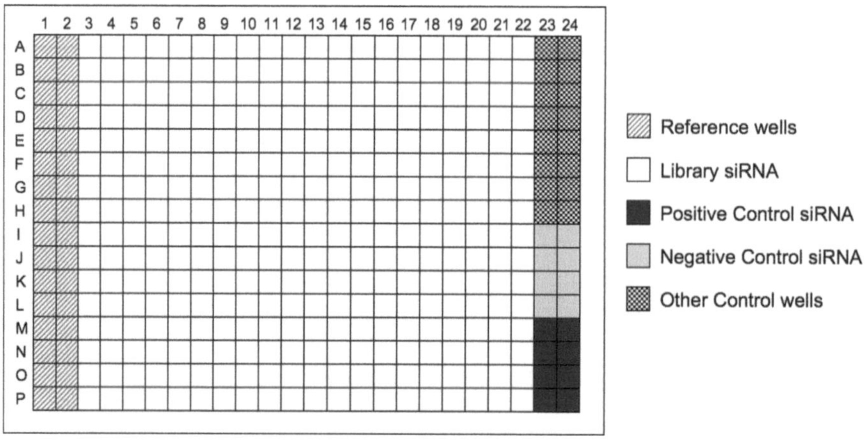

Fig. 2. Plate format for siRNA library screening. A 384-well daughter plate contains library siRNA as well as wells for positive and negative control siRNA and additional controls as needed for each specific assay. Reference wells are designated.

3. Transfer 2 μL of diluted siRNA from each daughter plate to replicate 384-well assay plates (see Note 5).
4. Seal daughter plates and store at –80°C until needed.

3.2. RNAi Library Screen

1. Remove frozen assay plates from freezer and thaw at room temperature. Centrifuge plates at 700 × *g* for 1 min to collect siRNA at the bottom of the wells.
2. In addition to library plates, prepare and label two additional clear 384-well plates as test plates.
3. Dilute Lipofectamine RNAiMAX in OptiMEM to a concentration of 3 μL lipid per mL of OptiMEM (see Note 6). Add diluted transfection reagent at a final volume of 20 μL/well to all plates (see Note 7). Allow transfection reagent to complex with siRNA at room temperature for at least 30 min.
4. Trypsinize HeLa cells and prepare cells at a concentration of 50,000 cells/mL diluted in DMEM containing FBS and L-glutamine (see Note 8). Add 20 μL of cell solution (final concentration of 1,000 cells/well) to all plates.
5. Allow the plates to sit at room temperature for approximately 15 min (see Note 9).
6. Incubate the plates in a 37 °C incubator with 5% CO_2.
7. After 72 h, remove the clear 384-well plates from the incubator and assess cell condition by microscope (see Note 10). Remove assay plates from the incubator and allow plates to equilibrate to room temperature.
8. Prepare Cell Titer Glo reagent and allow reagent to equilibrate to room temperature (see Note 11).

9. Add 25 μL of Cell Titer Glo to all white solid-bottom plates. Incubate the plate at room temperature for at least 1 h on a rotating platform. Read luminescence using a luminescence microplate reader (see Note 12).
10. Analyze data for each plate and normalize plate date for plate-to-plate comparison (see Note 13).

4. Notes

1. Libraries for siRNA screening can be obtained from several commercial sources and can be pre-designed or custom designed. Libraries can come as a pool of several (usually four) siRNA or as individual RNA duplexes. Some libraries come in a 96-well format and will have to be converted to 384-well format for daughter plate preparation and subsequent printing.
2. Additional clear bottom microplates are used as test plates to visually evaluate the cells throughout the assay.
3. Serum-free cell media can be substituted for OptiMEM for dilution of the transfection reagent. Adjustments to the serum concentration of the media used in cell addition may be needed.
4. The siRNA libraries are obtained as a 10 μM stock and we prepare a 1:30 dilution resulting in 333 nM daughter plate concentrations. Although this step can be done manually using multichannel pipets, it can be greatly expedited by the use of a liquid-handling system, such as the Biomek FX Workstation (Beckman Coulter; Brea, CA).
5. Printing of the assay plates from the daughter plate is usually done by dispensing 2 μL of 333 nM diluted siRNA to each assay plate. Up to 40 replicates can be made from each daughter plate. This results in approximately10 ng of siRNA/well in the assay plate.
6. This dilution results in a 6:1 lipid:siRNA (vol:wt) ratio. The optimal transfection reagent and ratio for the cell type used should be determined beforehand. This can be done by testing several commercially available transfection reagents at several lipid:siRNA ratios with control siRNAs including a non-silencing control and a lethal siRNA control. The optimal transfection reagent should show minimal effect on viability in wells treated with non-silencing siRNA (compared to no siRNA controls) and a greater than 90% decrease in viability in lethal siRNA-treated wells.

7. Addition of reagents to more than four 384-well plates is greatly expedited by the use of a semi-automated liquid-dispensing system such as the MicroFill Microplate Dispenser (BioTek; Winooski, VT).
8. The growth media should not include antibiotics as it might affect transfection efficiency and cell viability. The optimal cell number used in seeding the wells needs to be determined for each cell type and assay. During the addition of cells to the 384-well plates, it is important to keep the cells suspended and not allow them to settle. If using a semi-automated dispenser, lightly shake the cell suspension in between plates. If pipetting by hand, resuspend cells by pipetting up and down between plates. A semi-automated liquid-dispensing system is also helpful in this step.
9. Allowing the plates to equilibrate at room temperature will allow the cells to settle so as not to be affected by temperature currents when placed at 37°C. This results in better distribution of cells and can improve readout assays.
10. Assessing the condition of the cells ahead of time can save valuable reagents. Things to note include well contamination, empty wells, uneven volume in wells, and uneven cell growth. Cells in the test plate should look healthy and the confluency should be between 70 and 90%.
11. According to the manufacturer's protocol, an equal volume of Cell Titer Glo is recommended for stopping and reading out of the assay. We have found that a 1:2 (vol:vol) ratio of Cell Titer Glo:media yields comparable results, prevents high volumes in the assay wells, and also conserves reagent.
12. Allowing the Cell Titer Glo reagent and assay plates to equilibrate to room temperature is important as it prevents temperature differences in the wells from affecting the enzymatic activity of the reagent. If the reagent was added to plates directly removed from the incubator, the outer wells would adjust to room temperature quicker than the center wells, which could result in the plate data giving indications of an edge effect or deceased growth in the outer wells.
13. Several methods have been described to analyze siRNA screening data (22). It is important to calculate the Z′ factor for each plate to determine the robustness of the assay. Also, it is recommended that the siRNA screens be repeated at least once to obtain a biological replicate. This will greatly improve the data analysis and target hit selection by eliminating false-negatives. Furthermore, it allows for prioritization of target hits during the subsequent confirmation and validation steps.

References

1. Kittler R, Putz G, Pelletier L et al (2004) An endoribonuclease-prepared siRNA screen in human cells identifies genes essential for cell division. Nature 432:1036–1040
2. Henderson MC, Gonzales IM, Arora S et al (2011) High-throughput RNAi screening identifies a role for TNK1 in growth and survival of pancreatic cancer cells. Mol Cancer Res 9:724–732
3. Tyner JW, Walters DK, Willis SG et al (2008) RNAi screening of the tyrosine kinome identifies therapeutic targets in acute myeloid leukemia. Blood 111:2238–2245
4. Cole KA, Huggins J, Laquaglia M et al (2011) RNAi screen of the protein kinome identifies checkpoint kinase 1 (CHK1) as a therapeutic target in neuroblastoma. Proc Natl Acad Sci U S A 108:3336–3341
5. MacKeigan JP, Murphy LO, Blenis J (2005) Sensitized RNAi screen of human kinases and phosphatases identifies new regulators of apoptosis and chemoresistance. Nat Cell Biol 7:591–600
6. Azorsa DO, Gonzales IM, Basu GD et al (2009) Synthetic lethal RNAi screening identifies sensitizing targets for gemcitabine therapy in pancreatic cancer. J Transl Med 7:43
7. Morgan-Lappe SE, Tucker LA, Huang X et al (2007) Identification of Ras-related nuclear protein, targeting protein for xenopus kinesin-like protein 2, and stearoyl-CoA desaturase 1 as promising cancer targets from an RNAi-based screen. Cancer Res 67:4390–4398
8. Gilot D, Le Meur N, Giudicelli F et al (2011) RNAi-based screening identifies kinases interfering with dioxin-mediated up-regulation of CYP1A1 activity. PLoS One 6:e18261
9. Hsieh AC, Bo R, Manola J et al (2004) A library of siRNA duplexes targeting the phosphoinositide 3-kinase pathway: determinants of gene silencing for use in cell-based screens. Nucleic Acids Res 32:893–901
10. Berns K, Hijmans EM, Mullenders J et al (2004) A large-scale RNAi screen in human cells identifies new components of the p53 pathway. Nature 428:431–437
11. Bjorklund M, Taipale M, Varjosalo M et al (2006) Identification of pathways regulating cell size and cell-cycle progression by RNAi. Nature 439:1009–1013
12. Aza-Blanc P, Cooper CL, Wagner K et al (2003) Identification of modulators of TRAIL-induced apoptosis via RNAi-based phenotypic screening. Mol Cell 12:627–637
13. Zhao Y, Ding S (2007) A high-throughput siRNA library screen identifies osteogenic suppressors in human mesenchymal stem cells. Proc Natl Acad Sci USA 104:9673–9678
14. Simpson KJ, Selfors LM, Bui J et al (2008) Identification of genes that regulate epithelial cell migration using an siRNA screening approach. Nat Cell Biol 10:1027–1038
15. Iorns E, Lord CJ, Ashworth A (2009) Parallel RNAi and compound screens identify the PDK1 pathway as a target for tamoxifen sensitization. Biochem J 417:361–370
16. Whitehurst AW, Bodemann BO, Cardenas J et al (2007) Synthetic lethal screen identification of chemosensitizer loci in cancer cells. Nature 446:815–819
17. Arora S, Bisanz KM, Peralta LA et al (2010) RNAi screening of the kinome identifies modulators of cisplatin response in ovarian cancer cells. Gynecol Oncol 118:220–227
18. Bartz SR, Zhang Z, Burchard J et al (2006) Small interfering RNA screens reveal enhanced cisplatin cytotoxicity in tumor cells having both BRCA network and TP53 disruptions. Mol Cell Biol 26:9377–9386
19. Ziauddin J, Sabatini DM (2001) Microarrays of cells expressing defined cDNAs. Nature 411:107–110
20. Arora S, Beaudry C, Bisanz KM et al (2010) A high-content RNAi-screening assay to identify modulators of cholesterol accumulation in Niemann-Pick type C cells. Assay Drug Dev Technol 8:295–320
21. Arora S, Gonzales IM, Hagelstrom RT et al (2010) RNAi phenotype profiling of kinases identifies potential therapeutic targets in Ewing's sarcoma. Mol Cancer 9:218
22. Birmingham A, Selfors LM, Forster T et al (2009) Statistical methods for analysis of high-throughput RNA interference screens. Nat Methods 6:569–575

Chapter 7

Integration of RNAi and Small Molecule Screens to Identify Targets for Drug Development

Konstantinos Drosopoulos and Spiros Linardopoulos

Abstract

Cellular models for siRNA and small molecule high throughput screening have been widely used in the last decade to identify targets for drug discovery. As an example, we present a two-fold readout approach based on cell viability and multipolar phenotype. To maximize the discovery of potential targets and at the same time reduce the number of false positives in our dataset, we have combined focused and rationally designed custom siRNA libraries with small molecule inhibitor libraries. Here we describe a cellular model for centrosome amplification as an example of how to design and perform a multiple readout/multiple screening strategy.

Key words: High-throughput screening, Centrosome amplification, Centrosome clustering, High-content screening, siRNA screening, Target discovery

1. Introduction

Despite intensified efforts to identify new targets for drug development in cancer, in the last decades, only a very small fraction reached the preclinical development phase. The most common problems reported are drug selectivity and lack of understanding of the underlying mechanisms. The recent advances in RNA interference (RNAi) technology have permitted a wide adoption of high-throughput approaches for the discovery of new targets within their natural context. However, these strategies are still beset by selectivity problems due to siRNA dependent off-target effects, especially in mammalian cells (1). Network and protein connectivity analyses have contributed to shedding light on the potential therapeutic role of a particular target, but these are still dependent on high-quality input data in order to be useful (2). Several strategies have been suggested for improving the quality of data from high-throughput screens, including the use of isogenic

Jürgen Moll and Riccardo Colombo (eds.), *Target Identification and Validation in Drug Discovery: Methods and Protocols*, Methods in Molecular Biology, vol. 986, DOI 10.1007/978-1-62703-311-4_7,

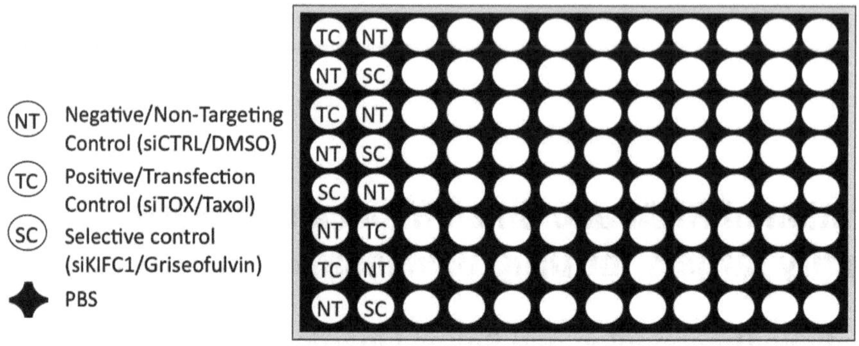

Fig. 1. Example of a 96-well plate with controls on columns 1 and 2. The rest of the plate is occupied by the samples (siRNA or small molecule). siRNA and small molecule libraries often come with columns 1 and 2 or 1 and 12 empty for the purpose of leaving space for the desired controls. Positioning the controls on columns 1 and 2 is preferable than the 1 and 12 option. Ideally, the controls should be dispersed throughout the plate if possible.

cell lines (cell lines of genetically identical background), use of multiple RNAi-oligos per gene and multiplexing viability data with high-throughput microscopy data (3). In order to improve the discovery of verifiable targets, facilitate identification of false positives, as well as understanding of the underlying mechanisms, we use a twofold readout approach, also limiting our assay to a specific phase of the cell cycle where possible. In addition, in order to further reduce screening artifacts and false positives, we utilize isogenic cell lines with or without centrosome amplification that are screened under identical conditions and siRNA transfection efficiency.

The centrosomes are microtubule organizing centers which play an essential role in organizing the mitotic spindle and maintaining spindle bipolarity in mammalian cells (4). Centrosome amplification is one of the most common abnormalities in cancer cells. In order to avoid the deleterious effects of multipolar mitotic spindles, cancer cells cluster supernumerary centrosomes in a bipolar fashion during mitosis, a process which is known to be "targetable" (5). Here we generate isogenic cells with and without centrosome amplification and screen a custom siRNA library representing all known microtubule-binding proteins with the rationale that centrosomes, which are microtubule-associated organelles, depend on microtubules and their associated proteins for their motility. In addition, we use a kinase siRNA library and a small molecule kinase inhibitor library to help us to identify the regulatory networks that lie upstream of the microtubule-associated effectors.

The siRNA screening is performed in duplicate in 96-well plates as shown in Fig. 1. The first replica is used for viability readout and the second for phenotypic readout where we look at spindle structure by using a high content screening platform. The small molecule library screening is also performed in duplicate with the first replica used for viability readout. In contrast to the siRNA libraries, small molecule libraries give us the opportunity to carefully time the treatment to include mostly one phase of the cell cycle, thus reducing the possibility for "noise" introduced by off-target inhibition of kinases that might be important in other phases of the cell cycle. For this reason, in the second replica, which we use for the phenotypic readout, cells are synchronized in S-phase by thymidine treatment and then released for 8 h before screening the library for 3 h, allowing us to limit the effects of the treatment to mitosis and, therefore, to reduce secondary effects. Using this approach, we identify a number of hit-genes that selectively kill cells with centrosome amplification validating at the same time both the underlying mechanism and the regulatory signaling involved in this process.

2. Materials

The nature of this approach requires specific optimization and setting up of the assay, depending on the cell model used. There is a variety of commercial lipids and siRNA controls that need to be tested for suitability for a particular cell line/cell model. It is desirable to achieve less than 10% toxicity of the lipid alone relative to growth media alone, less than 10% toxicity of the non-targeting (sequence does not match any known human mRNA) siRNA negative control relative to lipid alone, and more than 90% toxicity of the transfection efficiency control siTOX (highly toxic siRNA pool that will kill every transfected cell) relative to non-targeting siRNA negative control. In addition, cell plating should be optimized so that at the end of the assay cells are no more than 90–95% confluent for the viability assay and no more than 70–80% confluent for the phenotypic assay. It is important that every screen is set up individually by the person who is going to perform the screening since clonal variation of cell lines, available equipment, normal variation of different batches of transfection lipid and personal technique can have significant effects in a high-throughput screen even if an exact protocol is followed.

2.1. Generation of Isogenic Populations with or Without Centrosome Amplification

1. DLD1 colon adenocarcinoma cells grown in DMEM supplemented with 10% FBS.
2. Dihydro-cytochalasin B (DCB), an inhibitor of actin polymerization, required to promote cytokinesis failure.
3. Hoechst 33342 for live cell sorting and isolation of diploid/tetraploid populations.

2.2. siRNA Screening

1. Transfection reagent: DharmaFECT 2 (T-2002; Dharmacon).
2. Optimem I (Invitrogen).

3. siRNA controls (Dharmacon): Non-targeting siRNA pool (D-001206-13), siTOX Transfection Control (D-001500-01), and siKIFC1 (M-004958-02).
4. Dharmacon kinase-related siGENOME siRNA library.

2.3. Small Molecule Library Screening

1. DMSO for resuspension and dilution of small molecule library and for vehicle control.
2. Griseofulvin: used as a positive control at 5 μM since it is known to selectively kill cells with centrosome amplification.
3. Thymidine: Used at 4 mM for synchronizing the cells at G1.

2.4. Viability and Phenotypic Readout

1. Opaque white wall, clear bottom 96-well tissue culture plates to be used with a luminescent cell viability reagent suitable for high-throughput screening assays.
2. CellTiter-Glo Luminescent Cell Viability Assay (G7572, Promega).
3. Black wall, clear bottom 96-well tissue culture plates to be read by a suitable high content screening platform.
4. Ice-cold methanol/acetone 10:1.
5. Anti-Aurora-A (see Note 1) monoclonal antibody (610939, BD) and Alexa Fluor 488 Goat Anti-Mouse (A-11001, Molecular Probes).
6. Blocking solution: 5% FBS in PBS.
7. Antibody solution: 2% FBS in PBS.
8. DAPI solution: 1:10,000 DAPI (10 mg/ml stock) in PBS.
9. Black adhesive film to seal the plates before reading.

3. Methods

3.1. Generation of Isogenic Cell Lines

1. Treat DLD1 diploid cells with normal centrosome number for 16 h with 5 μM DCB and release for 96 h.
2. Prepare cell suspension of 500,000/ml and incubate with Hoechst 33342 (10 mg/ml stock) in PBS 1:5,000 at 37°C for 20 min.
3. Cell sort populations according to cell cycle profile. Isolate diploid populations from diploid G1 and tetraploid populations from tetraploid G2 (see Note 2).
4. Grow isolated populations for 1 week and verify DNA content by comparing isolated tetraploid populations to the original DLD1 cells. Repeat cell sorting if necessary.

3.2. siRNA Screening

All reagents must be at room temperature.

1. Split, pellet, and resuspend diploid, tetraploid, and centrosome amplified cells at appropriate density (2,000 cells/well final for viability and 12,000 cells/well final for phenotypic) in growth media (DMEM) without antibiotics.
2. Add 80 μl PBS in the space between the wells of the 96-well assay plates (see Note 3).
3. Distribute library and control siRNAs (Fig. 1) in V-bottom 96-well plates (one siRNA pool/well). Add appropriate amount of DharmaFECT 2 in Optimem (final concentration 0.2 μl per sample) using a 12 channel pipette and precision tips.
4. Gently pipette up and down 4–5 times and immediately distribute to flat-bottom 96-well plates.
5. Place the lid on the plate and incubate for 30 min at RT to allow the formation of siRNA–lipid complexes.
6. Add cell suspension on complexes by a single swift pipetting step and shake on an orbital shaker for 2–3 s at 400 rpm to equally distribute the cells throughout the well.
7. Place the plates in the incubator, avoiding stacking if possible (see Note 3).
8. After 5–7 h carefully add 200 μl of growth media, taking caution not to disturb the cells. This is done to minimize lipid toxicity.
9. (Optional) Next morning, remove half of the media in the plates using a multichannel pipette and carefully add an equal amount of fresh media to further dilute the lipid. This will help to further reduce lipid toxicity effects and at the same time minimize disturbance of cells that will likely happen if the full amount of growth media is replaced.
10. For the phenotypic readout, 48 h after the transfection, remove media and fix cells adding 150 μl of ice-cold methanol/acetone 10:1.
11. Keep the plates at –20°C for 30 min.
12. Leave for 5 min at RT before removing methanol/acetone and immediately add 5% FBS blocking solution.
13. Incubate for 30 min at RT.
14. Add 60 μl of antibody solution with Aurora-A antibody 1:1,000.
15. Keep the plates at 4°C for 16–24 h.
16. Wash 3× 5 min with PBS taking care not to leave the cells dry for more than 1 min maximum.
17. Incubate with fluorescent secondary antibody at 1:1,000 for 3 h at RT.

18. Wash 3 × 5 min with PBS using 1:10,000 DAPI in PBS for the second wash.
19. Seal the plate with black adhesive film and acquire images using a high-content platform, such as the *IN Cell Analyser* (GE Healthcare) or *Opera* (PerkinElmer) or an inverted fluorescence microscope.
20. For viability assay, 96 h after the transfection, remove growth media by flicking the plate sideways above a sink and add 100 μl of DMEM:Cell Titer Glo 1:1 solution.
21. Mix for 2–3 min on an orbital shaker and incubate for 20 min at RT.
22. Read using a luminescence plate reader.

3.3. Small Molecule Library Screening

1. Plate cells in 96-well flat-bottom tissue culture plates.
2. After 12–16 h treat phenotypic replicas (but not the viability ones) with 5 mM thymidine.
3. After 24 h release cells from thymidine.
4. 8 h after releasing cells from thymidine, dilute in growth media appropriate amount (see Note 4) of the small molecule solutions in V-bottom plates at 3× the final concentration.
5. Mix well (10 min) using an orbital shaker and proceed immediately to step 6.
6. Treat viability and phenotypic replicas with previously prepared small molecule solutions.
7. Continue from Subheading 3.2, step 10 with the difference that the phenotypic replicas are treated for 3 h instead of 48 h.

3.4. Analysis and Interpretation of the Results

1. For the analysis of viability data, we use the MAD method selecting over one standard deviation (6).
2. For the analysis of phenotypic results, we consider hits the si-RNAs or small molecules that induce a multipolar spindle phenotype at a frequency which is at least two times over the one observed in the control-treated centrosome amplified cells and at the same time do not induce the multipolar spindle phenotype in control cells (Fig. 2).
3. The common hits between the two different readouts represent our high confidence hits, which we validate further for their suitability to be used as targets for drug development for which both the target population and the underlying mechanisms involved, are known.
4. The results from the chemical library serve three purposes. To identify tools that can be used at later stages of drug development (i.e., for determining the target population) and for providing potential models for structure–activity relationship

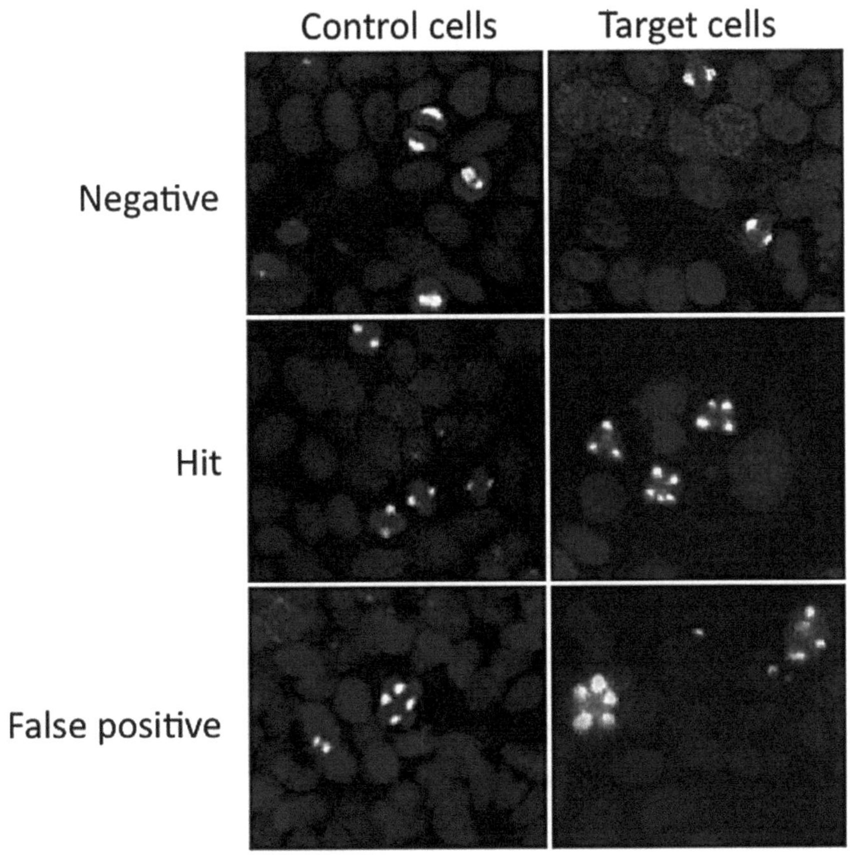

Fig. 2. Example images acquired with IN Cell Analyser 3000 (GE Healthcare). Control cells are 4 N DLD1 cells with no centrosome amplification, target cells are 4 N DLD1 cells with centrosome amplification. First set shows cells transfected with siControl non-targeting siRNA (Negative), second set shows cells transfected with siKIFC1 (Hit), and the third set shows cells transfected with siESPL1 (False positive) which is known to cause cytokinesis failure when knocked-down. Note that in the first set no multipolar spindles are observed, in the second set multipolar spindles are observed only in the target cells, and in the third set multipolar spindles are observed on both control and target cells.

(SAR). In addition, small molecules have the advantage of being suitable for dose response experiments that are helpful in validating borderline hits from the kinome siRNA libraries.

4. Notes

1. Anti-alpha or beta-tubulin can also be used. Aurora-A provides the advantage of easier identification of mitotic cells and spindle structure since it only localizes on the mitotic spindle and centrosomes during mitosis and not to microtubules in general. Also, it allows visualization of both spindle structure and centrosomes with a single antibody.

2. Cell suspension denser than 500,000 cells/ml is not recommended as it will make it harder to separate the different phases of the cell cycle. Also, avoid keeping the cells with Hoechst for more than 3 h to avoid toxicity. The procedure will give stable diploid and tetraploid populations due to the tendency of the cells that do not originally present centrosome amplification to reduce the centrosome number back to normal over time but maintaining 4 N DNA content. Cells with 4 N content and no centrosome amplification are used as a control cell line for the purpose of screening. 4 N cells with centrosome amplification are generated by 24–36 h incubation with DCB followed by a 24 h release. These cells will maintain abnormal centrosome numbers for a period of 6–10 days during which they can be used for screening.
3. When the plates are placed in the incubator, temperature and evaporation gradients start to form inside the plates which result in positional biases (edge effects) that introduce noise in the final data. The most affected samples are the ones in the periphery of the plate. Filling the spaces between the wells with PBS reduces the edge effect. Stacking the plates inside the incubator will also influence the evaporation gradient inside the plate, especially when the top plate of a stack is compared to the bottom plate of a stack. If the plates must be stacked then the plates that are to be compared directly should be in the same position on different stacks.
4. Different small molecules have different efficacies. A very potent inhibitor can inhibit its target at nanomolar concentrations and start being toxic at micromolar concentrations. Similarly, a less potent inhibitor might only work at micromolar concentrations. When screening small molecule libraries, it is best to do it at three different concentrations minimum (e.g., 100 nM to 1 μM to 10 μM) in order to maximize the useful data.

References

1. Westwick JK, Lamerdin JE (2011) Improving drug discovery with contextual assays and cellular systems analysis. Methods Mol Biol 756:61–73
2. Dezso Z, Nikolsky Y, Nikolskaya T et al (2009) Identifying disease-specific genes based on their topological significance in protein networks. BMC Syst Biol 23:3–36
3. Ashworth A, Bernards R (2010) Using functional genetics to understand breast cancer biology. Cold Spring Harb Perspect Biol 2:a003327
4. Nigg EA, Stearns T (2011) The centrosome cycle: centriole biogenesis, duplication and inherent asymmetries. Nat Cell Biol 13: 1154–1160
5. Kwon M, Godinho SA, Chandhok NS et al (2008) Mechanisms to suppress multipolar divisions in cancer cells with extra centrosomes. Genes Dev 22:2189–2203
6. Zhang XD, Yang XC, Chung N et al (2006) Robust statistical methods for hit selection in RNA interference high-throughput screening experiments. Pharmacogenomics 7:299–309

Chapter 8

CellProfiler and KNIME: Open Source Tools for High Content Screening

Martin Stöter, Antje Niederlein, Rico Barsacchi, Felix Meyenhofer, Holger Brandl, and Marc Bickle

Abstract

High content screening (HCS) has established itself in the world of the pharmaceutical industry as an essential tool for drug discovery and drug development. HCS is currently starting to enter the academic world and might become a widely used technology. Given the diversity of problems tackled in academic research, HCS could experience some profound changes in the future, mainly with more imaging modalities and smart microscopes being developed. One of the limitations in the establishment of HCS in academia is flexibility and cost. Flexibility is important to be able to adapt the HCS setup to accommodate the multiple different assays typical of academia. Many cost factors cannot be avoided, but the costs of the software packages necessary to analyze large datasets can be reduced by using Open Source software. We present and discuss the Open Source software CellProfiler for image analysis and KNIME for data analysis and data mining that provide software solutions which increase flexibility and keep costs low.

Key words: High content screening, Image processing, Statistics, Open Source, CellProfiler, KNIME, Distributed computing

1. Introduction

In the 1960s, digital image processing was first applied at the Jet Propulsion Laboratory in California to enhance images from the lunar surface sent back by the Ranger 7 moon probe (1, 2). In the 1970s, digital image processing became more widely applied mainly in fields such as medical imaging (3) or forensic science (4). In the late 1980s, the pharmaceutical industry developed methods to increase the number of natural extract to screen for biological activity leading to the field of high-throughput screening (HTS) (5–7). By the end of the 1990s, HTS and imaging were brought together and high content screening (HCS) (also called high content analysis (HCA) or high content imaging (HCI)) was born (8).

Jürgen Moll and Riccardo Colombo (eds.), *Target Identification and Validation in Drug Discovery: Methods and Protocols*, Methods in Molecular Biology, vol. 986, DOI 10.1007/978-1-62703-311-4_8,

The motivation for the pharmaceutical industry to use the analytical power of microscopy was to obtain early in the drug development process information about compound action in a cellular system. Typically, validated hits emerging from primary screening campaigns are tested in cell lines to assess gross toxicity, membrane permeability, and specificity. In this secondary screening phase, many primary hit compounds need to be analyzed and an increased throughput of microscopy analysis was desired. Therefore automated microscopy coupled to automated image analysis was developed to obtain stable user-friendly instruments by companies such as Cellomics.

As HCS was primarily a technology used by the pharmaceutical industry and only few academic laboratories were equipped with automated microscopes, the hardware and software providers were mainly dealing with industrial clients with access to large budgets compared to academia. The market was limited and the development costs were high. All these factors contributed to the high costs for instruments, software, maintenance contracts, and yearly licensing fees. Consequently, access to the technology was limited to institutions with the required budget to maintain an HCS infrastructure.

In the early 2000s academic institutions started founding screening centers mainly due to two events. First, with the availability of several sequenced genomes and the discovery of RNA interference (RNAi), the systematic analysis of biological pathways by gene silencing became feasible. Second, academic institutions became more involved in biomedical research and began screening for novel drugs, especially since the launch of the NIH roadmap in 2004 (http://commonfund.nih.gov/aboutroadmap.aspx) (9).

Some of the newly created screening centers were equipped with automated imagers and HCS entered the academic field. In general, these academic screening centers were well funded and could afford the prices demanded by vendors, perpetuating the tradition of cost-intensive screening.

In recent years, R&D budgets in many pharmaceutical companies have been reduced and many countries had to cut their research budgets because of the global financial crisis (10–13). Concomitantly, the demand for screening small chemical compounds and RNAi libraries, especially in Europe, has been growing in academia. Several national and European networks are being actively discussed to give wider access to existing screening infrastructures (14). These counteracting trends have led academia to become an important customer for HCS providers. This trend might be further accentuated in coming years, as HCS has the potential to become a standard research tool and not solely a screening tool. There are several reasons why we believe that automated microscopy will be a widely adopted technology in normal research laboratories in the future. Firstly, many common biological experiments would benefit from the exploration of hundreds of experimental conditions to fully analyze a biological question. For instance antibody staining and transfection optimization or time course experiments with various chemicals or stains are best carried out in multiwell plates. Often the parameter space is not fully explored, simply because the manual acquisition of images and manual image analysis would be too time consuming. The automation of this process with high content imagers solves that problem and allows more thorough investigation of experimental conditions. Secondly, the automation of data acquisition and data analysis eliminates human bias. The large amount of data generated does not allow visual inspections but requires automated image analysis and quantification of the data. Thus, automated microscopy tends to improve results as they are based on larger datasets that underwent rigorous statistical analysis rather than biased human interpretation. Thirdly, the miniaturization of assays results in significant time and cost savings.

For HCS to be fully accepted by academia, several conditions will need to be fulfilled. Current HCS instruments are closed black boxes and their expensive maintenance contracts do not allow any hardware or software modifications for adaptation to the diverse needs of academic research. Academic research is typically more diversified than pharmaceutical industry research and the instruments need to be more customizable than they are now. In addition, the image and data formats need to be accessible and open. In academia data is shared between collaborators and will be analyzed with various, partly custom-made software. Therefore the data needs to be accessible and open. Lastly, the yearly costs of maintenance contracts and licenses are particularly difficult to finance in academic research that relies heavily on grants. Grants typically do not cover licensing costs or if they do, when the grant runs out, new sources of funding must be found. In reality, those costs must generally be covered by institutional funds.

In order to reduce costs and increase the flexibility of HCS readers, it is to be expected that academic laboratories will assemble automated microscopes themselves. Hopefully this will be done with the Open Source hardware (OSHW) model, allowing other laboratories to reproduce the design (http://www.openhardware.org) (15, 16).

Another expensive factor of HCS is software. All HCS instruments have on board tools for image analysis and various level of sophistication for data analysis and viewing. Often image and data analysis is done after the acquisition of the screening plates in order not to slow down the microscope. Vendors therefore offer off-line licenses increasing the cost of software further. The image analysis software installed on instruments is often restricted in functionality to deliver easy-to-use software, although some vendors

Table 1
List of Open Source software for image analysis and statistics and their Web sites

Software name	Main application	GUI/scripting	Web site	Comment
Octave	Image analysis/ statistics	Scripting	http://www.gnu.org/ software/octave/	Similar to MATLAB
ImageJ	Image analysis	GUI/scripting	http://www.rsbweb.nih. gov/ij/	Java
Fiji	Image analysis	GUI/scripting	http://www.fiji.sc/wiki/ index.php/Fiji	Java
CellProfiler	Image analysis	GUI	http://www.cellprofiler.org	
BioImageXD	Image analysis	GUI	http://www.bioimagexd.net	For 3D images
R	Statistics	Scripting	http://www.r-project.org	Similar to S
Rapid miner	Statistics	GUI	http://www.rapid-i.com	
KNIME	Statistics	GUI	http://www.knime.org	
SciPy	Statistics	Scripting	http://www.scipy.org	Python programming
Shogun	Statistics	Scripting	http://www.shogun-toolbox.org	Integrated into many languages
Weka	Statistics	GUI	http://www.cs.waikato.ac. nz/ml/weka/	
µManager	Microscope control	GUI	http://www.micro-manager.org/	
Motmot	Image acquisition, storage, and analysis	GUI	http://www.code.astraw. com/projects/motmot	Python and C

have expert-mode tools allowing accessing many image analysis algorithms. For the diverse applications in academia, the push button applications found on the readers are in many case not sufficient and more advanced tools are required. Furthermore, academic laboratories need to publish their results in such a manner that other laboratories can reproduce their work. With commercial software this is very difficult, and reviewers cannot readily assess the used algorithm or software. For the statistical analysis of screening data, visualization of results and images and Web mining it is necessary to have more sophisticated software tools than typically found on HCS imagers. Many HCS vendors have entered partnerships with companies offering statistics software such as Pipeline Pilot, Genedata, or Spotfire. These software packages also have high licensing fees, increasing further the costs.

Fortunately, many powerful Open Source software are available both for image analysis and statistical analysis (Table 1). Some of these tools have user-friendly GUIs and do not require any scripting skills. Using Open Source software has three crucial advantages for academic laboratories: (1) flexibility to further develop the code within a community when needed, (2) reduced costs, and (3) the possibility to develop transparent analysis routines whose methods can be published. For screening facilities like ourselves, there is the added advantage that we can provide to our users the raw data, the processed data, and the analytical tools that were used to process it. The last point is particularly advantageous, as the collaborators are able to extend and modify easily the initial analysis on their own.

In this chapter we discuss two Open Source software we use routinely in our screening facility to analyze high content RNAi and chemical compound screens. With these software we are able to process several terabytes of data with relative ease.

2. Software

2.1. CellProfiler, CellProfiler Analyst

CellProfiler and CellProfiler Analyst are free Open Source software for automated image analysis, data visualization, and machine learning. Versions for Mac, Windows, and Linux are available and both software can be downloaded at http://www.cellprofiler.org. CellProfiler was developed by Anne Carpenter and Thouis Jones in the laboratory of David Sabatini at the Whitehead Institute for Biomedical Research and by Polina Golland at the CSAIL of the MIT.

CellProfiler is designed to analyze large amounts of images automatically (17, 18). Since the analysis of thousands of images is very computationally intensive, a version is available to run on a Linux cluster allowing the parallelization of the analysis. The software is written in the Python programming language and the source code can be downloaded and extended by the community.

CellProfiler has an intuitive user interface with point and click actions that allow assembling modules in a pipeline for processing images. Several features in CellProfiler make it an easy-to-use software. The Web site hosts many tutorials and examples of image processing pipelines together with the corresponding images. In the GUI of CellProfiler, every button or drop-down box for adjusting parameters has a help button, allowing to quickly look up its functionality. Once a pipeline has been assembled, CellProfiler has a test mode to verify the settings of the modules in the pipeline and adjust parameters. With these helpful features, it is possible for beginners to start producing quite complex image analysis pipelines in a very short time, provided they have basic knowledge in image processing.

CellProfiler provides many advanced image processing algorithms allowing developing image analysis solutions for a wide variety of assays in mammalian cells, tissue slices, and small organisms. There are several modules for preprocessing images such as calculating illumination correction functions and applying them to images or measuring image quality (19). Other modules detect objects in images using many different algorithms for distinguishing background from foreground and drawing separation lines between objects. There are several modules for performing various morphological operations for feature enhancement. Lastly, features such as intensity, texture, and morphology can be extracted either on a cell-by-cell basis or on an image-by-image basis and exported to CSV files, a MySQLite, or a MySQL database.

Visual control of the performance of the image analysis is essential in large-scale screens to verify the segmentation quality of the algorithms. CellProfiler generates overlays of the detected object outlines on images, for instance on RGB color merges of the various channels created within CellProfiler or directly on the original images. Another useful feature of the software is the ability to annotate images and to extract metadata either from the path or the file name itself. The handling of the data in the statistical analysis is made much easier and allows to quickly perform quality control checks of the images and the assay. When all the data and all the image file paths are exported to a MySQL database a properties file can be created. This properties file can then be read by CellProfiler Analyst to access the data in the database (20, 21).

CellProfiler Analyst is a software for rapidly analyzing and visualizing large image-based datasets exported to a MySQL database using the "ExportToDatabase" module in CellProfiler. The measurements extracted from the images or segmented objects can be visualized with four different types of plots: histograms, scatter plots, density plots, and parallel coordinates. All plots are interconnected and highlighting data points in one plot will highlight the same data points in the others. In this manner it is possible to rapidly explore relationships among conditions or among parameters. From this graphical representation data points or regions can be selected (gating) and either the images or the objects giving rise to the measurement can be displayed for visualization. This process is very important in large-scale imaging experiments, as it allows to quickly access data points standing out of the dataset and visually inspect the corresponding images. In this manner, novel phenotypes or artifacts stemming either from image analysis or plate preparation can be verified and sometimes be found as a distinct data cloud of points in a scatter plot. The software CellProfiler Analyst also has classifier tools and machine learning algorithms to classify cells. The user creates phenotype classes according to prior knowledge and trains the classifier. After a few iterations of testing and correcting, the software is able to classify whole screens and obtain enrichment scores for the various predefined phenotypes.

2.2. KNIME

KNIME is an Open Source platform for data mining developed by the Chair for Bioinformatics and Information Mining at the University of Konstanz (http://www.knime.org). The software can be applied for many different types of analysis such as data mining, cheminformatics, image analysis, text mining, deep sequencing, and many more. KNIME provides a wide set of functionalities for many different common tasks such as input/output, data processing, statistics, data mining, and visualization. As KNIME is a rich client application based on the Eclipse platform, it benefits from the Eclipse plug-in concept. This concept provides an interface for software developers to easily extend the functionality of KNIME. New tools developed in KNIME can be for private use or published as community contribution.

Especially in academic science, data analysis often starts as an exploratory creative process with evolving ideas of the data analysis flow and rapidly changing analysis parameters or conditions. Therefore, data analysis software has to be extremely flexible in order not to limit the exploration of data. Furthermore, it is important that the data analysis process is comprehensible and easily readable at all time points to ensure that scientists can share their approach with colleagues and to better prevent conceptual mistakes. A third requirement to data analysis software is the minimization of effort and time a scientist has to invest to implement various methods.

KNIME provides a user-friendly interface to visually create workflows allowing a step-by-step data analysis flow. A node—as a single entity of such a workflow—provides a very confined analysis step with a set of parameter configurations. Workflows can branch at any point, which allows easily implementing multiple approaches.

As KNIME is not focused on data analysis of high-throughput/high content experiments, it lacks common methods necessary for screen mining. We have therefore developed a set of KNIME nodes for various popular screen analysis methods that are available as community contribution in extension packages (http://www.tech.knime.org/community). The first extension is called "HCS Tools" (http://www.github.com/knime-mpicbg/HCS-Tools/wiki). The package includes readers for various file formats of image analysis software and plate readers, quality control metrics (Z-prime factors (Z′), strictly standardized mean difference (SSMD), coefficient of variance (CV)), common plate normalization methods (percent of control (POC), *z*-score, *B*-score), an IC50 calculation and plotting tool, a very powerful plate heatmap viewer, as well as plate utilities for easy handling of barcodes, wells, and plate layouts.

The second set of extensions is the Open Source scripting integration framework for KNIME (https://www.github.com/knime-mpicbg/knime-scripting/wiki). Even for software developers it is often a challenge to implement complicated statistic or data-mining methods. Instead of investing time and resources in such tasks, it is more efficient to exploit methods that were already developed in software for statistical computing such as R or Matlab. Although KNIME already implemented nodes to execute R-scripts, we devised our own integration to simplify the use of R scripts for biologists without any scripting background. We implemented a new open-source scripting integration framework for KNIME, which is based on templates. Its main purpose is to hide the script complexity behind a user-friendly graphical interface (GUI). Furthermore, our approach goes beyond the existing integration of R as it provides better and more flexible graphics support, flow variable support, and an easy-to-extend server-based script template repository.

The implementation is based on RGG (an R GUI generator http://www.rgg.r-forge.r-project.org/) to provide a graphical user interface instead of a scripting interface. It allows us to further integrate the scripting languages Python, Matlab, and Groovy. Using these tools we have developed many statistical methods and plotting functions in KNIME.

3. Workflow of an HCS Analysis

3.1. Assay Development

Assay development is an iterative process whereby experimental conditions of the assay are optimized to obtain the greatest, reliable separation between positive and negative controls. In contrast to homogeneous assays that measure mostly one or two parameters, HCS can measure a plethora of parameters, some of which are useful for describing the phenotype under study and some of which are not. To complicate matters further, both the image acquisition and the image analysis are often optimized in parallel during assay development to better measure the observed phenotypes. Thus, assay development in HCS involves a combination of optimization of experimental conditions, parameters, acquisition settings, and data analysis methods. To carry out this complex task CellProfiler and KNIME are ideally suited thanks to their flexibility.

3.2. Loading and Preprocessing Images

Our facility has several high content imagers such as the OPERA and Operetta from Perkin Elmer, the BD Pathway 855, and the Arrayscan from Cellomics. These platforms produce images in various formats since standards are lacking in the imaging community. CellProfiler can read all of the formats with a "LoadImages" module using the BIO-Formats library of LOCI. In this module, metadata such as plate name, date, well, or field can be extracted from the file name and the path using regular expressions. Once images are loaded, in a pipeline, they often require to be corrected for uneven illumination, which can be carried out with the "IlluminationCorrectionCalculate" and "IlluminationCorrectionApply" modules. Automated imagers focus automatically using either hardware or software focusing (or both for the BD Pathway). There is generally a proportion of out-of-focus images in large screens, especially when using high-magnification lenses. The "MeasureImageQuality" module of CellProfiler can be used to flag out of focus images. Alternatively, data from out-of-focus images can be removed in subsequent analysis steps (see Subheading 3.9), as they often represent outliers.

We analyze our images with CellProfiler installed on a Linux cluster to speed up this calculation-intensive process. The cluster is a 17-node cluster (1 master node, 16 compute nodes). Each node has 8 cores with 2.26 GHz and 48 GB of RAM. The master node shares 24 TB of disk space out to the compute nodes over NFS.

3.3. Segmenting Objects

The most crucial and computer-intensive step in image analysis is object segmentation. A common task in most image analysis routines is to detect nuclei as primary objects and cytoplasm of cells as secondary objects around the nuclei. We generally stain nuclei with Hoechst and the cytoplasm with HCS CellMask Blue (Life Technologies). Nuclei are detected using Otsu's three-class thresholding and fused objects are separated by finding local maxima followed by watershed. These methods are available in the "IdentifyPrimaryObjects" module. Next, using the "DetectSecondaryObjects" module, the stained cytoplasms are detected by a propagation algorithm using local image similarity and the nuclei as seeds. The local thresholds are again determined by using multi-threshold Otsu. A further common task in many biological assays is to detect tertiary objects such as organelles in the cytoplasm of cells. For small structures with high background intensities, a white top-hat filter is applied using the "EnhanceOrSuppressFeatures" module to increase the contrast. The tertiary objects are then detected, using the "IdentifyPrimaryObject" module with multi-threshold Otsu and watershed seeded from local maxima. Finally, to obtain tertiary object data on a cell-by-cell basis, the objects are linked to their parent cell with a "RelateObjects" module.

3.4. Extracting Parameters

After segmenting all the objects of interest, features are extracted for the quantitative phenotypic description. Several measurement modules are available in CellProfiler for both image-based measurements and object-based measurements. Using a "MeasureObjectIntensity" module, many aspects of the intensity of objects can be captured such as their integral, average, or median intensity. Further intensity properties are the maximum, minimum, upper, and lower quantile or the standard deviation of the intensity within an object. Depending on the nature of the object and the possible phenotypes, any of these intensity attributes can be used to describe the phenotype of interest. Other parameters describe the morphology

("MeasureObjectSizeShape"), texture ("MeasureTexture"), or spatial distribution of objects ("MeasureObjectNeighbors"). Extracting all these parameters can easily generate tables of several hundreds of columns. All these features can be used to describe the objects either for statistical analysis or directly within a pipeline to select objects for further processing and segmenting. For instance, in a mixed population of large, irregular cells and small round cells, it is possible to select with "FilterObject" module one population based on cellular feature to further segment tertiary objects. In general, we extract many parameters during initial image analysis to avoid a rerun of the analysis, since segmentation is a computationally intensive work. In this manner, it is possible to select relevant parameters later in the workflow of a screen analysis (see Subheading 3.8).

3.5. Control Images and Exporting Data

When analyzing large screens, it is impossible to visually control the quality of segmentation of all the images. A spot control needs to be carried out and control images showing the outlines of the segmented objects are very useful for this. Furthermore, it is paramount to visualize all the images of hit wells to ensure that no spurious segmentation artifact leads to the result. In CellProfiler it is possible to create color images of three- or four-channel images with the "GrayToColor" module. The outlines of each segmented object can be saved in the corresponding segmentation module and overlaid on the color merge image with the "OverlayOutlines" module. The image can then be saved with the "SaveImages" module in many different formats and the file name and path stored in the data output file. Thus, after identifying hits, the control images can be automatically retrieved within KNIME (see Subheading 3.13) for visualization. The last step of the analysis exports the data and associated metadata as comma-separated file (CSV file) using the "ExportToSpreadsheet" module or to a database using the "ExportToDatabase" module. If the image analysis of the entire screen is run on a cluster, the data is split into several jobs, each resulting in a separate CSV file; hence the screen results are distributed among several CSV files.

3.6. Importing Data into KNIME

To read the exported CSV result files of CellProfiler into KNIME, we first capture the path of all files with a "List Files" node. The list of paths is then connected to an "Iterate List of Files" node to load the data into a KNIME workflow. The barcode, plate row, and plate column metadata contained in the CSV files are used to associate a plate layout file (either a CSV file or a Microsoft Excel file) to an experimental condition for each line (lines representing either objects or images). This association can be carried out either with a "Joiner" or a dedicated "Join Layout" node developed by us. We have generated tables containing over 10 million lines and hundreds of columns. KNIME is able to carry out computations with files of this size on a normally equipped desktop computer. The only crucial factor is that enough free disk space is available to store temporary files.

3.7. Parameter Selection

Parameter selection is a crucial process in multiparametric screening and a full discussion on the subject is beyond the scope of this chapter. Briefly, parameters are sought that allow best separation between negative control and positive control. The parameters should describe various aspects of the phenotype and should therefore not be too much correlated to each other. Furthermore, the parameters should be robust.

A classical method to estimate the discrimination strength of parameters is the Z-prime factor (22). We have developed a KNIME node "Z-Primes" to calculate both robust Z-prime (with median and MAD) and non-robust Z-prime factors (mean and standard deviation). The node allows selecting several positive and negative controls, which is useful during assay development for determining the best controls to use during screening. In this manner, a matrix is obtained of Z-prime factors for all parameters and possible combination of positive and negative controls. Similarly the SSMD (23), which is another value describing the strength of a parameter, can be calculated using the "SSMD" KNIME node.

Lastly, we have developed two KNIME nodes to select parameters based on mutual information (24). The "Parameter Mutual Information" computes the mutual information matrix for all parameters. The "Group Mutual Information" computes the mutual information between two reference populations for a set of selected parameters. In this manner it is possible to select parameters and discover new phenotypes in a screen.

3.8. Binning of Population, z-Score Profiles

In some assays, only a subpopulation shows a phenotypic response and well averages are not able to discriminate between experimental conditions. Under these circumstances, we have found that dividing the negative control population into percentiles and comparing the population distribution in the corresponding parameter could be a useful analysis method for scoring hits. We have developed a KNIME node for this subpopulation analysis. The script partitions the sorted population of a negative control into n equal sized bins. Each bin covers a percentage of a parameter range. The script then determines the count of objects in each parameter bin for each well. Objects whose values lie either below or above the range of the negative control are attributed to either the lowest or the highest bin, respectively. The percentage of the population in each bin and each well is used to compute *z*-score values relative to the corresponding bin of the negative control wells (Fig. 1). This process is applied automatically for each parameter and each experimental condition in KNIME partitioning the population into a chosen number of bins. The final output of the node is a profile of

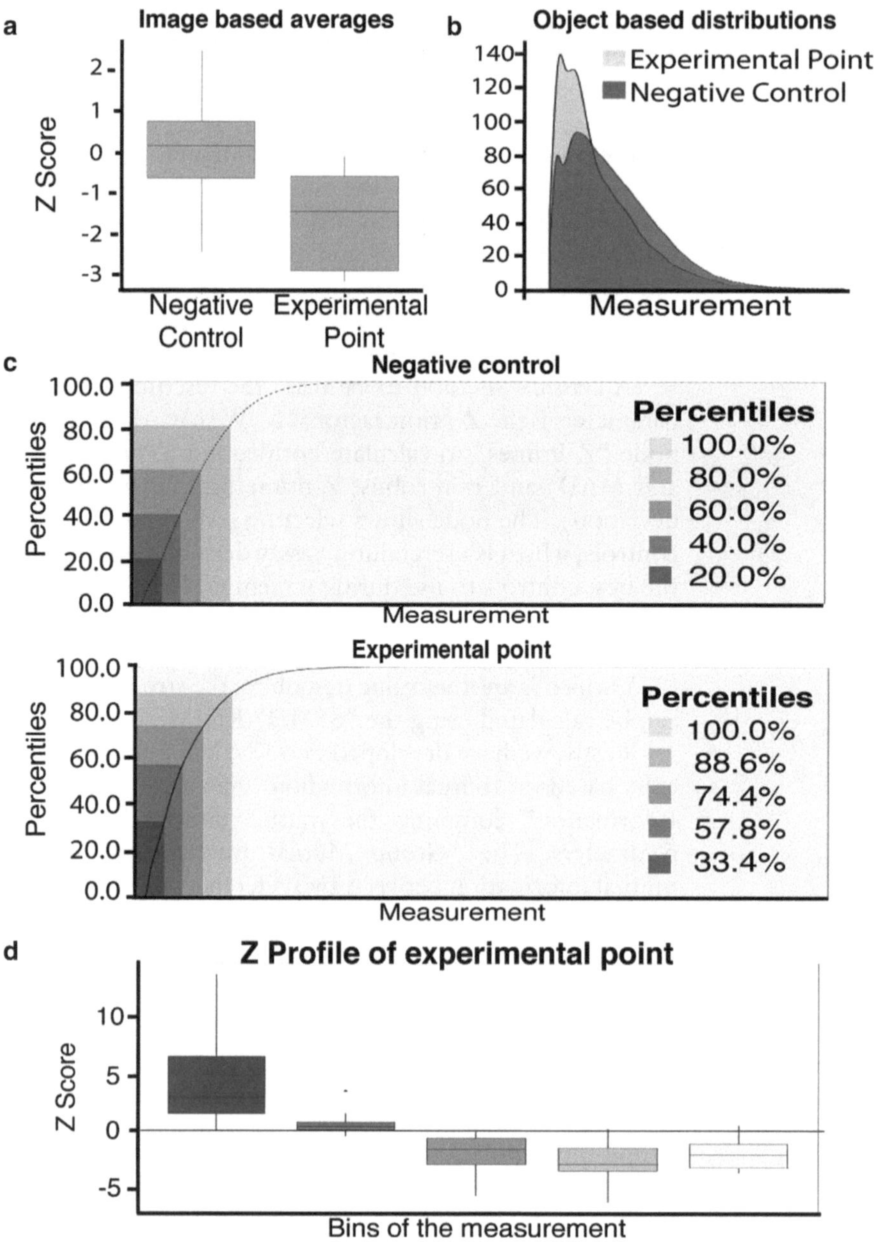

Fig. 1. Binning analysis allows detecting changes in subpopulations. Image-based averages of repeats of an experimental population might show low *z*-scores for a chosen parameter as shown by the boxplot in *panel A*. The distribution of object-based values of the same parameter in experimental populations might still show significant differences as indicated by the density histogram in *panel B* or the cumulative distribution function in *panel C*. The binning analysis subdivides the control population into a chosen number of equal sized bins (*panel C*, *upper graph*). The *upper* and *lower* parameter values of each bin divide all well populations. The population percentages in all bins are used for computing *z*-scores relative to the negative control wells. As shown in the boxplot of the percentiles in *panel D*, some subpopulations show significantly higher/lower *z*-score values than the average *z*-score in *panel A*, demonstrating the superior discriminative power of binning analysis.

n *z*-scores that we call "*z*-score profiles." Eliminating *z*-scores between +2 and −2 using a KNIME node named "Range Filter" reveals the remaining parameters whose mean is at least two standard deviations away from the control population and have some discriminatory power. In this manner it is possible to select parameters that score changes in population distributions.

3.9. Redundant Parameter Elimination

In order to eliminate parameters that are correlated to each other, we calculate their Pearson correlation coefficients (25). Linearly uncorrelated parameters have Pearson correlation coefficients close to zero and likely describe different aspects of the phenotype under study (exception for non-linearly correlated parameters which cannot be scored using Pearson's coefficient). We have developed an R template in KNIME to calculate Pearson correlation coefficients between parameters. Redundant parameters that yield Pearson correlation coefficients above 0.4 are eliminated. It is important to visually inspect the structure of the data using scatter matrices. A "Scatter Plot" and a "Scatter Matrix" node from KNIME exist that allow color-coding the controls for ease of viewing.

3.10. Robust Parameter Selection

Lastly, it is desirable that parameters are able to discriminate between positive and negative conditions in a variety of experimental conditions. In other words they should be robust and reproducible. For this purpose, the Pearson correlation coefficient between all experimental repeats using control wells is calculated. Robust parameters have high Pearson correlation coefficients (above 0.7) in pairwise comparisons of experimental repeats. For this analysis we have developed another R template in KNIME to calculate the Pearson correlation coefficient between experimental runs.

3.11. Outlier Removal

Removing outliers due to out-of-focus images, debris, or segmentation errors is an important step in analyzing large datasets, as they can introduce errors and increase erroneously the estimated variance. It is important to carry out this analysis within the same and not between experimental conditions, as this might effectively eliminate hits. For instance, outliers from the entire set of images of the negative control wells can be removed to correctly estimate the location and spread of that population. We have developed a "Remove Outliers" node in KNIME that allows selecting either of the two methods for determining outliers from a chosen population: standard deviation and box plot statistics. In the standard deviation method, objects or images whose parameter values lie more than 4 standard deviations from the mean of the chosen population can be considered outliers. In the box plot statistics method, objects or images lying outside the whiskers are removed. Useful parameters are for instance fluorescence intensity or surface of nuclei. Also images with very few or too many cells could be removed, since the segmentation tends to be unreliable in such images.

3.12. Plate Quality Control

During a screening campaign some plates and even entire batches can fail due to various reasons. To control the quality of plates, during a screen, we use the "Z-primes" and the "Multivariate Z-primes" nodes we have developed. Plates failing to fulfill a minimal criterion such as a Z-prime factor above 0.3 are eliminated from further analysis and scheduled for rescreening. A further useful node to verify the quality of plates is the "CV" node that calculates coefficients of variance for selected parameters.

Lastly, it is important to visualize plates to obtain a graphical overview of the screen. To this end we have developed a "Plate Viewer" to create heatmaps and an R template in KNIME to generate scatter plots of screening campaigns. These tools allow to visualize row and column artifacts and to compare the performance of various plate batches during a screening campaign.

3.13. Data Normalization

Due to plate-to-plate variations from different days or runs a normalizing step is necessary to render the data comparable across entire screens. We have developed several KNIME nodes for popular normalization methods in HTS such as POC, normalized percentage inhibition (NPI), standard score (*z*-score), and *B*-score (26). For all nodes, robust statistics, grouping, negative control, and parameters can be chosen. The method chosen for normalization is dependent on the screening results and the normality of the data. A full discussion on this issue is beyond the scope of this chapter and the reader is referred to excellent reviews (27, 28).

3.14. Hit Selection

Hit selection is the purpose of any screen and is therefore very important. Many different methods and criteria are applied in screening. Our method consists in calculating the separation of experimental wells from control in multiparametric space using either the Mahalanobis distance (29) or multivariate Z factor. We have implemented an R template in KNIME to calculate the Mahalanobis distance and a KNIME node for the multivariate Z-prime factor. Prior to applying this analysis it is critical to verify that the data are approximately normally distributed. We have developed an R template in KNIME to produce quantile–quantile plots and compute a Shapiro–Wilk test for normality. We have also developed an R template for Box Cox power transform in KNIME to attempt modifying data to become more normally distributed. To select a threshold when applying Mahalanobis distance, we either calculate the Chi square p value for the 99 percentile or set the threshold above the highest Mahalanobis distance of the negative control.

3.15. Clustering

Classification of hits into different phenotype classes is a debated issue in the field. Many clustering algorithms are implemented in KNIME and a discussion about the choice of algorithm to use goes beyond the scope of this chapter. However, a clustering approach can yield insights into mode of action of hits and also help discover novel phenotypes.

Our approach consists of first projecting the vectors onto a unity hypersphere. In this manner, the amplitudes of the phenotypes are ignored and only the directions of the vectors are of importance for clustering. To normalize the multiparametric vectors to a length of one we have developed a "Vector Length Normalization" node. We then calculate the cosine or Euclidian distances of the length-normalized vectors with the "Distance Matrix Calculate" node. The cosine distance reflects the angle between the vectors whereas the Euclidian distance measures the distance amongst points on the hypersphere at radius one. To obtain a feel for the likely number of clusters in the dataset we carry out a hierarchical clustering with the "Hierarchical Clustering" node. The result of the clustering is examined with the "Hierarchical Cluster View" node. Deciding on the number of clusters in the dataset is difficult and subjective. Furthermore, different clustering approaches yield different results and we therefore apply additional clustering algorithms to explore our screening data and compare the results. Another clustering method is k-Means clustering. Using the "k-Means" node in KMINE for the length-normalized vectors and loop variables for the number k of clusters, the best number can easily be determined by following the propagation of positive controls for different values for k. After classification of the data, hits are selected by applying a threshold on the amplitude of the phenotype as judged by the Mahalanobis distance (Subheading 3.12).

Deciding on an appropriate analysis is at this point subjective, but additional information such as the average Pearson correlation of profiles in a cluster or simply how positive controls cluster can be useful for judging the result of any clustering method.

3.16. QC of Hit Selection

It is important to verify the results of the hit selection. Hits can be visualized as a heatmap in the "Plate Viewer" node to determine whether certain well positions, such as edge rows or edge columns, have a higher number of hits. In the clustering approach we verify that negative control wells are found in all clusters, indicating correct length normalization and clustering.

It is also paramount to visualize the images of the identified hits. The path information and file name information of the control images generated by CellProfiler can be used to this end. In KNIME all the images of the hits can be opened using the "Picture Chooser" node. In this manner, all the control images with the corresponding object outlines can be quickly visualized.

Selected hits are rescreened and, for validated hits, a dose–response relationship is generated. We have developed a "Dose Response" node in KNIME to plot dose–response curves and to calculate IC50 values.

3.17. Screen Annotation

Information of our chemical and siRNA libraries is stored in a PostgreSQL database. We annotate our validated hits using a "DatabaseReader" and a "Joiner" node to obtain either chemical structures for chemical screens or GeneID for RNAi screens. KNIME has many tools for cheminformatics, can visualize the molecule structures, and has tools to retrieve data from external public databases via Web queries to further annotate hits, allowing clustering either by chemical substructures or GO terms.

4. Conclusion

With the availability of CellProfiler and KNIME, two very powerful Open Source software tools have been created to analyze high content screens. These tools allow creating very complex analysis workflows, yet they can be handled by noncomputer scientists. A further advantage of using these software tools is that our analysis workflows become standardized, in spite of using several different automated imaging platforms. The imaging community has still not managed to agree on standards, which makes the comparison of data across laboratories difficult or impossible. By using CellProfiler and KNIME, we have established a de facto standard for internal use and for our users.

Several conditions need to be fulfilled for image analysis and statistical software to be really useful for analyzing large datasets. For image analysis software, it is paramount that it can be run robustly on large computer clusters since calculating complex image analysis requires significant amount of processing power. The software must also have good visualization tools to allow verifying the performance of the segmentation. Image quality controls need to be available to correct for uneven illumination and out-of-focus images. All of these requirements are found in CellProfiler. For the statistical software, it is important that it can handle large datasets. Especially when analyzing screens at the individual cell level, the software must deal with millions of rows and hundreds of columns. For analyzing screens, special statistical methods are employed and they need to be readily available. Visualization of the data at each single data processing step is also extremely important for handling the data properly and gaining insights into the data. The implementation of new statistical methods and many visualization tools and their flexible extension is of great importance. KNIME with its plug-in concept and various extensions integrates all these requirements.

Lastly, both CellProfiler and KNIME are easily extendable Open Source software that both have a highly active and fast-growing community since the release of their first versions in 2005 and 2006, respectively, a key for further development and implementation of new functions and state-of-the-art methods. Together CellProfiler and KNIME offer the possibility to apply advanced analysis tools to HCS data without paying expensive licensing fees. Especially in a time of reduced research budget, it is an important aspect for academic institutes interested in acquiring high content imagers. If the costs of running high content imagers will decrease, automated imaging will become more widely applied and subsequently new applications and modalities will be developed. With academia entering the field of automated imaging, smart microscopy with advance imaging modalities in automated mode implementing feedback loops between microscope, image analysis, and control of acquisition mode might become more widely applied. Thus many applications will become automated in the future, freeing up time for researchers to carry out other tasks (30).

Acknowledgment

This work was supported by the Max Planck Gesellschaft.

References

1. Johnston AR, Powell RV (1970) Optics at the Jet Propulsion Laboratory. Appl Opt 9:271–275
2. Harmon LD, Knowlton KC (1969) Picture processing by computer. Science 164:19–29
3. Lipkin LE, Lipkin BS (1975) Computers in the clinical pathologic laboratory: chemistry and image processing. Annu Rev Biophys Bioeng 4:529–577
4. Blackwell RJ, Crisci WA (1975) Digital image processing technology and its application in forensic sciences. J Forensic Sci 20:17
5. Archer JR (2004) History evolution, and trends in compound management for high throughput screening. ASSAY Drug Dev Technol 2: 675–681
6. Newman DJ, Cragg GM, Snader KM (2003) Natural products as sources of new drugs over the period 1981–2002. J Nat Prod 66:1022–1037
7. Ortholand J-Y, Ganesan A (2004) Natural products and combinatorial chemistry: back to the future. Curr Opin Chem Biol 8:271–280
8. Giuliano KA, DeBiasio RL, Dunlay RT et al (1997) High-content screening: a new approach to easing key bottlenecks in the drug discovery process. J Biomol Screen 2:249
9. Verkman AS (2004) Drug discovery in academia. Am J Physiol: Cell Physiol 286: C465–C474
10. Cressey D (2011) Drug-maker plans to cut jobs and spending as industry shies away from drug discovery. Nature 470:154
11. Holt R (2011) Dueling visions for science. Science 333:1549
12. Gulledge J (2011) Debt crisis: crunch time for US science. Nature 477:155–156
13. Hunter P (2010) Facing the credit crunch. EMBO Rep 11:924–926
14. Roy A, McDonald PR, Sittampalam S, Chaguturu R (2010) Open access high throughput drug discovery in the public domain: a mount everest in the making. Curr Pharmaceut Biotechnol 11:764–778
15. D'Ausilio A (2012) Arduino: a low-cost multi purpose lab equipment. Behav Res Meth. 44(2):305–313
16. Santos AF, Zaltsman AB, Martin RC et al (2008) Angiogenesis: an improved in vitro biological system and automated image-based workflow to aid identification and characterization of angiogenesis and angiogenic modulators. ASSAY Drug Devel Technol 6: 693–710
17. Carpenter A, Jones T, Lamprecht M et al (2006) Cell Profiler: image analysis software for identifying and quantifying cell phenotypes. Genome Biol 7:R100

18. Kamentsky L, Jones TR, Fraser A et al (2011) Improved structure, function and compatibility for CellProfiler: modular high-throughput image analysis software. Bioinformatics 27: 1179–1180
19. Bray M-A, Fraser AN, Hasaka TP et al (2011) Workflow and metrics for image quality control in large-scale high-content screens. J Biomol Screen 17:135–143
20. Jones R, Carpenter E, Lamprecht R et al (2009) Scoring diverse cellular morphologies in image-based screens with iterative feedback and machine learning. PNAS 106:1826–1831
21. Jones T, Kang I, Wheeler D et al (2008) CellProfiler analyst: data exploration and analysis software for complex image-based screens. BMC Bioinformatics 9:482
22. Zhang J-H, Chung TDY, Oldenburg KR (1999) A simple statistical parameter for use in evaluation and validation of high throughput screening assays. J Biomol Screen 4:67–73
23. Zhang X. D, (2007) A pair of new statistical parameters for quality control in RNA interference high-throughput screening assays, Genomics 89:552–561
24. Moddemeijer R (1999) A statistic to estimate the variance of the histogram based mutual information estimator based on dependent pairs of observations. Signal Process 75:51–63
25. Pearson K (1896) Mathematical contributions to the theory of evolution. III. Regression, heredity and panmixia. Phil Trans Roy Soc London Ser A 187:253–318
26. Brideau C, Gunter B, Pikounis B et al (2003) Improved statistical methods for hit selection in high-throughput screening. J Biomol Screen 8:634–647
27. Birmingham A, Selfors LM, Forster T et al (2009) Statistical methods for analysis of high-throughput RNA interference screens. Nat Meth 6:569–575
28. Shun TY, Lazo JS, Sharlow ER et al (2011) Identifying actives from HTS data sets. J Biomol Screen 16:1–14
29. Mahalanobis PC (1936) On the generalised distance in statistics. Proc Natl Inst Sci India 2:49–55
30. Conrad C, Gerlich DW (2010) Automated microscopy for high-content RNAi screening. JCB 188:453–461

Chapter 9

PARP Inhibition as a Prototype for Synthetic Lethal Screens

Xuesong Liu

Abstract

Although DNA damaging chemotherapy and radiation therapy remain the main stay of current treatments for cancer patient, these therapies usually have toxic side effect and narrow therapeutic window. One of the challenges in cancer drug discovery is how to identify drugs that selectively kill cancer cells while leaving the normal cell intact. Recently, synthetic lethality has been applied to cancer drug discovery in various settings, and has become a promising approach for identifying novel agents for the treatment of cancer. A prototypical example is the synthetic lethal interaction between PARP inhibition and BRCA deficiency. PARP inhibitors represent the most advanced clinical agents targeting specifically DNA repair mechanisms in cancer therapy. In this chapter, I will review the molecular mechanism for this synthetic lethality and the clinical applications for PARP inhibitors. I will also discuss the formats of synthetic lethal screens, current progress on the utilization of these screens, and some of the advantages and challenges of synthetic lethal screens in cancer drug discovery.

Key words: Synthetic lethal screen, DNA repair, PARP, BRCA, BER, HR

1. Introduction

The revolution of biomedical research in the twentieth century began with Watson and Crick's discovery of the DNA double helix and continues to this day. The identification of DNA as genetic information transformed cancer research from a mere descriptive science to that entailing the elucidation of molecular mechanisms underlying the disease. Since the discovery of the first oncogene, Src, by Michael Bishop and Harold Varmus in the 1970s (1), our knowledge of cancer has exploded due to advances on many frontiers of cancer research, and so has the effort to discover and develop anticancer drugs, which has a direct impact on the improvement of the quality of life for cancer patients.

Jürgen Moll and Riccardo Colombo (eds.), *Target Identification and Validation in Drug Discovery: Methods and Protocols*, Methods in Molecular Biology, vol. 986, DOI 10.1007/978-1-62703-311-4_9,

One of the challenges with cancer drug discovery is how to identify agents that kill cancer cells specifically while leaving normal cells intact. Although DNA damaging chemotherapy and radiation therapy remain the mainstay of current cancer treatment, these therapies usually cause severe side effects. Since most of them were selected solely on their ability to kill rapidly dividing cells in culture and in animal tumor models without substantial molecular basis for cancer cell specificity, it is not surprising that these drugs have a relatively narrow therapeutic window.

One of the hall marks of cancer is that certain cancer cells display a phenotype of "oncogene addiction," where the initiation and maintenance of a particular type of tumor cell is dependent on the activation of a specific oncogene (2). Typical examples of such oncogenes include EGFR in non-small cell lung cancer (NSCLC), HER2 in breast cancer, BCR-ABL in chronic myeloid leukemia (CML), and EML4-ALK in NSCLC. These oncogenes frequently activate signaling pathways that promote cancer growth and survival through, for example, over-expression or gain of function mutation. The initiation and maintenance of tumorigenesis for these cancer cells are dependent on these addictive oncogenes. Drugs that target specifically addictive oncogenes have been developed and shown great clinical benefit in treating subgroups of cancer patients who harbor the respective driver oncogenes. For example, EGFR drives cell proliferation and survival in cells containing either EGFR amplification or activating mutations. Drugs approved by the FDA as targeted therapy for NSCLC patients with EGFR amplification or activation mutation in the clinic include both small molecule inhibitors (gefitinib, erlotinib, and lapatinib) and biologics (cetuximab and panitumumab). Other approved targeted agents include for example Bcr-Abl inhibitors (imatinib, nilotinib, and dasatinib), anti-Her2 mAb (trastuzumab), or ALK inhibitor (crizotinib). However, not all tumor cells harbor a well-characterized driver oncogene and tumor cells are usually heterogeneous. Since there are a large number of genetic alterations present in a given tumor sample, it is difficult to differentiate a driver mutation from the plethora of coincidental passenger mutations. Of those that are well characterized, not all are currently amenable to targeted therapy. Direct targeting of driver oncogenes including K-Ras, c-Myc, and many oncogenic transcription factors are currently not feasible since such targets have been pharmacologically intractable with current approaches to develop small molecule inhibitors. Novel strategies to target tumors harboring such oncogenes, as well as tumors that do not display well-characterized oncogenes, are therefore in urgent need to improve survival in patients affected with such cancers.

2. Synthetic Lethality

2.1. An Attractive Approach for Cancer Drug Discovery

Recently, synthetic lethality has been applied to cancer drug discovery, hoping to discover and develop novel cancer therapies that specifically kill cancer cells and spare normal cells. The concept of synthetic lethality was first described in lower organisms such as yeast or drosophila (3–5). Two genes are considered synthetic lethal if mutations of both genes cause cell death while mutation of either one alone has little effect on cell viability. The synthetic lethal genes can function in related pathways or along a single pathway, where simultaneous mutations of both genes, rather than just one, severely affect the function of pathways that are critical to cell survival. Cancer cells arise from normal cells through the accumulation of a series of genetic lesions. These tumor cell-specific genetic lesions can be used to exploit their synthetic lethal partner gene whose loss of function or inhibition with pharmacological agents will in principle result in selective tumor cell death while leaving normal cells intact. Thus synthetic lethality provides a rational approach to the development of cancer cell-selective targeted therapeutics with an improved therapeutic window (5).

Cancer cells are usually derived from normal cells through the activation of oncogenes and loss of tumor suppressor genes. Many addictive oncogenes are not currently amenable to direct pharmacological inhibition, and there are limited options to restore the loss of tumor suppressor genes. While gene therapy has been proposed as an option to restore tumor suppressor gene function and thereby inhibit tumor growth, this technology remains in its infancy and the approach has yet to be clinically validated.

Synthetic lethality has been applied as a promising strategy for these pharmacologically evasive oncogenes as well as tumor suppressor genes (5, 6). The concept is to identify a drug that selectively kills tumor cells through a mechanism of synthetic lethality or to find a synthetic lethal partner gene whose protein product can be exploited in targeted therapy. A prototypical example is the synthetic lethal interaction between PARP inhibition and BRCA deficiency. PARP inhibitors represent the most advanced clinical agents targeting specifically DNA repair mechanisms in cancer therapy. In this chapter, I review the molecular mechanism for this synthetic lethality and the clinical applications for PARP inhibitors. I also review the formats of synthetic lethal screens and current progress on the utilization of these screens. In addition, I discuss some of the advantages and challenges of these synthetic lethal screens.

2.2. PARP Inhibition Is Synthetic Lethal in Cancer Cells Deficient in BRCA1/2

Mammalian DNA is under constant attack by exogenous insults including ionizing radiation and carcinogenic compounds as well as endogenous stresses such as reactive oxygen species (ROS) generated through cellular metabolism. Single strand breaks (SSB) and

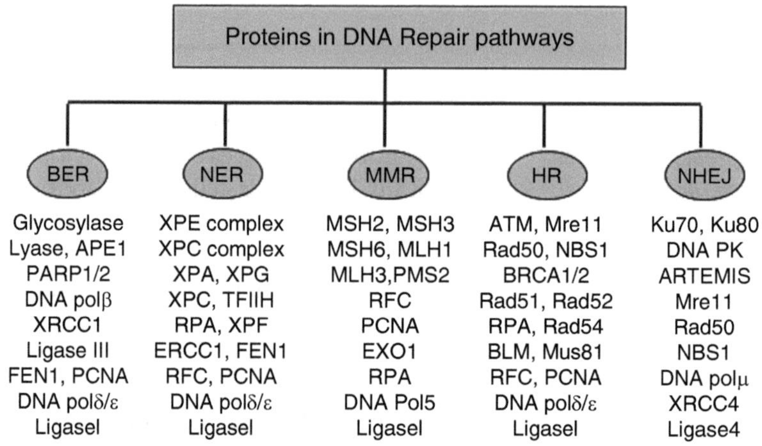

Fig. 1. Proteins in DNA repair pathways. DNA repair proteins are listed for each of the following pathways: BER (Base Excision Repair), NER (Nucleotide Excision Repair), MMR (Mismatch Repair), HR (Homologous Recombination), and NHEJ (Nonhomologous End Joining). PARP1/2 and BRCA1/2 are relevant in BER and HR pathways, respectively.

double strand breaks (DSB) are the major forms of DNA damage generated by both exogenous and endogenous stress in mammalian cells. It has been estimated that about 10,000 spontaneous SSB events occur inside a human cells every day (7). If not repaired, the damaged DNA leads to mutation and can cause genetic instability. In order to maintain genome integrity, mammalian cells are equipped with multiple DNA repair pathways to deal with various stress-induced DNA damage.

DNA repair pathways can be divided into those that respond to SSB and those that respond to DSB. SSB repair pathways include base excision repair (BER), mismatch repair (MMR), and nucleotide excision repair (NER). DSB repair pathways include nonhomologous end joining (NHEJ) and homologous recombination (HR). The proteins involved in these DNA repair pathways are shown in Fig. 1.

Poly (ADP-ribose) polymerase-1 (PARP-1) is the founding member of a family of proteins that share a catalytic PARP homology domain and are characterized by their ability to poly (ADP-ribosyl)ate their protein substrates (8–10). Of the PARP family members identified to date, PARP-1 and PARP-2 are unique in exhibiting stimulation of catalytic activity in response to DNA damage (11). Activation of PARP-1 and PARP-2 is an immediate eukaryotic cellular response to DNA damage induced by a variety of stimuli including ionizing radiation, alkylating agents, and oxidants (9). Activated PARP-1 and PARP-2 bind to DNA strand breaks, and covalently attach poly(ADP-ribose) to nuclear proteins including PARP-1 itself, histones, and transcription factors (8). Both PARP-1 and PARP-2 are essential for the repair of single strand DNA breaks through base excision repair (BER) (8).

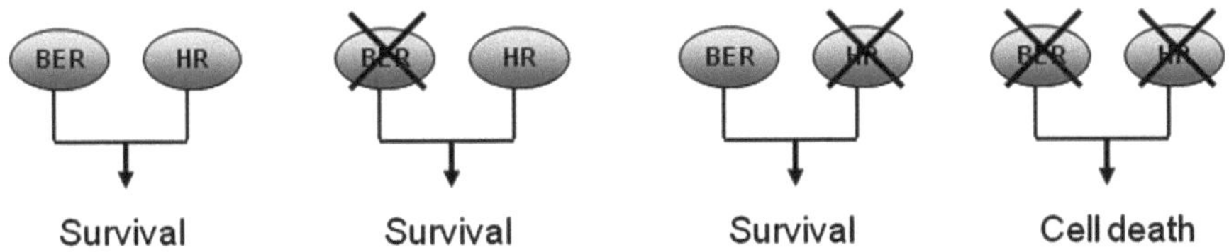

Fig. 2. Synthetic lethal interaction between BER and HR. BER and HR are DNA repair pathways for SSB and DSB, respectively. Inhibition of BER or HR alone does not cause cell death, while simultaneous inhibition of BER and HR results in cell death.

In response to DSB, the Mre11/Rad50/NBS1 (MRN) complex is recruited to the DNA damage site triggering the activation of ATM, ATR, and DNA-PK family members to phosphorylate Ser139 on the carboxyl tail of H2AX. Phosphorylated H2AX (γH2AX) plays an important role in the recruitment of DNA repair proteins such as BRCA1, BRCA2, MDC1, and 53BP1, followed by recruitment of Rad51. Rad51 is a central DNA recombinase which catalyzes homologous pairing and strand exchange in the HR pathway. BRCA1 plays critical roles in the cell cycle checkpoint and recruitment of other DNA repair proteins in the HR pathway, while BRCA2 promotes the assembly of Rad51 filaments on ssDNA generated in the HR DNA repair process. Cancer cells with deletion of either BRCA1 or BRCA2 are defective in HR repair.

BRCA1 and BRCA2 are frequently mutated in a variety of human cancers. Breast cancer is the most common malignancy in women. A total of more than a million women will be diagnosed with breast cancer worldwide with more than 400,000 deaths annually. The germ line mutations of BRCA1/2 are associated with hereditary breast cancer which account for 5–7% of all breast cancer cases (12, 13). Ovarian cancer is the leading cause of death from gynecological cancer with more than 140,000 deaths annually worldwide. The germ line and somatic BRCA1/2 mutations have been reported to occur in approximately 10% of ovarian cancer patients (14, 15). BRCA1/2 mutations have also been shown to occur frequently in Ashkenazi Jewish families with pancreatic cancer (16).

The synthetic lethal effect between PARP inhibition and BRCA1/2 deficiency was first described in 2005 by Farmer et al and Bryant et al in preclinical studies (17, 18). BRCA1/2 deficient cell lines are more sensitive to PARP inhibition than BRCA1/2 proficient cell lines in colony formation assays and in xenograft models (17, 18). BER and HR are DNA repair pathways that carry out the repair of SSB and DSB, respectively. Inhibition of either pathway alone is not lethal, but loss of both pathways causes cell death (Fig. 2). Mechanistically, PARP inhibition blocks the activity of BER and leads to DNA SSB accumulation inside cells. When cells enter the next round of DNA replication, these DNA SSBs are converted to DNA DSBs which can be efficiently repaired through

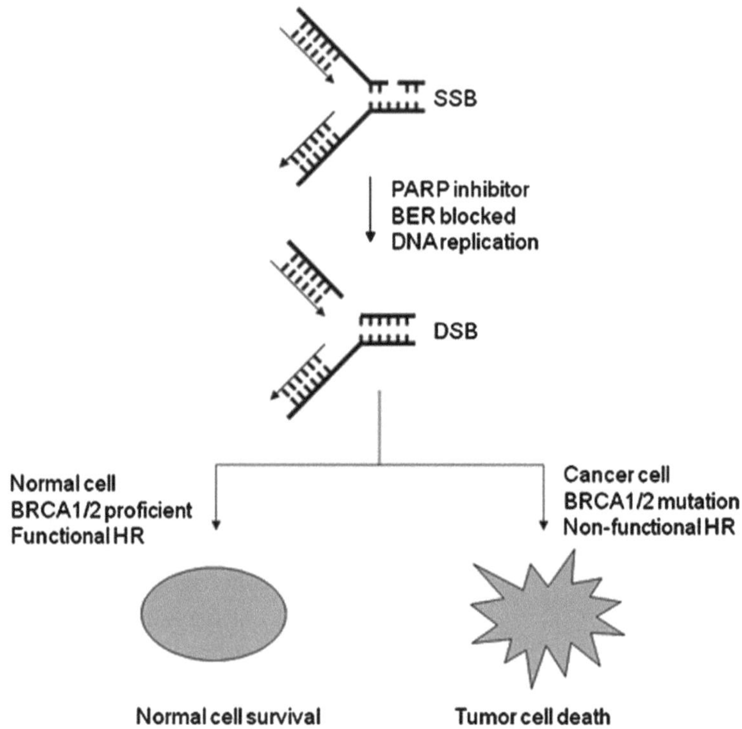

Fig. 3. Mechanism of synthetic lethality between PARP inhibition and BRCA deficiency. PARP inhibition blocks the activity of BER and leads to DNA SSB accumulation inside cells. When cells enter the next round of DNA replication, these DNA SSBs are converted to DNA DSBs which can be efficiently repaired through the HR pathway in normal cells. BRCA deficient cancer cells are unable to repair the DNA DSBs that accumulate following PARP inhibition. This eventually causes toxicity and cancer cell death.

the HR pathway. Synthetic lethality between PARP inhibition and BRCA deficiency arises, because BRCA deficient cancer cells are unable to repair the DNA DSBs that accumulate following PARP inhibition. This eventually causes toxicity and cancer cell death (Fig. 3). On the other hand, normal cells with at least one copy of wild-type BRCA have sufficient HR DNA repair capacity and are able to repair DNA DSBs that accumulate following PARP inhibition (Fig. 3).

A number of PARP inhibitors have entered clinical trials over the last several years to investigate the effectiveness of these agents in various oncology settings. Most of these small molecule inhibitors are based on a nicotinamide-like pharmacophore, owing to their mechanism as competitive binders to the NAD^+ pocket within PARP-1/2 (19). In contrast, one of these compounds, iniparib, is reported to be a pro-drug whose nitroso metabolite, 4-iodo-3-nitrosobenzamide, selectively kills tumor cells by oxidizing the zinc finger of PARP-1 resulting in ejection of zinc and inhibition of PARP activity (20). Although iniparib has been claimed to be a PARP inhibitor and is already in Phase-III clinical trial in combination with gemcitabine and carboplatin, we have recently shown that iniparib reacts nonspecifically with many cysteine-containing proteins and is not a bona fide PARP inhibitor (21). The primary mechanism of action for iniparib is likely not via inhibition of PARP activity, which will have important implications for the clinical development of this compound and selective PARP inhibitors.

Recent data from clinical trials of some of the NAD^+-competitive PARP inhibitors showed promising synthetic lethal effect in cancer patients with BRCA1/2 deficiency. Representatives of these potent PARP inhibitors include veliparib, olaparib, and MK-4827. Results from two open-labeled phase-II studies of olaparib have been reported in germ line BRCA1/2 mutation carriers with advanced breast (ICEBERG1) and ovarian (ICEBERG2) cancers (22, 23). At a maximal tolerated dose (MTD) of 400 mg twice daily, an objective response rate (ORR) of 41% (11 of 27) was reported in the ICEBERG1 study in breast cancer patients with BRCA1/2-mutation. Among these 11 patients, one patient had a complete response (CR) and 10 patients had partial response (PR) using Response Evaluation Criteria In Solid Tumors (RECIST). In the ICEBERG2 study, an ORR of 33% (11 of 33) was achieved for ovarian cancer patients who received a dose of 400 mg twice daily. Another phase-II study showed an ORR of 41% (7 of 17) for ovarian cancer patients with BRCA1/2 mutations at 400 mg twice a day dose of olaparib (24), although caution has to be taken that olaparib may not be better than the current standard of care in these ovarian cancer patients with BRCA1/2 mutation (25). Promising antitumor activity has been reported in both BRCA-deficient and sporadic cancers in a phase-I trial for MK-4827 (26). In addition to the single agent activity for PARP inhibitors, combination of DNA damaging agents also represents an attractive treatment strategy for BRCA deficient cancers. Combination of veliparib with TMZ showed promising antitumor activity in BRCA-deficient patients in clinical trials (27).

The synthetic lethal effect for PARP inhibitors goes beyond germ line or sporadic BRCA1/2 mutations. PARP inhibition has also been shown to be synthetic lethal with compromised HR DNA repair activity in cancer cells with wild-type BRCA1 and BRCA2 (so-called "BRCAness"). BRCAness describes a phenotype for certain cancer cells that display similarity to BRCA1/2 deficiency in terms of their HR DNA repair activity. BRCAness can result from many conditions. Inactivation mutation or deletion of DNA repair proteins in HR pathway will generate BRCAness phenotypes. Moreover, post-translational modification or epigenetic regulation also plays a key role here. For example, epigenetic regulation of the BRCA1 genes, especially through promoter methylation, induces BRCAness by reducing the expression of BRCA proteins and renders cancer cells sensitive to PARP inhibition (28, 29).

Down-regulation of BRCA protein or interference of Rad51 foci formation has been reported for a variety of conditions including inhibition of CDKl (30), mild hyperthermia (31), hypoxia (32), loss of PTEN (33, 34), loss of PALB2 (35), or over-expression of EMSY (36). The BRCAness phenotype induced under all these conditions makes cancer cells more sensitive to PARP inhibition. The key is to develop a functional assay for BRCAness to enable stratification of cancer patients for clinical trials of PARP inhibitors.

3. Methods: Formats of Synthetic Lethal Screens

In order to identify synthetic lethal partners or agents for tumor driver oncogenes or tumor suppressor genes, synthetic lethal screens have been applied to cancer drug discovery in various settings. Usually an isogenic pair of cell lines which differ only on a tumor suppressor gene or oncogene are employed in the screens. The goal here is to identify agents or synthetic lethal partners whose inhibition will kill these cancer cells that carry the driver oncogene or lose the tumor suppressor gene, but have little effect on the wild-type control cells. The beauty of this screen is that the wild-type control cells function as a filter to triage these agents that kill cells nonspecifically. Here I review two formats of synthetic lethal screens.

3.1. Synthetic Lethal Screen of Small Molecule Compound Library

Hartwell et al. were the first to employ synthetic lethal screening of a small molecule library consisting of FDA-approved chemotherapeutic agents to identify drugs that can kill yeast cells defective in DNA repair (37). They also suggested that this approach can be applied to other tumor suppressor genes that are not considered to be ideal drug targets (37). Currently, synthetic lethal screening is applied in human cancer drug discovery, not only for these tumor suppressor genes, but also for tumor driver oncogenes (5). A number of synthetic lethal screens of small-molecula compound libraries have been conducted using isogenic human cell lines with cancer-relevant genetic alterations. VHL is a tumor suppressor gene whose loss of function is associated with a variety of tumors, especially in the development of renal cell carcinoma (RCC). Turcotte et al. conducted a synthetic lethal screen of a small molecule library of 64,000 compounds and identified STF-62246, which selectively induced autophagy in cancer cells defective in VHL (38).

Activating Ras mutations have been found in more than 25% of human tumors. However, a drug-like small molecule inhibitor for Ras protein is not currently available. Several synthetic lethal screens have been carried out using isogenic cell lines for Ras. Torrance et al. conducted a synthetic lethal screen of a library of about 30,000 compounds and identified some novel hits including triphenyltetrazolium and a sulphinylcytidine derivative (SCD) that kill colon carcinoma cells with mutant K-Ras (39). Shaw et al. carried out a similar screen using isogenic MEF cells harboring a conditional oncogenic K-Ras^{G12D} allele. They identified a class of compounds represented by lanperisone that selectively kill K-Ras^{G12D} expressing cells through the induction of oxidative stress (40). In a separate screen, Dolma et al. screened chemical libraries using isogenic cell lines for H-Ras (41). This screen identified erastin as a compound that selectively kills tumorigenic cells harboring oncogenic H-Ras. One of the targets that erastin interact with is the mitochondrial voltage-dependent anion channels (VDACs) (41). Thus, it has been proposed that VDACs are a synthetic lethal partner of H-Ras. It has also been proposed that inhibition of VDACs by erastin results in mitochondria dysfunction, release of ROS, and oxidative cell death in cancer cells with oncogenic H-Ras (41). However, the molecular mechanism on how erastin induces oxidative cell death is not well-understood.

It is of great promise that synthetic lethal screening of chemical libraries can generate potential tools for target identification, validation, lead compounds for further optimization, or even potential drug candidates. For some compounds identified from synthetic lethal screening, the respective protein targets are well characterized. However, for a majority of hits, the pharmacological targets are still unknown. Hence, the challenge is how to identify the synthetic lethal target with which a compound interacts. For example, the targets of triphenyltetrazolium and lanperisone, synthetic lethal compounds for K-Ras, remain elusive. A common approach for target identification involves generation of affinity resins with the compounds and pull-down assays for potential protein targets to be determined by mass spectrometry. This method had been applied to identify the target, VDACs, for erastin (41). However, it is not always straightforward to identify the synthetic lethal target through affinity pull-down assay and mass spectrometry, due to reasons such as the availability of SAR information for the compounds, availability of positions in the compound to be coupled to a solid resin, the abundance of the target protein, and the specificity of the compound to its target.

3.2. Synthetic Lethal Screen of shRNA/siRNA Libraries

shRNA and siRNA are powerful tools for silencing genes in cancer cells. The employment of shRNA/siRNA to knock down its corresponding gene generates a multitude of opportunities in cancer target discovery, especially in the screening for the synthetic lethal partner of a known oncogene or tumor suppressor gene. A number of screens have been carried out using shRNA/siRNA libraries for a list of oncogenes and tumor suppressor genes. The shRNA/siRNA library could consist of target genes from the whole genome or consist of a focused library for a protein family such as target genes for all protein kinases.

Several synthetic lethal screens of shRNA/siRNA libraries have been conducted for oncogenic Ras using isogenic cell lines. Luo et al. undertook a genome-wide shRNA screen for K-Ras mutated cells and identified synthetic lethal pathways involving mitotic kinase PLK1, APC/C, and the proteasome (42). Similar screens have identified other potential target pathways in K-Ras mutant cell lines, including the noncanonical IκB kinase TBK1 and a serine/threonine kinase STK33 (43). In addition, a synthetic lethal interaction between K-Ras and CDK4 has been reported (44). These studies indicate that there could be multiple pathways that are synthetic lethal with mutant K-Ras. High-throughput screening of shRNAs for 88 kinases has been carried out to identify synthetic lethal kinases for VHL. The shRNA hits from this screen include CDK6, Met, and MEK1, whose knockdown by shRNA selectively inhibits the viability of VHL deficient renal cell carcinoma cells (45). Proteins of the mismatch repair pathway (MMR) such as MLH1, MSH2, and MSH6 are frequently mutated in colon cancer carcinoma. In order to identify synthetic lethal genes for MMR deficiency, Martin et al. carried out a screen of shRNA from a kinase library, and identified PTEN-induced putative kinase 1 (PINK1) as synthetic lethal partner (46). Inhibition of PINK1 by shRNA induced elevation of ROS and oxidative DNA damage in cancer cells with MMR deficiency, but not in cells with proficient MMR (46).

All the synthetic lethal screens of siRNA/shRNA library mentioned above utilize a pair of isogenic cells for a known genetic alteration. In fact, large scale synthetic lethal screens can be conducted for more comprehensive study using a panel of cancer cell lines with well-characterized genetic background. As a first attempt, Brough et al. carried out a siRNA screen for a large panel of breast cancer cell lines to identify protein kinase targets that are required for cell survival (47). The panel of breast cancer cell lines have been profiled using transcript microarray, array-based comparative genomic hybridization (aCGH) and gene mutation (47). As positive controls, PIK3CA and ERBB2 are identified as oncogenic drivers for breast cancer cells with PIK3CA mutation and ERBB2 amplification, respectively. Moreover, novel protein kinase targets, TTK and ADCK2, have been identified as potential synthetic lethal partners for PTEN mutation and ER expression in breast cancer cells, respectively (47). Similar screens can be conducted for other solid cancer cell line panels from lung, colon, prostate, and other tumor types.

Synthetic lethal screening of a shRNA/siRNA library will directly result in the identification of the synthetic lethal target for the corresponding oncogene or tumor suppressor gene. Ideally, the identified target is tractable to pharmacological inhibition and can be used to screen small molecule compound libraries to identify potential drugs or candidates for optimization into lead compounds.

However, caution has to be taken about the results from siRNA/shRNA synthetic lethal screen. First, siRNA/shRNA often has off-target activity. Hence, the results from a siRNA/shRNA screen have to be validated through extensive biochemistry and molecular biology studies. For example, STK33 was initially identified and reported as a synthetic lethal partner for K-Ras mutation (43). However, this synthetic lethal interaction has since been called into question. Babij et al. carried out a comprehensive study using both siRNA and small molecule inhibitors of STK33, and they showed that inhibition of STK33 had no effect on K-Ras signaling or the viability of cancer cells with K-Ras mutation (48). This result suggests that the STK33 shRNA used in the initial shRNA screen may have had off-target activity. It maybe the unidentified off-target activity, but not STK33, that showed synthetic lethal interaction with K-Ras mutation, although it is also possible that the siRNA/shRNA has context-dependent effects (49). One approach commonly used to validate siRNA/shRNA screening results is to carry out a rescue experiment where the effect of shRNA/siRNA can be overcome by re-expression of the target protein using a mutated cDNA construct that is resistant to the siRNA/shRNA (Fig. 4). If the shRNA/siRNA is located in the 5′-UTR or 3′-UTR, a rescue construct with the coding sequence should work (Fig. 4a). If the shRNA/siRNA is located in the coding sequence, site-directed mutagenesis in the rescue construct can be carried out by introducing silent mutations at each residue corresponding to a codon's wobble position in the corresponding cDNA sequence targeted by the shRNA/siRNA so that the mutated construct still expresses the wild-type protein (Fig. 4b). If the phenotype of the shRNA/siRNA can be rescued, the effect of shRNA/siRNA is specific. Otherwise, the shRNA/siRNA is likely to have off-target activity (Fig. 4c). Second, it has been well documented that the biological phenotype may be different between knockdown of a protein and pharmacological inhibition of its activity (50). A lot of proteins have scaffolding functions unrelated to their enzymatic activity. Experiments can be designed to test whether the enzymatic activity of an identified target through siRNA/shRNA screen is critical for its function. In this case, a catalytically dead mutant will be generated from the rescue construct that are resistant to siRNA/shRNA treatment. If the enzymatic activity is important for function of the identified target, the expression of the catalytically dead mutant protein should not be able to rescue the effect of siRNA/shRNA. The two forms of synthetic lethal screens described here are not mutually exclusive. In fact, they can be complimentary to each other. Each screen may generate its own hits that are of interest to warrant follow-up studies.

In addition, synthetic lethal screens can also be used to identify potential biomarkers for a known drug or compound. In a recent siRNA screen, Wiltshire et al. carried out a study to identify genes

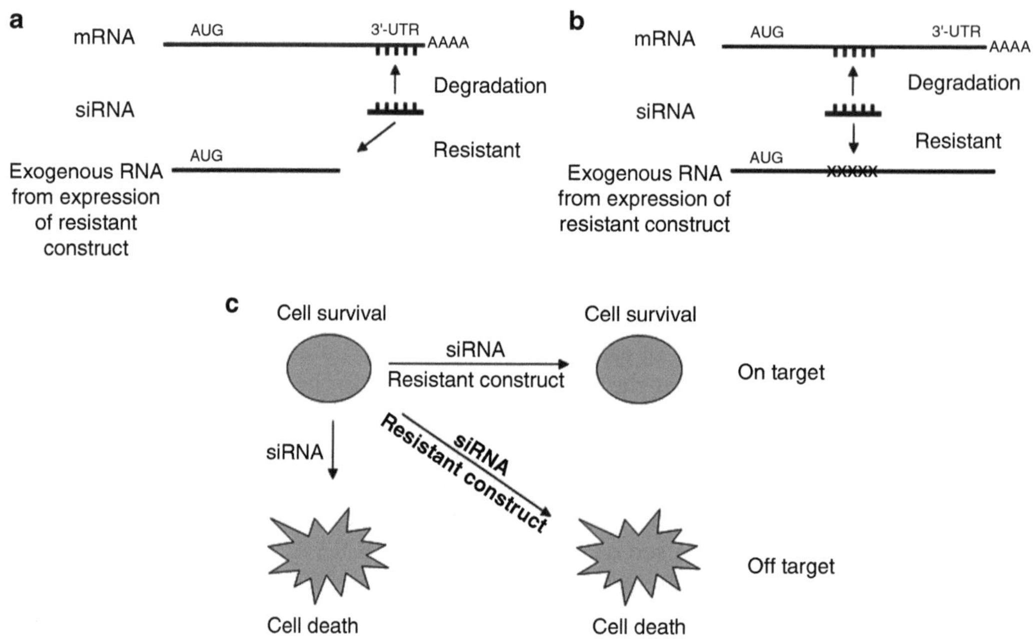

Fig. 4. Rescue experiment to test the specificity of siRNA/shRNA. (**a**) If the siRNA/shRNA targeting sequence is located in 3′-UTR, a rescue construct containing the coding sequence should be resistant to the siRNA-induced degradation. (**b**) If the shRNA/siRNA is located in the coding sequence, a rescue construct can be made through site-directed mutagenesis by introducing silent mutations at each residue corresponding to a codon's wobble position in the corresponding cDNA sequence targeted by the shRNA/siRNA. The resulting construct should be resistant to the siRNA/shRNA, while it still expresses wild-type protein. (**c**) If siRNA/shRNA treatment results in cell death, and expression of the siRNA/shRNA-resistant construct keeps the cells from dying in the presence of siRNA, the siRNA is on target. Otherwise, the siRNA has off-target activity that induces cell death. The knockdown of the endogenous protein by siRNA/shRNA and the expression of exogenous protein by rescue construct should be verified by western blot analysis.

that when silenced cause synthetic lethality with PARP inhibitor AZD2281 (51). The screen identified ubiquitin specific protease USP11 which plays a role in HR DNA repair through its ability to regulate foci formation for Rad51 and 53BP1 (51).

As novel approaches for identifying tumor-specific therapies, synthetic lethal screens have been applied widely in cancer drug discovery, especially for cancer cells with known loss of function mutations of tumor suppressor genes or gain of function mutations of driver oncogenes whose protein products are not amenable for small molecule inhibition. PARP inhibition represents a prototype for the successful employment of synthetic lethality in cancer drug discovery. PARP inhibitors also represent a novel class of targeted therapies for DNA repair. Significant antitumor activity has been reported for PARP inhibitors in BRCA-deficient cancer patients in the clinic. These proof-of-concept studies demonstrate the feasibility of leveraging synthetic lethality as a promising strategy for anticancer drug discovery and development.

Acknowledgments

I would like to thank Drs. Alexander Shoemaker, Eric Johnson, and Gui-dong Zhu for critically reading the manuscript.

References

1. Martin GS (2004) The road to Src. Oncogene 23:7910–7917
2. Bernard WI, Joe A (2008) Oncogene addiction. Cancer Res 68:3077–3080
3. Leonard G (1993) Synthetic enhancement in gene interaction: a genetic tool come of age. Trends Genet 9:362–366
4. Lucchesi JC (1968) Synthetic lethality and semi-lethality among functionally related mutations of *Drosophila melanogaster*. Genetics 59:37–44
5. Kaelin WG (2005) The concept of synthetic lethality in the context of anticancer therapy. Nat Rev Cancer 5:689–698
6. Kaelin WG (2009) Synthetic lethality: a framework for the development of wiser cancer therapeutics. Genome Med 1:99
7. Norbury CJ, Hickson ID (2001) Cellular responses to DNA damage. Annu Rev Pharmacol Toxicol 41:367–401
8. Schreiber V, Dantzer F, Ame JC et al (2006) PolyADP-ribose: novel functions for an old molecule. Nat Rev Mol Cell Biol 7:517–528
9. Amé JC, Spenlehauer C, de Murcia G (2004) The PARP superfamily. BioEssays 26:882–893
10. Helleday T, Petermann E, Lundin C et al (2008) DNA repair pathways as targets for cancer therapy. Nat Rev Cancer 8:193–204
11. Amé JC, Rolli V, Schreiber V et al (1999) PARP-2, a novel mammalian DNA damage-dependent poly(ADP-ribose) polymerase. J Biol Chem 274:17860–17868
12. Moynahan ME, Pierce AJ, Jasin M (2001) BRCA2 is required for homology-directed repair of chromosomal breaks. Mol Cell 7:263–272
13. Moynahan ME, Chiu JW, Koller BH et al (1999) Brca1 controls homology-directed DNA repair. Mol Cell 4:511–518
14. Hennessy B, Timms KM, Carey MS et al (2010) Somatic mutations in BRCA1 and BRCA2 could expand the number of patients that benefit from poly ADP ribose polymerase inhibitors in ovarian cancer. J Clin Oncol 28:3570–3576
15. Kwon JS, Daniels MS, Sun CC et al (2010) Preventing future cancers by testing women with ovarian cancer for BRCA mutations. J Clin Oncol 28:675–682
16. Stadler ZK, Salo ME, Patil SM, Pietanza M et al (2012) Prevalence of BRCA1 and BRCA2 mutations in Ashkenazi Jewish families with breast and pancreatic cancer. Cancer Res 118:493–499
17. Farmer H, McCabe N, Lord CJ et al (2005) Targeting the DNA repair defect in BRCA mutant cells as a therapeutic strategy. Nature 434:917–921
18. Bryant HE, Schultz N, Thomas HD et al (2005) Specific killing of BRCA2-deficient tumours with inhibitors of poly ADP-ribose polymerase. Nature 434:913–917
19. Penning TD (2010) Small-molecule PARP modulators-current status and future therapeutic potential. Cur Opin Drug Discov Dev 13:577–586
20. Mendeleyev J, Kirsten E, Hakam A et al (1995) Potential chemotherapeutic activity of 4-iodo-3-nitrobenzamide: metabolic reduction to the 3-nitroso derivative and induction of cell death in tumor cells in culture. Biochem Pharmacol 50:705–714
21. Liu X, Shi Y, Maag DX et al (2012) Iniparib nonselectively modifies cysteine-containing proteins in tumor cells and is not a bona fide PARP inhibitor. Clin Cancer Res 18:510–523
22. Tutt A, Robson M, Garber JE et al (2010) Oral polyADP-ribose polymerase inhibitor olaparib in patients with BRCA1 or BRCA2 mutations and advanced breast cancer: a proof-of- concept trial. Lancet 376:235–244
23. Audeh MW, Carmichael J, Penson RT et al (2010) Oral polyADP-ribose polymerase inhibitor olaparib in patients with BRCA1 or BRCA2 mutations and recurrent ovarian cancer: a proof-of- concept trial. Lancet 376:245–251

24. Gelmon KA, Tischkowitz M, Mackay H et al (2011) Olaparib in patients with recurrent high-grade serous or poorly differentiated ovarian carcinoma or triple-negative breast cancer: a phase 2, multicentre, open-label, non-randomised study. Lancet Oncol 12:852–861
25. Kaye SB, Lubinski J, Matulonis U et al (2012) Phase II, open-label, randomized, multicenter study comparing the efficacy and safety of olaparib, a poly (ADP-ribose) polymerase inhibitor, and pegylated liposomal doxorubicin in patients with BRCA1 or BRCA2 mutations and recurrent ovarian cancer. J Clin Oncol 30:372–379
26. Schelman WR, Sandhu SK, Moreno GV et al (2011) First-in-human trial of a poly(ADP)-ribose polymerase (PARP) inhibitor MK-4827 in advanced cancer patients with antitumor activity in BRCA-deficient tumors and sporadic ovarian cancers (soc). J Clin Oncol 29:abstr 3102
27. Isakoff S, Overmoyer B, Tung N et al (2010) A phase II trial of the PARP inhibitor veliparib (ABT888) and temozolomide for metastatic breast cancer. J Clin Oncol 28:abstr 1019
28. Ibragimova I, Cairns P (2011) Assays for hypermethylation of the BRCA1 gene promoter in tumor cells to predict sensitivity to PARP-inhibitor therapy. Methods Mol Biol 780:277–291
29. Drew Y, Mulligan EA, Vong WT et al (2011) Therapeutic potential of polyADP-ribose polymerase inhibitor AG014699 in human cancers with mutated or methylated BRCA1 or BRCA2. J Nat Cancer Ins 103:334–346
30. Johnson N, Li Y, Chen ZE et al (2011) Compromised CDK1 activity sensitizes BRCA-proficient cancers to PARP inhibition. Nat Med 17:875–882
31. Krawczyk PM, Eppink B, Essers J et al (2011) Mild hyperthermia inhibits homologous recombination, induces BRCA2 degradation, and sensitizes cancer cells to poly ADP-ribose polymerase-1 inhibition. Proc Natl Acad Sci USA 108:9851–9856
32. Chan N, Pires IM, Bencokova Z et al (2010) Contextual synthetic lethality of cancer cell kill based on the tumor microenvironment. Cancer Res 70:8045–8054
33. Shen WH, Balajee AS, Wang J et al (2007) Essential role for nuclear PTEN in maintaining chromosomal integrity. Cell 128:157–170
34. Mendes P, Ana M, Martin SA et al (2009) Synthetic lethal targeting of PTEN mutant cells with PARP inhibitors. EMBO Mol Med 1:315–322
35. Buisson R, Dion C, Anne M et al (2010) Cooperation of breast cancer proteins PALB2 and piccolo BRCA2 in stimulating homologous recombination. Nat Struct Mol Biol 17:1247–1254
36. Rodriguez C, Hughes DL, Vallès H et al (2004) Amplification of the BRCA2 pathway gene EMSY in sporadic breast cancer is related to negative outcome. Clin Cancer Res 10:5785–5791
37. Hartwell LH, Szankasi P, Roberts CJ et al (1997) Integrating genetic approaches into the discovery of anticancer drugs. Science 278:1064–1068
38. Turcotte S, Chan DA, Sutphin PD et al (2008) A molecule targeting VHL-deficient renal cell carcinoma that induces autophagy. Cancer Cell 14:90–102
39. Torrance CJ, Agrawal V, Vogelstein B et al (2001) Use of isogenic human cancer cells for high-throughput screening and drug discovery. Nat Biotechnol 19:940–945
40. Shaw AT, Winslow MM, Magendantz M et al (2011) Selective killing of K-ras mutant cancer cells by small molecule inducers of oxidative stress. Proc Natl Acad Sci USA 108:8773–8778
41. Dolma S, Lessnick SL, Hahn WC et al (2003) Identification of genotype-selective antitumor agents using synthetic lethal chemical screening in engineered human tumor cells. Cancer Cell 3:285–296
42. Luo J, Emanuele MJ, Li D et al (2009) A genome-wide RNAi screen identifies multiple synthetic lethal interactions with the Ras oncogene. Cell 137:835–848
43. Scholl C, Fröhling S, Dunn IF et al (2009) Synthetic lethal interaction between oncogenic KRAS dependency and STK33 suppression in human cancer cells. Cell 137:821–834
44. Puyol M, Martín A, Dubus P et al (2010) A synthetic lethal interaction between K-Ras oncogenes and Cdk4 unveils a therapeutic strategy for non-small cell lung carcinoma. Cancer Cell 18:63–73
45. Bommi RA, Almeciga I, Sawyer J et al (2008) Kinase requirements in human cells: III altered kinase requirements in VHL−/− cancer cells detected in a pilot synthetic lethal screen. Proc Natl Acad Sci USA 105:16484–16489
46. Martin SA, Hewish M, Sims D et al (2011) Parallel high-throughput RNA interference screens identify PINK1 as a potential therapeutic target for the treatment of DNA mismatch repair-deficient cancers. Cancer Res 71:1836–1848

47. Brough R, Frankum JR, Sims D et al (2011) Functional viability profiles of breast cancer. Cancer Discov 1:260–273
48. Babij C, Zhang Y, Kurzeja RJ et al (2011) STK33 kinase activity is nonessential in KRAS-dependent cancer cells. Cancer Res 71: 5818–5826
49. Frohling S, Scholl C (2011) STK33 kinase is not essential in KRAS-dependent cells-letter. Cancer Res 71:7716
50. Weiss WA, Taylor SS, Shokat KM (2007) Recognizing and exploiting differences between RNAi and small-molecule inhibitors. Nat Chem Biol 3:739–744
51. Wiltshire TD, Lovejoy CA, Wang T et al (2010) Sensitivity to polyADP-ribose polymerase PARP inhibition identifies ubiquitin-specific peptidase 11 USP11 as a regulator of DNA double-strand break repair. J Biol Chem 285:14565–14571

Part IV

Protein-Focused Technologies

Chapter 10

Structure-Based Target Druggability Assessment

Jean-Yves Trosset and Nicolas Vodovar

Abstract

The focus of this chapter is on the important concepts behind the in silico techniques that are used today to assess target druggability. The first step of the assessment consists of finding cavity space in the protein using 2D and/or 3D topological concepts. These concepts underlie the geometry and energy-based pocketfinder algorithms. Analysis pursues on the physico-chemical complementarity between the binding site and the drug like molecule. Geometrical and molecular flexibility aspect are also included in this assessment. The presence of hot interaction spots are shown to be particularly important for targeting protein-protein interactions. Finally, binding site promiscuity can be assessed by large scale structural comparison with other targets. Common chemical features amongst protein cavities can predict potential cross-reactivity with unwanted targets.

Key words: Drug targets, Hot spots, Druggability, Flo-QXP, Pocket finder, Structure superimposition, Target promiscuity, Protein cavity

1. Introduction

A "druggable" target is typically a receptor that interacts with a drug, i.e., a chemical compound or a protein that will trigger a biological response which will influence the evolution of the disease. In the context of this chapter, only proteins are considered as a target and small molecular weight molecules as a drug.

Identification of a disease relevant target is a challenge on its own. Typically this involves a significant number of information research (1–9) and validation studies to prove the influence of this specific target for the particular disease. Once the relevance of the target is found, the next step is to find a reliable biochemical or cellular assay to identify interacting compounds. With both a validated target and an efficient bioassay in hands, drug discovery can start, even without structural information. However, the increasing numbers of experimental tridimensional structures have provided insight on the way drugs interacts with protein targets.

Jürgen Moll and Riccardo Colombo (eds.), *Target Identification and Validation in Drug Discovery: Methods and Protocols*, Methods in Molecular Biology, vol. 986, DOI 10.1007/978-1-62703-311-4_10,

This understanding has contributed to the development of structure based drug design techniques which have been adapted to assess the structural druggability of a target, i.e., its capacity to bind a small drug-like molecule. Structure based druggability assessment can be carried out independently of the biological relevance of the target, or the existence of a suitable screening assay for high-throughput screening (HTS).

Various aspects of target druggability can be assessed in silico (9–11). The first one is to check for the presence of a protein cavity that can be triggered by a small molecule. The second step is to check whether the physicochemical properties of this cavity can complement with the properties of a drug-like molecule through the existence of a suitable pharmacophore site at the surface of the binding site. The shape and existence of micro-cavities or narrow clefts that can become strong interacting hot spots (12–16) for the ligand is the third aspect of the assessment of the binding pocket. The particular case of protein–protein interactions (PPI) is discussed in this context. A fourth aspect of the druggability assessment concerns the binding site flexibility (17). Finding a cavity in a rigid target supposes that the cavity already exists. Certain pockets like inducible allosteric sites are revealed after protein conformational changes. With the existence of multiple X-ray conformers for a specific target, it is possible to assess the relative variability of certain residues within the binding site. Certain techniques that assess the plasticity of the binding site are presented.

The last issue of the assessment concerns the promiscuity of the target. The degree of structural similarity with other proteins of the organism will condition the selectivity profile of the drug. High similarity with unwanted targets might jeopardize the development of a drug for that specific target. The cross-reactivity with unwanted targets can be identified using in silico structural comparison techniques (18). The impact of the target promiscuity on the drug selectivity profile is discussed in this last section.

2. Algorithms Used for Druggability Assessments

2.1. Finding Protein Cavities

Geometry-based approach: from a geometrical point of view, a cavity is a concave empty space that can be described using 2D (surface) or 3D shape descriptors (19–21). We consider three regions in the protein environment: the protein bulk, the bulk solvent and the cavity space. The protein bulk is the space filled by the protein atoms. The bulk solvent is the space "outside" the protein which differentiates from the space "inside" the protein which defines the cavity where the drug-like molecule is supposed to bind. The identification of protein pockets by numerical methods suppose the capacity to discriminate first the protein bulk from the rest which is relatively easy, and second to differentiate the bulk solvent from the cavity space which is more difficult and involve some 2D and 3D topological concepts that are used alternatively or in combination to distinguish these three regions.

2D Approach: in this case, the delineation of the three regions is made through the calculation of a close surface surrounding either the protein alone or the protein including the cavities. A close surface defines an inner and outer part of the protein. The inside region depends on the way the surface is calculated and in particular whether the interstice and micro-cavities are included in the interior of the surface. This surface is calculated by rolling a spherical probe on the protein atoms and by estimating for each of them the surface area accessible to this probe (see Note 1)—the smaller the probe, the smaller the interstices accessible by the probe. For example, a small probe will be able to enter all interstices and cover the complete solvent accessible van der Waals surface of the protein (excluding overlap with protein atom). On the contrary, a very large probe will not be able to enter any of the protein interstices. This probe will describe a convex envelop of the protein including both the protein bulk as well as the cavities space. This surface would exclude nicely the bulk solvent from the rest (protein bulk + cavities). This surface can be considered as an upper limit surface that leaves the bulk solvent "outside" the surface. This surface is somehow equivalent to the convex hull of the set of protein atoms (see Note 2). The region outside the convex hull is exclusively made of bulk solvent. The reverse is not true: the inside region of the convex hull does not define exclusively the protein atoms and cavities space. Unfortunately, it may also include some bulk solvent area. The reason is that the protein envelop is not convex (see Note 2). Bended proteins like beta-catenin (22) for example, have a large solvent region between the two protein lobes that cannot be considered as a cavity (see Fig. 1). Other cavities like the RNA/DNA binding region of the polymerases are much larger than the average volume of small molecular weight compounds. In the absence of DNA/RNA, this cavity is filled by water that can be considered as part of the bulk solvent. For such type of targets, the convex hull is not a good protein envelop. It includes a large region of bulk solvent (23). Various surfaces can be calculated depending of the size of the probe. They can exclude the bulk solvent when using a large probe and exclude also the protein cavities as the probe get smaller. Alpha-shape theory is an attempt to capture the imbrication of the various hierarchical levels into a single mathematical framework (24). It provides a natural envelop that is in between the convex hull envelop of the protein and the protein surface that exclude all micro-cavities. Protein cavities surface are in between these two limits. In alpha-shape theory, the hierarchy of the various molecular surfaces is controlled by the size of the probe (1/alpha). For each alpha, there is a corresponding

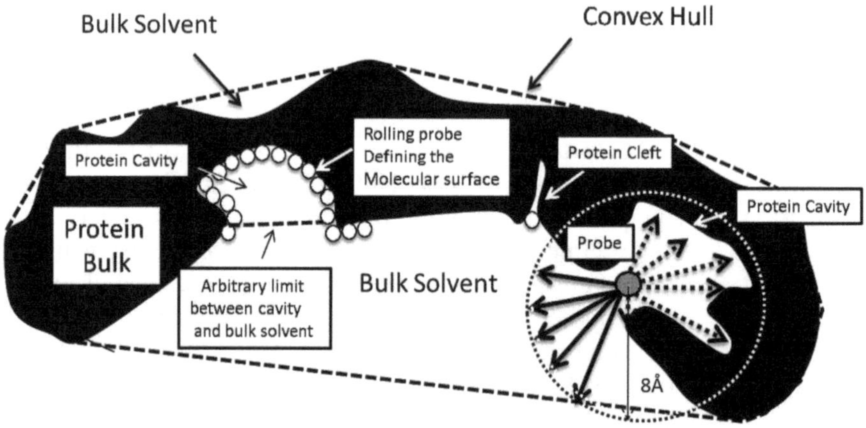

Fig. 1. Geometry concepts for analyzing protein cavity. The protein bulk is represented in *black*, probes as *little spheres*. The convex Hull of the protein is represented in *dash line*. The plain vectors emerging from the probe in grey are pointing towards the bulk solvent, whereas the dash vectors will encounter protein atoms within a radius of 8 Å (radius of the influence *circle* in *dot line*). The probe in front of the cleft defines the degree of precision in representing the molecular surface.

surface of the protein that envelops the cavities for which the entrance is smaller than the probe size (see white probe at the entrance of the cleft in Fig. 1). Alpha-shape theory is a tool to describe the surface of objects like proteins using a single continuous alpha parameter. There is no mathematical consideration concerning the best alpha parameter to use. Topological concepts are still necessary to choose the suitable protein surface that will delineate adequately the bulk solvent from protein cavities (24). This hierarchical concept is discussed in detail by Kawabata et al. and implemented in pocket finder algorithms (25, 26).

3D Approach: here we consider the 3D environment of a probe. We investigate whether this probe is part of the protein bulk or within an empty space. It is relatively easy to see whether a probe is in the vicinity or overlapping with a protein atom. The difficulty is to distinguish between the cavity space and the bulk solvent. 3D neighborhood of a probe can easily be estimated numerically by immersing the protein into a 3D grid. This 3D grid is made of nodes located at a center of a unit cube of a given size (usually between 0.25 and 1 Å). Each node can be considered as a reference point to which we can assign a spherical probe of a given size. The default value is often chosen to be that of a solvent water molecule (radius = 1.4 Å). Once the probe is attached to a node, we can calculate various parameters such as the local atomic density, the energetic field or partial free energy contribution of all the protein atoms or any other physical or geometrical local parameters at that grid position (27). The calculated information is then stored at the corresponding node (see Note 3). The center of the node can either be inside a protein atom or outside in the empty space.

A simple pocket finder algorithm would consist to identify the nodes that are not overlapping with protein atoms and delineate the cluster of nodes in the empty space of suitable size to be considered as a pocket.

As mentioned above, this empty space can itself be divided into bulk solvent and protein cavities. It is very easy to localize the high density region (protein bulk), but the real challenge is to distinguish between bulk solvent and protein cavity. In the 3D framework, the concept used to make this distinction is atom density. A solvent molecule in a cavity is surrounded by a lot of protein atoms, certainly more than when it is in the bulk solvent or close to a flat protein surface. The calculation of the number of surrounding protein atoms at an "empty" node of the 3D grid (node in the empty space) will tell whether the node is in a protein cavity or outside in the bulk solvent. This approach has been used by Weisel et al. Their PocketPicker algorithm (28) defines the degree of buriedness as the number of protein atoms which lies within a sphere of 8 Å centered at the probe. A region with high buriedness index corresponds to a cavity region inside the protein. Such an approach is very pertinent to find micro-cavities or clefts that are surrounded by a lot of protein atoms. However, when the volume of the protein cavity increases or when the cavity broadly open to solvent, the distinction between cavities and bulk solvent becomes more difficult. Cavity can always be defined theoretically as an empty space that is predominantly surrounded by protein atoms at a given shell radius. An equal exposition towards protein atoms or towards the bulk solvent corresponds to a "flat" protein surface. This can be estimated easily using a vector based argument. Looking at all the directions around a given probe (node of 3D grid), a linear search can be carried out in each direction to check for the presence of a protein atom within the circle of influence (the radius of which is defined by the user). For a uniform distribution of vectors emerging from each node, two types of vectors will be identified: those pointing towards protein atoms (dashed vectors in Fig. 1) and those pointing towards the bulk solvent (plain vectors in Fig. 1). The relative percentage of these two types of vectors will tell whether the probe is in the bulk solvent or inside the cavity. For example, the probe in the center of the dotted circle (on the right side of Fig. 1) is localized at the entrance of the cavity: 50 % of the vectors are pointing towards the bulk solvent, the rest towards the protein bulk. This vector-based approach has been implemented into the program VICE by Tripathi and Kellogg (29).

These 2D and 3D concepts have been implemented into geometry-based pocket finding algorithms (21, 25, 28, 30–41) that are listed in Table 1.

Energy-based approach: The concept of the energy-based pocket finding algorithms is to describe the atomic density at a neighborhood of a probe using potential functions similar to the one used to

Table 1
Geometry-based pocket finder algorithms available online

Cavity finder algorithm	Web references
APROPOS (Automatic PROtein Pocket Search) (30)	http://www.csb.yale.edu/userguides/datamanip/apropos/manual.html
Binding-response (31)	http://www.mackerell.umaryland.edu/CADD/CADD_bindingresponse.html
CASTp (32)	http://www.sts.bioengr.uic.edu/castp/
CAVER (33)	http://www.loschmidt.chemi.muni.cz/caver/
Fpocket (34)	http://www.sourceforge.net/projects/fpocket/
GHECOM (Probe-based HECOMi finder) (25)	http://www.biunit.naist.jp/ghecom/
McVol (35)	http://www.bisb.uni-bayreuth.de/People/ullmann/mcvol/mcvol.html
PASS (putative active sites with spheres) (36)	http://www.ccl.net/cca/software/UNIX/pass/index.shtml
PocketDepth (37)	http://www.proline.physics.iisc.ernet.in/pocketdepth/
PocketPicker (28)	http://www.gecco.org.chemie.uni-frankfurt.de/pocketpicker/
Screen (surface cavity recognition and evaluation) (38)	http://www.interface.bioc.columbia.edu/screen/
SplitPocket (39)	http://www.pocket.uchicago.edu/
SURFNET (40)	http://www.biochem.ucl.ac.uk/_roman/surfnet/surfnet.html
LigSitecsc (41)	http://www.projects.biotec.tu-dresden.de/cgi-bin/index.php
VOIDOO (21)	http://www.xray.bmc.uu.se/usf/voidoo.html

calculate energy of molecular complexes. Various functions are suitable to estimate the presence of an atom at a given distance from a reference point. Depending on the degree of influence we want to assign to a protein atom, short range or long range potential functions can be designed. The step function for example mimic the hard-sphere geometrical view of atomic interactions: the value of the function is infinite or constant if two atoms overlap and zero otherwise. With this infinite short range function, the probe cannot "feel" any protein atoms unless they touch the probe. Its first order derivative is also not continuous and therefore not adapted for gradient-based searching algorithms. The sigmoidal function is a continuous differentiable version of the step function. It includes a zone of influence where the probe can still feel surrounding protein atoms even it does not overlap with the probe. Finally, the third

Table 2
Energy-based pocket finder algorithms available online

Cavity finder energy-based algorithms	Web references
SiteMap (44)	http://www.schrodinger.com/
ICM-PocketFinder (45)	http://www.molsoft.com/
Q-SiteFinder (46)	http://www.modelling.leeds.ac.uk/qsitefinder/
SITEHOUND (47)	http://www.bsbbsinai.org/SHserver/SiteHound/download.html
AutoLigand (43)	http://www.mgltools.scripps.edu/downloads
GRID (27)	http://www.moldiscovery.com/soft_grid.php
Surflex–Protomol (48)	http://www.biopharmics.com

most common function is the hydrophobic van der Waals potential energy which is implemented in most protein force fields. The van der Waals energy function shows an infinite positive energy asymptote when the center of the van der Waal spheres of two atoms overlap and a negative energy minimum when the atoms are at optimal distance from each other. At longer interatomic distance, the energy tends to zero. Such a function is therefore suitable to check whether a solvent probe is in the protein bulk, close to the molecular surface or far from any protein atom. The van der Waals energy will be positive in the first case, negative in the second and close to zero in the third (42). Note that the same difficulty as in the 2D approach remains: the energy value of the probe could not differentiate whether the probe is within the bulk solvent or located in the middle of a large cavity space. In both cases, the probe is far away from any protein atoms.

Using an energy-based approach it is possible to implement the conformational sampling strategies that are implemented in the protein modeling package to sample the probe in the protein environment. Once the "cavity field" is implemented into a 3D-grid, standard sampling techniques such as Molecular Dynamics (33) or Monte Carlo (35) can be used to search for protein cavities. Full force field as it is implemented in Auto-Ligand (43) could help in characterizing the hydrophobic or electrostatics properties of the surface. Energy-based algorithms available online (27, 43–48) are listed in Table 2.

To conclude this section, we remind that the main purpose of the protein finding algorithms is to localize the cavities within the protein and estimate its size (see Note 4). They do not give information on the druggability of the pocket itself. This is the object of the next section.

2.2. Druggability of Binding Site

2.2.1. Physicochemical Properties

The Lipinski's rule of five (RO5) connects the physicochemical properties of a drug with its pharmacokinetics properties (49). Molecules of reasonable size (<500 MW), that are able to cross the lipophilic membrane (CLogP > 5) while capable to interact with polar solvent (number of H-bond donors > 5; sum of N's and O's > 10) will be characterized as drug-like molecules (49, 50). The physicochemical properties of a druggable pocket should be the mirror image of the physical properties of the drug-like molecule itself (51, 52). The concept of druggable pocket was born from this analogy. The complementary properties of the pocket therefore reflect the Lipinski's rule of five of drug-likeness. Qualitatively, we should expect an adequate proportion of H-bond donors and acceptors as well as hydrophobic patches to accommodate both the polar and unpolar character of the drug. Graphical interfaces are valuable tools to inspect the proportion of polar versus hydrophobic area of the protein molecular surface (see Note 5). Druggability assessment can be made by visual inspection of the cavity with an adequate molecular surface. The relative number of H-bond donors and acceptors as well as the relative proportion of polar versus hydrophobic protein atoms at the surface of the cavity can easily be visualized using color code surface: a given color for polar atoms (positive and negatively charged) and a third color for neutral atoms like aliphatic or aromatic carbons (see Note 5). For an expert modeler, this is certainly the most precise assessment.

Quantitatively, physicochemical features can be coded into in silico descriptors that can be used by a machine learning algorithm. Such target druggability prediction models (10) are useful to screen the whole PDB and to estimate the size of the druggable genome. These models supposed a pre-classification of the binding pockets into "druggable" and "non-druggable" target sets. For undruggable targets, it is always tempting to define them as such because of the absence of known ligands. This does not mean of course that the target is undruggable as it was demonstrated after the discovery of PPI targets a decade ago (13).

Here there are two choices: either we want a good statistics, i.e., a good discrimination rate, and only very well defined targets such as enzyme entering in the druggable set or we are willing to "fish" exotic targets such as PPI at the risk of decreasing the statistical performance with characterized targets. So, care should be taken when using statistical target druggability predictors.

2.2.2. Shape Properties

Not only are the physicochemical properties important to assess pocket druggability but also the geometry. The shape of the binding pocket and the existence of micro-cavities that could interact strongly with certain chemical groups of the drug can be crucial for suitable interactions with a small molecule. The shape of the binding site is always best described by visual inspection. It is indeed always difficult to code into a program all the geometrical aspect of

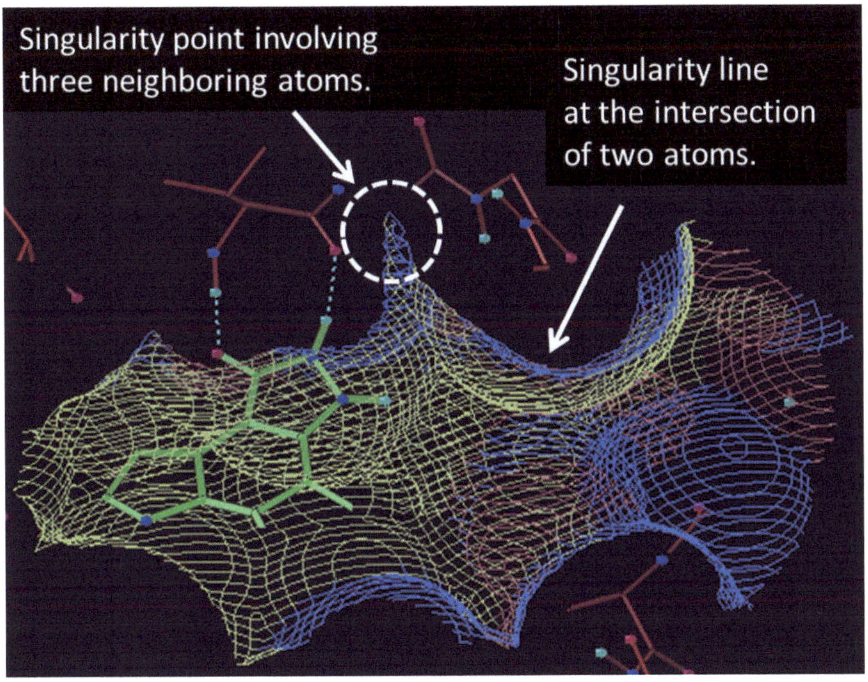

Fig. 2. QSURF protein surface from FLO-QXP around ligand in the hinge region of the ATP binding site of a kinase. The scaffold makes two hydrogen bonds with the protein backbone (hinge). The micro-cavity on the right of the ligand (*dot circle*) is created by three atoms in the vicinity, the lines of singularity in the molecular surface results of the intersection between two protein atoms. The three color code of the protein surface corresponds to the hydrophobic area (*yellow*) or negative and positive polar areas (*blue* and *red*, respectively).

a molecule that the eye can see. If statistical automated methods are useful for large scale assessment of targets, visual inspection of the shape of the cavity using an adequate surface rendering is perfectly adapted when investigating a specific target of our choice.

As seen above, atoms located at the contact energy surface, defined in the program FLO-QXP (see Note 5), give major contribution to the hydrophobic interaction (or contact energy) as compared to the atoms lying outside of the surface, i.e., within the bulk of the protein cavity. What ultimately counts for assessing the druggability of a target is the amount of protein surface that can wrap around the drug. In technical terms, this is the amount of van der Waals contact between the drug and the protein.

Large spherical cavities, like the ones observed for example in RNA or DNA polymerases (53) are not very druggable pockets. The large cavity that hosts the double strands of RNA or DNA is well exposed to solvent and not really suitable for binding a small molecule. Narrow cavities offer more van der Waals contact and are in this sense more druggable. They indeed offer a better surface–volume ratio which implies a larger number of clefts or micro-cavities that can be used by the ligand to optimize the hydrophobic interactions. A typical example for such a type of surface is found in kinase ATP binding sites of protein kinases (Fig. 2). With a flat interface, each ligand atom contributes twice to the contact energy, from the face "below" and from the one "above". Inducible pockets that specifically adapt the small molecules optimize this contact energy. A good example is given in the case of imatinib binding in the allosteric pocket of Bcr-Abl kinase (54).

Optimum atom ligand efficiency (55–57) is obtained when the atom is inside a micro-cavity (dot circle in Fig. 2). We define such micro-cavities as hot spots. They usually correspond to singularities of the molecular surface (see Note 6). The presence of such hot spots is characteristics of highly druggable targets, especially for PPI (see below). It is important to have graphical tools that can identify them. The QSURF program (58) has the advantage that it does not smooth out the surface and therefore does not remove these surface singularities. The hot spots are therefore easily identified with the FLO-QXP package. These singularities are the most informative elements when assessing target druggability and are particularly important for the inhibition of PPI as we see in the following section.

2.2.3. Protein–Protein Interactions

The basic concept behind the inhibition of PPI with a small molecule is competition: David against Goliath, i.e., a small ligand competing with a big protein. The number of interactions is what makes the difference. For example, the competition between a kinase inhibitor and ATP is more or less equilibrated if not advantageous for the drug: the ATP is only a weak binder (K_d of the order of μM). It is enough for the kinase inhibitor to trigger additional interactions equivalent to one or two strong H-bonds to compete efficiently against ATP. Basically, most kinase inhibitors mimic ATP: a hydrophobic scaffold with two or three H-bonds interacting with the hinge of the kinase. Such a ligand would not compete with a protein: the size of the ligand would have to be equivalent to estimated 30–50 ATPs. Therefore, a different strategy is necessary to identify useful inhibitors.

Although, a small molecule cannot accommodate the large protein interface, yet the disruption of such interfaces by small drug-like molecules is possible (13, 59, 60). Particular features of the protein interface should be considered to make this possible (61). As seen above, the presence of hot spots, i.e., the presence of micro-cavities in which ligand atoms can interact very strongly is one of the most important criteria (62–67).

However, the presence of more than one anchoring point is important in order for the small molecule to be correctly "locked" within the binding site. When the ligand is well anchored in various hot spot positions, the more difficult it will be for a large competitor like protein, RNA or DNA strand to displace this ligand. A PPI inhibitor can therefore be compared to a spider shaped molecule: the "body" of the spider is a chemical moiety that connects the different "arms," i.e., chemical side chains that interact with

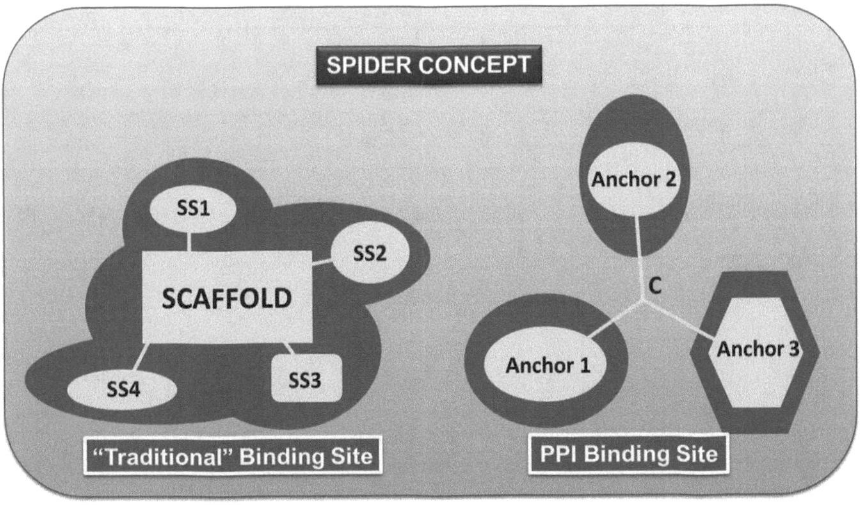

Fig. 3. "Traditional" binding site model compared to the "Spider" model of PPI. In traditional binding sites, the scaffold is part of the cavity and gives major contribution to the ligand binding affinity whereas at the protein–protein interface, the scaffold is more open to bulk solvent and plays the role of a connector ("C") of highly interacting chemical groups anchoring to hot spots of the cavity of one of the protein partner.

the anchors (Fig. 3). What contributes to the binding potency are mainly the "arms" of the molecule, not necessarily the "body" (i.e., the scaffold) as it is often the case in traditional binding sites like enzyme.

Once the scaffold is selected, the affinity and the pharmacological properties of the compound are optimized using the decorating side-chains (SSi). In the PPI inhibitors, the chemical groups anchored to hot spots are contributing most to the binding energy. The connector "C" plays the role of a linker between the different anchors. It does not necessarily have a strong role in terms of affinity as the scaffold does in classical drugs.

An example of PPI inhibitors is shown in Fig. 4. The inhibitor PNU-74654 of the protein–protein complex β-catenin-Tcf4 is binding in the middle of a large groove of the PPI interface. This 350 nM inhibitor has been found by virtual screening using the above described spider concept (62). It disrupts the β-catenin/Tcf4 interaction by targeting two anchoring points (marked by surrounded small circles in the upper panel of Fig. 4). Right- and left-hand cavities correspond to the groove of α-helices 23 and 30 of the armadillo repeat, respectively. The one from the right is hosting a benzene moiety that substitutes for the D16 aspartic side chain of Tcf4. On the left hand side of the upper panel, the methyl on the furan moiety is contributing most of the activity of the compound. The same compound without this methyl group is about ten-fold less potent.

To illustrate the spider concept, the central benzamide of the PNU-74654 inhibitor of β-catenin can be considered as a linker

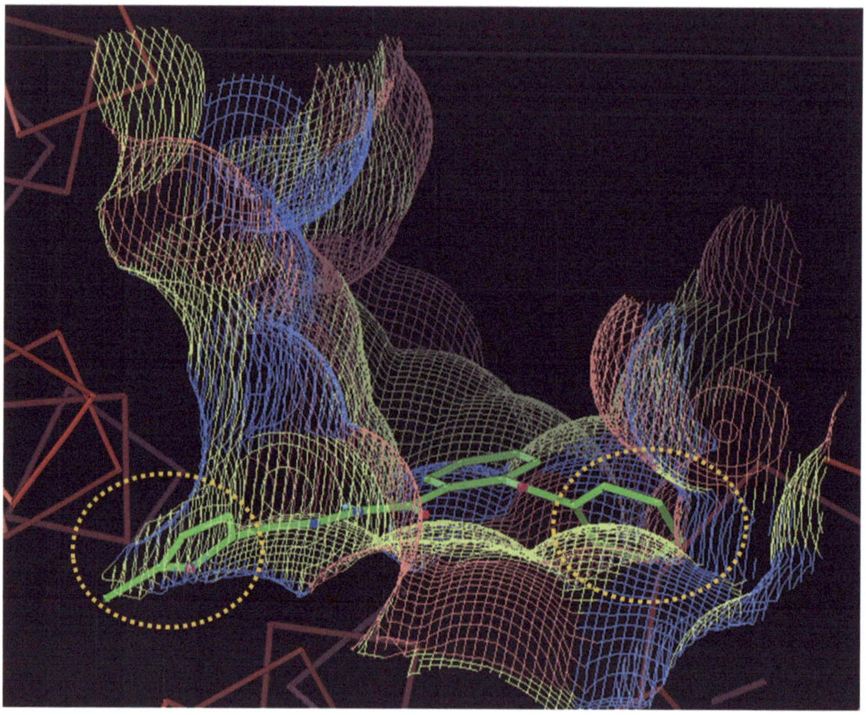

Fig. 4. Protein contact surface representation of the beta-catenin/Tcf4 protein complex interface near Tcf4 (D16). The QSURF color-coded surface highlight two protein micro-cavities that anchor the ligand PNU-74654 at two opposite sites (*orange dot circle*) involving a methyl-furan moiety and a benzene ring.

that connects two anchoring points: the methyl-furan and the benzene ring that target the two hot spots of the protein receptor.

The spider type molecules are a general characteristic of the inhibitors of the PPI (13, 15, 60). It replaces the scaffold-containing molecules of the traditional binding sites. The design of PPI inhibitors is mainly focused on the anchoring points of the protein. The challenge is to find a molecule that can target these hot spots simultaneously. Targeting these anchoring points is believed to be the key for successful design of PPI inhibitors.

2.2.4. Flexibility of Binding Sites

The intuitive concept discussed in this section is the ligand instability due to the protein conformational flexibility. From a dynamical point of view, the ligand interacts with its receptor through a type of "breathing" that can be modeled by calculating the normal modes of the protein ligand complex. The normal modes of the protein correspond to concerted motions of the atoms that arises at large scale such as domains motions, protein loops movements and at lower scale as well such as residues side chains rearrangement.

Those internal movements reflect also the function of the receptor, like the closing/opening of a catalytic region to accept the substrate and release the product of an enzymatic reaction for example. Drugs usually interfere with these particular regions of the receptor that are usually localized at the hinge between movable domains and are therefore less perturbed by the internal motions of the receptor. Kinase hinge binding regions correspond to these zones of minimal movement amplitude. This hinge is at the intersection between the N and C-terminal domains of the kinase which move with respect to each other. The drugs are therefore in synchrony with the internal motions of the receptor.

The study of the flexibility of the receptor binding pocket is meant to see whether sheering movement of domains, protein loop movements can make certain region of the binding site unstable and make the binding with a ligand difficult. Thanks to the presence of various crystal structures for a same target with different ligands, it is possible to study the conformational variation of these protein binding pockets (68).

When several X-ray structures are known for the same target, it is possible to detect the parts of the molecular surface of the binding site that move when transferring from one protein conformation to the other. The morphing technique is a simple way to analyze the variability of the binding site.

The morphing describes the transition from one conformer to another using a predefined scheme of transformation. If the 3D coordinates of protein A and B are represented by X_A and X_B (an $n \times 3$ matrix where n is the number of atoms that are common to both proteins), then a linear morphing from protein A to protein B is a transformation that is defined by:

$$X(t) = X_A + t(X_B - X_A) = tX_B + (1-t)X_A \quad \text{with } t = [0,\ldots,1]$$

where t is a parameter that runs from 0 to 1 which corresponds to the protein conformation A and B, respectively. It is possible to visualize the intermediate structures $X(t)$ at every step t. Since all the intermediate structures are displayed sequentially like in a movie, the parameter t can be assimilated with "time" even if this morphing does not involve dynamical equations. The morphing technique can be applied to the protein surface as well. Differences in the shape and the physicochemical properties of the cavity surface inform on the hot spots that are not affected by the relative protein motions. Interaction with these hot spots will not dependent of the change of binding site conformation. The morphing technique can be applied to the molecular surface as implemented in FLO-QXP. The moving molecular surface helps at identifying the flexible part of the protein binding site.

Prediction of the conformational change of the protein under binding is a challenge. New structural parameters based on high resolution X-Ray structures have been presented recently (69). Such studies open the door to a more extended analysis of conformational changes of protein cavity when binding with different ligands.

2.3. Target Promiscuity Assessment

Difficulty in developing a drug for a given target is not only related to the druggability of the target itself but also on the structural promiscuity of the target. Drugs that target one member of a large protein family, like kinases, are likely to cross-react with other family members due to the high degree of structural similarity within this family. If therapeutically promising multi-target effects can be identified (70, 71) high structural similarity with unwanted targets will represent a handicap for the drug development. The similarity of the binding sites could make the design of selective inhibitors difficult. The effective assessment of the structural landscape of the primary target binding site is therefore one of the ultimate steps in the target druggability assessment.

One way to discover whether a compound can bind to different targets is to compare their binding sites (72–77). Partial information can be obtained from sequence multi-alignment. When the degree of conservation between the two sequences is sufficiently high, identical amino acids in the sequence will likely correspond to identical binding site structures (78). The contrary is not true. A change of amino acids in the sequence may or may not influence the binding mode of the inhibitor, depending whether the residue side-chain is pointing towards the protein cavity.

Finding a mapping between the atoms of the equivalent proteins is a preliminary step before the structural superposition. There are many ways of finding a mapping between a pair of macromolecular targets. I will describe two of them in the following section.

2.3.1. Structural Alignment Based on 3D Grid Force Field

The field around the binding site can be calculated using various types of energy terms, including electrostatic, hydrophobic, hydrogen bonding, etc. This approach has been implemented into the GRID program (27). Calculation of the Molecular Interaction Fields (MIFs) has been used by Zamora et al. to study ADME properties of drugs (79, 80). The field potentials can be calculated for each protein and are used for comparing binding sites. Recently, a GRID potential approach has been implemented to compare protein binding sites based on their GRID molecular interaction fields (81). The structural similarity between a pair of proteins is studied by comparing the auto- and cross-correlation functions of the various MIF's of the two grids, respectively. Structure superimposition can also be obtained by using the fast Fourier transformation of the correlation functions as it is applied in the rigid body protein–protein docking problem (82–84) or by a global Monte Carlo minimization optimizer with a least square strategy on the two grids of the corresponding proteins (75).

2.3.2. Structural Alignment Based on Pharmacophore

Another approach consists of identifying chemical pharmacophore features that interact directly with the bound ligand. They can be summarized as hydrophobic centers, H-bond donors and acceptors, positive and negative charges, aromatic centers, etc. These surface

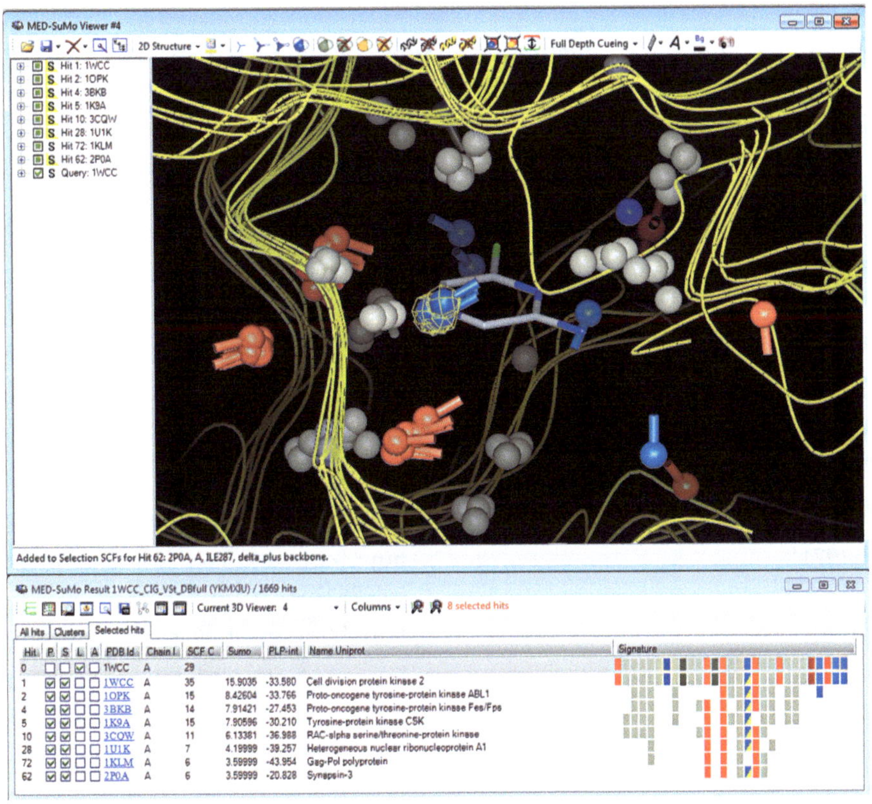

Fig. 5. Structural superposition of binding sites using Med-Sumo™. The Surface Chemical Features (SCF) are used to superimpose the protein binding sites (*top panel*). The SCF are represented in color code bar (*bottom panel*) and the SUMO score measure the quality of the 3D structural superimposition that is calculated using a bit-wise matching algorithm of the color code-bar fingerprint (*bottom panel*).

chemical features (SCF) are determined on the whole protein surface, or on a chosen cavity. They form a 3D graph that reflects the shape and the size of the cavity of interest. Such surface chemical features (SCF) can be calculated for all target structures of the protein data bank (http://www.rcsb.org) as well as a query protein structure of interest. Triplets and quadruplets of SCF are then scanned against the pre-calculated SCF PDB database. This approach has been implemented in the commercial software Med-SuMo (85), an extension of the original free version SuMo (86). It can be used either to find cavities from a query protein using a structural protein database (87), or to study the promiscuity of a target as shown below.

The SCF of the protein binding sites can be represented with colored bar-codes for which each color bit represents a 3D SCF pharmacophore feature. Binding sites with the highest number of matches between these triplets or quadruplets of SCF have the highest similarity with the query binding site (Fig. 5). The larger

Table 3
Summary of available tools evaluating binding site similarities

Binding site similarity algorithms	Web references
CavBase (89)	http://www.relibase.ccdc.cam.ac.uk
CPASS (comparison of protein active site structures) (91)	http://www.bionmr-c1.unl.edu/CPASS_OV/
eF-seek (92)	http://www.ef-site.hgc.jp/eF-seek/top.do
FINDSITE (93)	http://www.cssb.biology.gatech.edu/skolnick/
MultiBind (94)	http://www.bioinfo3d.cs.tau.ac.il/MultiBind/
PROSURFER (PROteinSURFaceExploreR) (95)	http://www.tsurumi.yokohama-cu.ac.jp/
Query3d (96)	http://www.pdbfun.uniroma2.it/
SiteAlign (97)	http://www.bioinfo-pharma.u-strasbg.fr/template/
SiteBase (98)	http://www.modelling.leeds.ac.uk/sb
SiteEngine (101)	http://www.bioinfo3d.cs.tau.ac.il/SiteEngine/
SuMo (86)	http://www.sumo-pbil.ibcp.fr
@TOME-2 (100)	http://www.abcis.cbs.cnrs.fr/AT2

the set of SCF's that match between the two proteins, the larger the similarity SuMo score.

Such binding sites comparison is a powerful tool to assess the specificity of the target, especially when it is part of a large protein family, like kinases for example. Large scale comparison has been carried out on the full PDB or on important protein families such as the purine-like target family (88). This software is used in drug discovery to reveal potential off-targets or for drug repurposing (85).

Methods for evaluating binding site similarities and related databases have been reviewed recently (18). Table 3 lists available online structural comparative methods (86, 89–101).

It is worth to mention that binding site comparison relies on the entries of the structural protein database. The increasing number of available crystal structures will facilitate the identification of similar binding sites and potential off-targets and is most suitable for large protein families like kinases or proteins with purine containing binding sites (93).

3. Conclusion

There are many ways to modulate the action of a receptor from ions, to small molecules to PPI. If the concept of target druggability depends on the size of the partner it also depends on the mode of action of the drug itself. Druggable pockets might have different characteristics depending on whether we search for an inhibitor, an activator, a compound competing with a small molecule or a protein. For small molecular weight inhibitors, sufficiently strong interactions with the protein receptor are necessary for the inhibitor to reach the efficacy threshold. On the contrary, for activators, partial activation might be sufficient to reach the desired biological effect (102). From a structural point of view, we can suppose that a sufficiently large binding site is necessary to accommodate drug-like inhibitors (MW > 150). On the contrary, a smaller binding site can be sufficient for a partial activating effect if we suppose that a small molecule can achieve the necessary interaction. The size and the physicochemical properties of a druggable pocket might therefore not follow the same criteria whether we are looking for an inhibitor or an activator of an enzyme, a membrane receptor or a PPI.

Structure-based druggability assessment consists first in finding protein cavities. A combination of 2D and/or 3D concepts is used to isolate these empty cavities inside the protein and differentiate them from surrounding bulk solvent. This distinction becomes more difficult as cavities become larger and broadly open to solvent. Once a cavity is identified, physicochemical analysis of the cavity surface will determine whether a drug-like small molecule could complement the surrounding protein residues. Analysis focuses on the proportion of the H-bond donor/acceptors and hydrophobic areas in the binding pocket. In the case of PPI druggability assessment, the presence of small pockets can serve as anchoring points to interact with two or more chemical groups of the molecule. Previous studies have shown that the presence of such micro-pockets is an important criteria to assess the druggability of PPI (62).

It is important to note that structural assessments are usually made on rigid protein structures. This supposes that the binding pocket is already present from the experimentally observed structure. If this is usually the case for substrate binding sites, allosteric sites however are rarely observed in ligand receptor crystal structures. Some druggable targets might still remain invisible from such a type of assessment. Allosteric sites are indeed discovered serendipitously by determining the crystal structure of a new chemical entity after observed activity (103). The discovery of such allosteric sites is one of today's challenges. If such allosteric sites are not easily identifiable, it is still possible to look for indices that could infer the presence of such inducible pockets. Allosteric pockets might indeed emerge from a nucleus of low atomic density within the protein. These nuclei are easily identified by cavity finder algorithms using smaller probe's size parameter on a rigid protein structure obtained either from X-ray or from a Molecular Dynamics trajectory.

Future developments on structure based target druggability assessment include the development of new parameters to describe the protein conformational changes that occurs upon binding. A first step towards this goal has been presented recently (69). This study presents a new geometrical force field that can be used to calculate a high resolution-like protein structure from an input low resolution experimental structure. Such structural parameters will help reproducing at high resolution the protein conformational changes that occurs upon binding different ligands.

Genome-wide druggability assessment is another today's challenge. Complete or partial similarities between known and unknown binding sites could help in discovering new binding sites (10, 88). Large scale protein druggability assessment will not only help in determining the multiple targets of a given drug but also to choose adequate chemistry needed for targeting a selected subset of a large protein family. The genome-wide approach of structural target druggability assessment could then serve as a template of a poly-pharmacological target assessment. As the concept of one drug, one target is certainly reductive the consideration of full-genomic transversal data will help at understanding the genome wide pharmacological profile of a drug.

4. Notes

1. When using pocket finding programs, investigation of the probe size parameter is recommended. With increasing size of the probe, the hierarchy of the various pockets will be more apparent. On the other hand, with small probe size values, micro-cavities will appear. These micro-clefts might correspond to hot spots where strong interactions might take place with some atoms of the drug. It is left to the judgment of the user to discern whether the micro-pockets that appear correspond to errors in protein coordinates (which are often less than 0.3 Å), or to a real protein cavity. With much caution, this parameter can be used to look for the existence of potential inducible binding sites.
2. The convex hull of a set of points is the volume surrounding these points such that any segment between any two of these points stay inside the volume. For proteins, we might relax this definition to include "inside" the protein volume, the closed protein cavities that are not open to bulk solvent. A more adequate

definition for protein would be: any segment between any two points at the *surface* of the protein should stay inside the protein. Note that the protein example in Fig. 1 has been chosen such that the convex hull is not a good discrimination between protein bulk and cavity space (even with the new adapted definition mentioned above. The large solvent area between the two lobes is considered as part of the bulk solvent. The cavity where drugs will interact corresponds to small clefts or hot spots within this protein–protein interface. This schematic representation of the protein mimics the 3D structure of β-catenin (22). Note that docking packages have most of the ingredients for designing energy-based cavity finding algorithms: energy force field, techniques for sampling the conformational space (Molecular Dynamics, Monte Carlo, systematic search) as well as the calculation of protein molecular surface around the ligand etc.

3. The nodes represent a discrete sample of the protein space. Any kind of information can be stored on each node of a 3D grid such as the number of surrounding protein atoms, the van der Waals contribution of the all protein atoms at that point, the local atomic density, etc. It is possible to get an estimation of those quantities everywhere on the 3D grid using a linear or non-linear interpolation scheme like B-splines. The B-Splines interpolation scheme has the advantage to ensure the continuity of the function when crossing the cubic element. It is also possible to request the continuity of all the derivatives of this function. This allows some gradient-based optimization techniques to be implemented to find energy-optimum positions within the 3D grid. Such optima correspond to interesting hot spots of the protein cavity (region of lower energy). The B-Spline interpolation scheme has been adapted by Oberlin et al. (104) for the estimation of van der Waals energies within 3D-grids and implemented into the protein modeling/docking package PRODOCK (105–107). The efficiency of the approach has been demonstrated for the case of the rigid docking problem (108).
4. Note that it is possible for certain protein modeling software (especially docking) to display the surface of buried or closed protein cavities by locating a ligand inside this pocket. Indeed, most graphical interface of docking software can display the protein surface around a docked ligand. For a pocket finding algorithm, the ligand should play the role of a probe. A protein cavity finding algorithm has been implemented into FLO-QXP by automatizing this concept: the protein is immersed into a 3D grid for which each node, considered as a single atom, is defined as part of a big ligand (the whole 3D grid). The QSURF program from FLO-QXP package (58) is calculating the

solvent accessible surface of the protein around each grid node. All types of close cavities are then displayed for visual investigation. This has the advantage to not only locate the cavities but also to understand the shape and physical properties of this cavity.

5. When visualizing the interactions of the drug receptor complex, it is very convenient to use the solvent accessible surface. In this case, the optimal van der Waals energy between a ligand atom and a protein atom corresponds very closely to the solvent accessible surface of the protein. Maximum energy contact is therefore obtained when a ligand atom is on the solvent accessible protein surface. Visually it is very easy to estimate whether an atom is on the surface or outside. The QSURF representation protein surface of FLO-QXP (58) is perfectly adapted for interactive drug design for this reason. The optimum inter-atomic distance that provides the best correlation with experimental drug potency and the contact energy term is about 1.4 Å. This is also the size of the water probe used to calculate the accessible surface area. This contact surface has been designed by C. McMartin and R. Bohacek for intuitive drug design (109). This surface is very similar to the solvent accessible surface.
6. There are basically two interesting singularities: on the intersecting line when two spheres overlap (contact energy surface of two protein atoms) or on the intersecting point when three spheres overlap. Of course, for this singularity the three atoms have to depart from each other to leave space for an atom of the ligand to stay in between. This little spherical hole of a size of an atom or chemical group like $-CH_3$ or $-SO_3$ represents a micro-cavity defined as a hot spot. These surface singularities are usually never represented in protein modeling packages. For esthetical constraints, the graphical software tends to smooth out the protein surface from all types of singularities (micro-cavities included). One of the main advantages of the molecular surface calculated by QSURF in FLO-QXP is to display the singularities of the surface. This is very informative as they correspond usually to positions in the protein where ligand atoms get maximum efficiency.

Acknowledgment

The author thanks Dr. Fabrice Moriaud for providing picture 5 on the MedSuMo software.

References

1. Sakharkar MK, Sakharkar KR (2007) Targetability of human disease genes. Curr Drug Disc Technol 4:48–58
2. Wyatt PG, Gilbert IH, Read KD et al (2011) Target validation: linking target and chemical properties to desired product profile. Curr Topics Med Chem 11:1275–1283
3. Taboureau O, Nielsen SK, Audouze K et al (2011) ChemProt: a disease chemical biology database. Nucleic Acids Res 39:D367–D372
4. Bender A, Young DW, Jenkins JL et al (2007) Chemogenomic data analysis: prediction of small-molecule targets and the advent of biological fingerprint. Comb Chem High Throughput Screen 10:719–731
5. Dixon SJ, Stockwell BR (2009) Identifying druggable disease-modifying gene products. Curr Opinion Chem Biol 13:549–555
6. Caffrey CR, Rohwer A, Oellien F et al (2009) A comparative chemogenomics strategy to predict potential drug targets in the metazoan pathogen, Schistosomamansoni. PloS One 4:e4413
7. Bredel M, Jacoby E (2004) Chemogenomics: an emerging strategy for rapid target and drug discovery. Nat Rev Gen 5:262–275
8. Li Q, Cheng T, Wang Y et al (2010) PubChem as a public resource for drug discovery. Drug Discov Today 15:1052–1057
9. Fauman EB, Rai BK, Huang ES (2011) Structure-based druggability assessment-identifying suitable targets for small molecule therapeutics. Curr Opin Chem Biol 15:463–468
10. Schmidtke P, Barril X (2010) Understanding and predicting druggability. A high-throughput method for detection of drug binding sites. J Med Chem 53:5858–5867
11. Huang N, Jacobson MP (2010) Binding-site assessment by virtual fragment screening. PloS One 5:e10109
12. Fuller JC, Burgoyne NJ, Jackson RM (2009) Predicting druggable binding sites at the protein–protein interface. Drug Discov Today 14:155–161
13. Sugaya N, Furuya T (2011) Dr. PIAS: an integrative system for assessing the druggability of protein–protein interactions. BMC Bioinformatics 9:12–50
14. Villoutreix BO, Bastard K, Sperandio O et al (2008) In silico-in vitro screening of protein–protein interactions: towards the next generation of therapeutics. Curr Pharm Biotechnol 9:103–122
15. Sperandio O, Reynès CH, Camproux AC et al (2010) Rationalizing the chemical space of protein–protein interaction inhibitors. Drug Discov Today 15:220–229
16. Panjkovich A, Daura X (2010) Assessing the structural conservation of protein pockets to study functional and allosteric sites: implications for drug discovery. BMC Struct Biol 10:9–33
17. Chène P (2008) Challenges in design of biochemical assays for the identification of small molecules to target multiple conformations of protein kinases. Drug Discov Today 13: 522–529
18. Pérot S, Sperandio O, Miteva MA et al (2010) Druggable pockets and binding site centric chemical space: a paradigm shift in drug discovery. Drug Discov Today 15:656–667
19. Morris RRJ, Najmanovich RRJ, Kahraman A et al (2005) Real spherical harmonic expansion coefficients as 3D shape descriptors for protein binding pocket and ligand comparisons. Bioinformatics (Oxford, England) 21: 2347–2355
20. Mak L, Grandison S, Morris RJ (2008) An extension of spherical harmonics to region-based rotationally invariant descriptors for molecular shape description and comparison. J Mol Graph Model 26:1035–1045
21. Kleywegt GJ, Jones TA (1994) Detection, delineation, measurement and display of cavities in macromolecular structures. Acta Crystallogr D Biol Crystallogr 50:178–185
22. Poy F, Lepourcelet M, Shivdasani RA et al (2001) Structure of a human Tcf4-beta-catenin complex. Nat Struct Biol 8: 1053–1057
23. Tagami S, Sekine S-I, Kumarevel T et al (2010) Crystal structure of bacterial RNA polymerase bound with a transcription inhibitor protein. Nature 468:978–982
24. Liang J, Edelsbrunner H, Woodward C (1998) Anatomy of protein pockets and cavities: measurement of binding site geometry and implications for ligand design. Protein Sci 7:1884–1897
25. Kawabata T, Go N (2007) Detection of pockets on protein surfaces using small and large probe spheres to find putative ligand binding sites. Bioinformatics 529:516–529
26. Kawabata T (2010) Detection of multiscale pockets on protein surfaces using mathematical morphology. Proteins 78:1195–1211
27. Goodford PJ (1985) A computational procedure for determining energetically favorable binding sites on biologically important macromolecules. J Med Chem 28:849–857

28. Weisel M, Proschak E, Schneider G (2007) PocketPicker: analysis of ligand binding-sites with shape descriptors. Chem Central J 1:7–24
29. Tripathi A, Kellogg GE (2010) A novel and efficient tool for locating and characterizing protein cavities and binding sites. Proteins 78:825–842
30. Peters KP, Fauck J, Frömmel C (1996) The automatic search for ligand binding sites in proteins of known three-dimensional structure using only geometric criteria. J Mol Biol 256:201–213
31. Zhong S, MacKerell AD (2007) Binding response: a descriptor for selecting ligand binding site on protein surfaces. J Chem Inf Model 47:2303–2315
32. Binkowski TA, Naghibzadeh S, Liang J (2003) CASTp: computed atlas of surface topography of proteins. Bioinformatics 31:3352–3355
33. Petrek M, Otyepka M, Banás P et al (2006) CAVER: a new tool to explore routes from protein clefts, pockets and cavities. BMC Bioinformatics 7:316–325
34. Guilloux VL, Schmidtke P, Tuffery P (2009) Fpocket: an open source platform for ligand pocket detection. BMC Bioinformatics 11:1–11
35. Till MS, Ullmann GM (2010) McVol—a program for calculating protein volumes and identifying cavities by a Monte Carlo algorithm. J Mol Modeling 16:419–429
36. Brady GP, Stouten PF (2000) Fast prediction and visualization of protein binding pockets with PASS. J Comput Aided Mol Des 14: 383–401
37. Kalidas Y, Chandra N (2008) PocketDepth: a new depth based algorithm for identification of ligand binding sites in proteins. J Struct Biol 161:31–42
38. Nayal M, Honig B (2006) On the nature of cavities on protein surfaces: application to the identification of drug-binding sites. Proteins: Struct Funct Bioinformatics 63:892–906
39. Tseng YY, Dupree C, Chen ZJ et al (2009) SplitPocket: identification of protein functional surfaces and characterization of their spatial patterns. Nucleic Acids Res 37:W384–W389
40. Laskowski RA (1995) SURFNET: a program for visualizing molecular surfaces, cavities, and intermolecular interactions. J Mol Graphics 13:307–308
41. Huang B, Schroeder M (2006) LIGSITE csc: predicting ligand binding sites using the Connolly surface and degree of conservation. BMC Structural Biol 11:1–11
42. Kitchen DB, Decornez H, Furr JR et al (2004) Docking and scoring in virtual screening for drug discovery: methods and applications. Nat Rev Drug Discov 3:935–949
43. Harris R, Olson AJ, Goodsell DS (2008) Automated prediction of ligand-binding sites in proteins. Proteins 70:1506–1517
44. Halgren T (2007) New method for fast and accurate binding-site identification and analysis. Chem Biol Drug Design 69:146–148
45. An J, Totrov M, Abagyan R et al (2005) Pocketome via comprehensive identification and classification of ligand binding envelopes. Mol Cell Proteomics 4:752–761
46. Laurie ATR, Jackson RM (2005) Structural bioinformatics Q-SiteFinder: an energy-based method for the prediction of protein—ligand binding sites. Bioinformatics 21:1908–1916
47. Hernandez M, Ghersi D, Sanchez R (2009) SITEHOUND-web: a server for ligand binding site identification in protein structures. Nucleic Acids Res 37:W413–W416
48. Ruppert J, Welch W, Jain AN (1997) Automatic identification and representation of protein binding sites for molecular docking. Protein Sci 6:524–533
49. Lipinski C (2004) Lead- and drug-like compounds: the rule-of-five revolution. Drug Discov Today: Technologies 1:337–341
50. Zhang M-Q, Wilkinson B (2007) Drug discovery beyond the "rule-of-five". Curr Opinion Biotech 18:478–488
51. Hopkins AL, Groom CR (2002) The druggable genome. Nat Rev Drug Discov 1:727–730
52. Costa PR, Acencio ML, Lemke N (2010) A machine learning approach for genome-wide prediction of morbid and druggable human genes based on systems-level data. BMC Genomics 5(Suppl 11):S9
53. Lesburg CA, Cable MB, Ferrari E et al (1999) Crystal structure of the RNA-dependent RNA polymerase from hepatitis C virus reveals a fully encircled active site. Nat Struct Mol Biol 6:937–943
54. Salah E, Ugochukwu E, Barr AJ et al (2011) Crystal structures of ABL-related gene (ABL2) in complex with imatinib, tozasertib (VX-680), and a type I inhibitor of the triazolecarbothioamide class. J Med Chem 54:2359–2367
55. Abad-Zapatero C, Metz JT (2005) Ligand efficiency indices as guideposts for drug discovery. Drug Discov Today 10:464–469
56. Ghersi D, Sanchez R (2009) Improving accuracy and efficiency of blind protein–ligand docking by focusing on predicted binding sites. Proteins 74:417–24

57. Hopkins AL, Groom CR, Alex A (2004) Ligand efficiency: a useful metric for lead selection. Drug Discov Today 9:430–431
58. Bohacek RS, McMartin C (1992) Definition and display of steric, hydrophobic, and hydrogen bonding properties of ligand binding sites in proteins using Lee and Richards accessible surface: validation of a high-resolution graphical tool for drug design. J Med Chem 35: 1671–1684
59. Licata L, Briganti L, Peluso D et al (2011) MINT, the molecular interaction database: 2012 update. Nucleic Acids Res 40: D857–D861
60. Bourgeas R, Basse M-J, Morelli X et al (2010) Atomic analysis of protein–protein interfaces with known inhibitors: the 2P2I database. PloS One 5:e9598
61. Kozakov D, Hall DR, Chuang GY et al (2011) Structural conservation of druggable hot spots in protein–protein interfaces. Proc Natl Acad Sci USA 108:13528–13533
62. Trosset J-Y, Dalvit C, Knapp S et al (2006) Inhibition of protein–protein interactions: the discovery of druglike beta-catenin inhibitors by combining virtual and biophysical screening. Proteins 64:60–67
63. Fasolini M, Wu X, Flocco M et al (2003) Hot spots in Tcf4 for the interaction with beta-catenin. J Biol Chem 278:21092–21098
64. Clackson T, Wells JA (1995) A hot spot of binding energy in a hormone-receptor interface. Science 267:383–386
65. Kuttner YY, Engel S (2011) Protein hot spots—the islands of stability. J Mol Biol 415:419–428
66. Wanner J, Fry DC, Peng Z et al (2011) Druggability assessment of protein–protein interfaces. Future Med Chem 3:2021–2038
67. Metz A, Pfleger C, Kopitz H et al (2011) Hot spots and transient pockets: predicting the determinants of small-molecule binding to a protein–protein interface. J Chem Inf Model 52:120–133
68. Henrich S, Salo-Ahen OMH, Huang B et al. (2010) Computational approaches to identifying and characterizing protein binding sites for ligand design. J Mol Recognit 23:209–219.
69. McMartin C (2012) A geometry force field which converts low-resolution X-ray models to structures with properties found at ultra high resolution. Protein Sci 21:75–83
70. Knight ZA, Lin H, Shokat KM (2010) Targeting the cancer kinome through polypharmacology. Nat Rev Cancer 10:130–137
71. Hopkins AL (2008) Network pharmacology: the next paradigm in drug discovery. Nat Chem Biol 4:682–690
72. Chikhi R, Sael L, Kihara D (2010) Real-time ligand binding pocket database search using local surface descriptors. Proteins 78: 2007–2028
73. Das S, Kokardekar A, Breneman CM (2009) Rapid comparison of protein binding site surfaces with property encoded shape distributions. J Chem Inf Model 49:2863–2872
74. Ivanisenko VA, Pintus SS, Grigorovich DA et al (2004) PDBSiteScan: a program for searching for active, binding and posttranslational modification sites in the 3D structures of proteins. Nucleic Acids Res 32:W549–W554
75. Totrov M (2011) Ligand binding site superposition and comparison based on Atomic Property Fields: identification of distant homologues, convergent evolution and PDB-wide clustering of binding sites. BMC Bioinformatics 12:S35
76. Gold ND, Jackson RM (2005) A searchable database for comparing protein–ligand binding sites for the analysis of structure-function relationships. J Chem Inf Model 46:736–742
77. Schmidtke P, Souaille C, Estienne F et al (2010) Large-scale comparison of four binding site detection algorithms. J Chem Inf Model 50:2191–2200
78. Feldman HJ, Labute P (2010) Pocket similarity: are alpha carbons enough? J Chem Inf Model 50:1466–1475
79. Fontaine F, Pastor M, Zamora I et al (2005) Anchor-GRIND: filling the gap between standard 3D QSAR and the GRid-INdependentdescriptors. J Med Chem 48: 2687–2694
80. Crivori P, Zamora I, Speed B et al (2004) Model based on GRID-derived descriptors for estimating CYP3A4 enzyme stability of potential drug candidates. J Comput Aided Mol Des 18:155–166
81. Baroni M, Cruciani G, Sciabola S et al (2007) A common reference framework for analyzing/comparing proteins and ligands. Fingerprints for ligands and proteins (FLAP): theory and application. J Chem Inf Model 47:279–294
82. Ritchie DW, Venkatraman V (2010) Ultra-fast FFT protein docking on graphics processors. Bioinformatics 26:2398–2405
83. Katchalski-Katzir E (1992) Molecular surface recognition: determination of geometric fit between proteins and their ligands by correlation techniques. Proc Natl Acad Sci 89:2195–2199
84. Brenke R, Kozakov D, Chuang G-Y et al (2009) Fragment-based identification of druggable "hot spots" of proteins using Fourier

domain correlation techniques. Bioinformatics 25:621–627
85. Doppelt-Azeroual O, Moriaud F, Adcock SA et al (2009) A review of MED-SuMo applications. Infect Disord Drug Targets 9:344–357
86. Jambon M, Andrieu O, Combet C et al (2005) The SuMo server: 3D search for protein functional sites. Bioinformatics 21: 3929–3930
87. Moriaud F, Doppelt-Azeroual O, Martin L et al (2009) Computational fragment-based approach at PDB scale by protein local similarity. J Chem Inf Model 49:280–294
88. Doppelt-Azeroual O, Delfaud F, Moriaud F et al (2010) Fast and automated functional classification with MED-SuMo: an application on purine-binding proteins. Protein Sci 19:847–867
89. Schmitt S, Kuhn D, Klebe G (2002) A new method to detect related function among proteins independent of sequence and fold homology. J Mol Biol 323:387–406
90. Bartlett S, Beddard GS, Jackson RM et al (2005) Comparison of the ATP binding sites of protein kinases using conformationally diverse bisindolylmaleimides. Methods 127:11699–11708
91. Powers R, Copeland JC, Germer K et al (2006) Comparison of protein active site structures for functional annotation of proteins and drug design. Proteins 65:124–135
92. Kinoshita K, Furui J, Nakamura H (2002) Identification of protein functions from a molecular surface database, eF-site. J Struct Funct Genomics 2:9–22
93. Skolnick J, Brylinski M (2009) FINDSITE: a combined evolution/structure-based approach to protein function prediction. Brief Bioinform 10:378–391
94. Shulman-Peleg A, Shatsky M, Nussinov R et al (2008) MultiBind and MAPPIS: webservers for multiple alignment of protein 3D-binding sites and their interactions. Nucleic Acids Res 36:W260
95. Minai R, Matsuo Y, Onuki H et al (2008) Method for comparing the structures of protein ligand-binding sites and application for predicting protein–drug interactions. Proteins: Struct Funct Bioinformatics 72: 367–381
96. Ausiello G, Via A, Helmer-Citterich M (2005) Query3d: a new method for high-throughput analysis of functional residues in protein structures. BMC Bioinformatics 4(Suppl 6):S5
97. Schalon C, Surgand JS, Kellenberger E et al (2008) A simple and fuzzy method to align and compare druggable ligand-binding sites. Proteins: Struct Funct Bioinformatics 71:1755–1778
98. Brakoulias A, Jackson RM (2004) Towards a structural classification of phosphate binding sites in protein–nucleotide complexes: an automated all-against-all structural comparison using geometric matching. Proteins: Struct Funct Bioinformatics 56:250–260
99. Tomei L, Roussel A, Incitti I et al (1999) Crystal structure of the RNA-dependent RNA polymerase of hepatitis C virus. Proc Natl Acad Sci 23:13034–13039
100. Pons J-L, Labesse G (2009) @TOME-2: a new pipeline for comparative modeling of protein–ligand complexes. Nucleic Acids Res 37:W485–W491
101. Shulman-Peleg A, Nussinov R, Wolfson HJ (2004) Recognition of functional sites in protein structures. J Mol Biol 339:607–633
102. Zorn J, Wells J (2010) Turning enzymes on with small molecules. Nat Chem Biol 6:179–188
103. Hantschel O, Grebien F, Superti-furga G (2011) Targeting allosteric regulatory "Drugging the Undruggable" modules in oncoproteins. Cancer Res 2:829–830
104. Oberlin D, Scheraga HA (1998) B-spline method for energy minimization in grid-based molecular mechanics calculations. J Comp Chem 19:71–85
105. Trosset J-Y, Scheraga HA (1999) Prodock: software package for protein modeling and docking. J Comp Chem 20:412–427
106. Trosset J-Y, Scheraga HA (1999) Flexible docking simulations: scaled collective variable Monte Carlo minimization approach using Bezier splines, and comparison with a standard Monte Carlo algorithm. J Comp Chem 20:244–252
107. Roisman LC, Piehler J, Trosset JY et al (2001) Structure of the interferon-receptor complex determined by distance constraints from double-mutant cycles and flexible docking. Proc Natl Acad Sci U S A 98:13231–13236
108. Trosset JY, Scheraga HA (1998) Reaching the global minimum in docking simulations: a Monte Carlo energy minimization approach using Bezier splines. Proc Natl Acad Sci USA 95:8011–8015
109. McMartin C, Bohacek RS (1997) QXP: powerful, rapid computer algorithms for structure-based drug design. J Comput Aided Mol Des 11:333–344

Chapter 11

Validating Pharmacological Disruption of Protein–Protein Interactions by Acceptor Photobleaching FRET Imaging

Janos Roszik, Gábor Tóth, János Szöllősi, and György Vereb

Abstract

Proteins are the major targets of drug discovery and many of the new drugs are designed to exert their effect by disrupting protein-protein interactions. Validation of the inhibition of molecular interactions is generally done by biochemical methods, however, these are often not feasible when the interaction is not stable enough. Fluorescence resonance energy transfer (FRET) is an excellent tool for determining direct molecular interactions between two molecules in the cell membrane or inside cells in their natural state. Although originally established as a flow cytometric approach, FRET has been adapted for microscopy, allowing for analysis of sub-cellular co-localization at the single cell level. In this chapter, we provide theoretical introduction to the phenomenon of FRET, and a protocol - including labeling techniques, measurement, and evaluation of microscopy images - of the simplest microscopic FRET approach, acceptor photobleaching FRET. This technique is generally usable for studying protein interactions and requires only a standard confocal laser scanning microscope. To demonstrate the value of image based FRET for testing pharmacological disruption of protein-protein interactions, we show how inhibition of the heterodimerization of ErbB2 and ErbB1 by the humanized monoclonal antibody pertuzumab can be validated using this technique.

Key words: Förster resonance energy transfer, Image cytometry, Protein interactions, Pertuzumab, ErbB2, ErbB1

1. Introduction

Pharmacological disruption of protein–protein interactions is one of the important emerging tools of today's medicine. When a new compound that targets a protein interaction in a crucial pathway related to a disease is developed, its mode of action needs to be validated to prove that it exerts the desired effect through inhibiting the interaction of its targets. Although several biochemical methods are available, these are not always applicable, for example, when the interaction is not stable enough, i.e., it is not preserved

Jürgen Moll and Riccardo Colombo (eds.), *Target Identification and Validation in Drug Discovery: Methods and Protocols*, Methods in Molecular Biology, vol. 986, DOI 10.1007/978-1-62703-311-4_11,

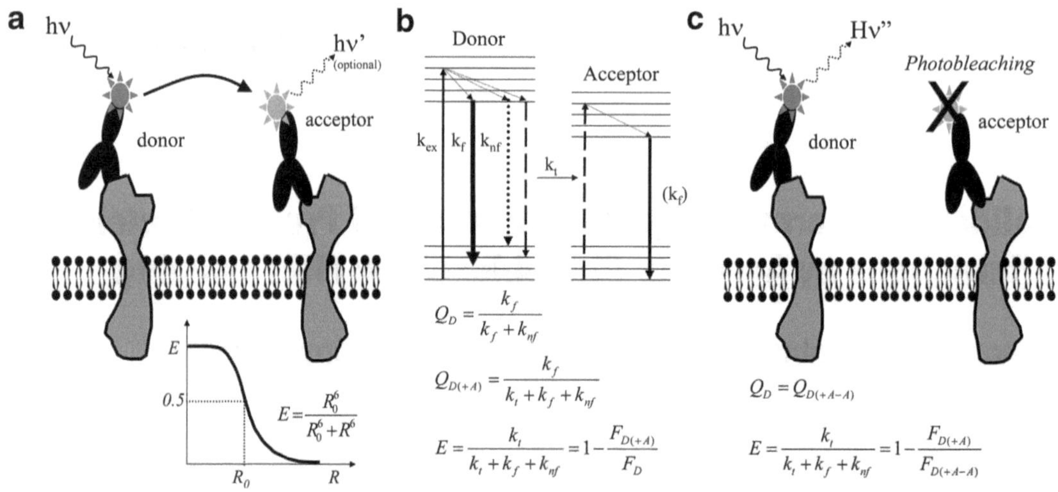

Fig. 1. Illustration of the Förster resonance energy transfer phenomenon. (**a**) FRET occurs when an excited fluorescent molecule, called donor, transfers energy to another molecule called acceptor by a dipole–dipole resonance energy transfer mechanism. The acceptor may emit this energy as a photon, provided it is fluorescent. The requirements for this event are the following: (1) The emission spectrum of the donor has to overlap with the excitation spectrum of the acceptor. (2) The emission dipole vector of the donor and the absorption dipole vector of the acceptor need to be close to parallel. (3) The distance between the donor and acceptor molecules has to be between 1–10 nm. The efficiency of FRET, E, is the function of the inverse 6th power of donor–acceptor separation and thus FRET occurs only over short R distances in the range of 1–10 nm, distributed around R0, which marks 50% efficiency for a given donor–acceptor pair. The requirement for molecular proximity makes the phenomenon ideal for detecting association or aggregation of various macromolecules even in functioning cells. (**b**) The Jablonski diagram shows that FRET is an additional de-excitation pathway (rate constant = *kt*) for the excited donor that otherwise can relax by radiative (*kf*) and non-fluorescent (*knf*) processes. Henceforth, the Q_D quantum efficiency of the donor decreases in the presence of FRET ($Q_{D(+A)}$). It follows that measuring donor fluorescence, a parameter proportional to *Q*, in the presence and absence of acceptor is an easy way to assess FRET efficiency. (**c**) Photobleaching of the acceptor relieves the donor from quenching and increases its quantum efficiency, with a consequent increase of its fluorescence. The strategy of determining first the donor fluorescence on the double-labeled (donor + acceptor) sample, then photobleaching the acceptor, and consecutively measuring, on the same sample, the de-quenched donor fluorescence (proportional to $Q_{D(+A-A)}$) allows for a self-controlled determination of FRET efficiencies.

after applying biochemical techniques. A further drawback of biochemical approaches is that the cells are destroyed during the process, and no information is provided on the subcellular localization of events induced by the drug. Förster resonance energy transfer (FRET, also known as fluorescence resonance energy transfer) (1) is an excellent and widely used tool for determining direct molecular interactions between two molecules on the surface or inside cells in their natural state. Furthermore, FRET methods have been adapted for confocal microscopy allowing the analysis of subcellular localization at the single cell level.

FRET is a non-radiative transfer of energy from an excited donor fluorophore to an acceptor fluorophore which needs to be at a distance of ~1–10 nm for conventional fluorescent tags (2–5) (see Fig. 1a). Energy transfer is characterized by the FRET efficiency *E*, which is the ratio of excited-state donor molecules relaxing by FRET to the total number of excited donors. The simplest way to determine E is to measure donor fluorescence in the absence (F_D) and in the presence ($F_{D(+A)}$) of acceptors (see Fig. 1b):

$$E = 1 - \frac{F_{D(+A)}}{F_D} \tag{1}$$

The rate of energy transfer is dependent on the negative sixth power of the distance R between the donor and the acceptor, and R_0 is the distance where $E=0.5$, that is, where there is a 50% chance that the energy of the excited donor will be transferred to the acceptor. Since the efficiency of FRET depends on the inverse sixth power of the distance between these fluorophores, it is not just a tool for determining molecular interactions, but it can also be used as a spectroscopic ruler (6). It is of much interest in many biological systems to know if and how specific molecules interact, and how this process is altered by pharmaceutical intervention. When a drug disrupts the association of proteins, FRET can be applied to confirm this mechanism. The relatively short range over which FRET occurs makes it an excellent tool for the visualization of these molecular interactions or even conformational changes. The occurrence of FRET has profound consequences on the fluorescence properties of both the donor and the acceptor. An additional process to dissipate the energy captured by the excited fluorophore is introduced in the donor, which decreases the fluorescence lifetime and the quantum efficiency of the donor, rendering it less fluorescent. The decrease in donor fluorescence (often termed donor quenching) can be one of the most easily measured spectroscopic characteristics that indicates the occurrence of FRET. The de-quenching of the donor upon photodestruction of the acceptor results in an increase of the donor fluorescence, which is proportional to the FRET efficiency E:

$$E = 1 - \frac{F_{\text{Donor(quenched\_by\_acceptor)}}}{F_{\text{Donor(de-quenched)}}} \tag{2}$$

Additionally, since the acceptor is excited as a result of FRET, those acceptors that are fluorescent will emit photons (proportional to their quantum efficiency) also when FRET occurs. This is called sensitized emission and can also be a good measure of FRET (see Fig. 1). To quantitate FRET efficiency in practice, several approaches have been evolved so far. In flow cytometric FRET (7), we can obtain cell-averaged statistics for large cell populations, while the subcellular details can be investigated with various microscopic approaches. Jares-Erijman and Jovin have classified 22 different approaches that can be used to quantify energy transfer (8). Most of them are based on donor quenching and/or acceptor sensitization, and a few on measuring emission anisotropy of either the donor or the acceptor. Some of these methods can be combined to extend the information content of the measurement, for example two-sided FRET (9) involves both acceptor depletion (10) and donor photobleaching rate measurements (11) on the same sample consecutively to assess the association state of two molecular pairs out of three distinct molecules in a correlated manner. Other combination techniques, like the three color FRET (12) and variations on the direct measurements of fluorescence lifetimes in microscopic imaging (FLIM) (13, 14), have also been developed and applied in the assessment of protein interactions.

In this chapter, we demonstrate the implementation of the acceptor photobleaching FRET method, which is the simplest, robust image cytometric FRET technique that can be performed using any confocal microscopes (15). It involves measuring the donor fluorescence in the presence of the acceptor, and then repeating the measurement after having photobleached the acceptor fluorophore. The result is independent of donor and acceptor concentration and stoichiometry and requires relatively simple image mathematics.

In order to obtain reliable results, it is of basic importance to have a sound knowledge of this technique to avoid the various pitfalls. The most common issue is the correction for spectral cross talk/bleed-through which cannot be neglected; however, it can be minimized by choosing appropriate fluorophores and emission filters. Here we describe in detail the application of this method for confocal microscopy, so that it can easily be applied to validating the pharmacological disruption of protein–protein interactions using basic instrumentation, and provide all the necessary information to make its application precise and fruitful.

Our model system for demonstrating this application is a potentially important interaction within the ErbB family of receptor tyrosine kinases—that of ErbB1 (EGFR) and ErbB2. Overexpression and interactions of ErbB2 have been shown to be of basic importance in development, progression, and metastasis formation of tumors (16, 17). Several humanized antibodies have been developed to target ErbB2, including trastuzumab, the first therapeutic antibody against solid tumors (17). Dimerization and activation of ErbB2 is inhibited by Hsp90 (18); however, Hsp90 inhibitors can be used to enhance internalization and downregulation of ErbB2 in ErbB2 overexpressing tumors that are resistant to trastuzumab antibody therapy (19). Another possibility in targeting ErB2, especially on trastuzumab resistant tumors, is the use of another humanized monoclonal antibody, pertuzumab, which inhibits the association of ErbB2 with other ErbB family members, and thus the formation of active receptor tyrosine kinase heterodimers (20). One important partner for such heterodimerization is ErbB1, which, upon binding EGF, can transactivate ErbB2, and further enhance tumor proliferation in an EGF containing microenvironment. Pertuzumab has been found clinically beneficial for the treatment of such tumors (21). In this chapter, we show how this effect can be validated in vitro using FRET imaging.

Table 1
The most frequently used donor–acceptor fluorophore pairs

Donor (excitation, emission nm)	Acceptor (excitation, emission nm)
Fluorescein (494, 521), AlexaFluor488 (495, 519), Cy2 (492, 510), Atto488 (501, 523)	Rhodamine (550, 570), AlexaFluor546 (556, 573), AlexaFluor555 (555, 565), Cy3 (554, 658), Atto550 (554, 576)
Cy3 (554, 658), AlexaFluor546 (556, 573), AlexaFluor555 (555, 565), Atto550 (554, 576), Rhodamine (550, 570), Phycoerythrin (480-565, 578)	Cy5 (650, 665), AlexaFluor633 (621, 639), AlexaFluor647 (650, 668), Atto647N (644, 669), Allophycocyanin (650, 660)
CFP and variants, such as Cerulean (433, 475)	YFP and variants, such as Venus (515, 528)
eGFP (488, 509)	tagRFP (555, 584), mStrawberry (574, 596)

2. Materials

2.1. Cells

Any kind of cells, adherent and also non-adherent, can be used grown on or attached to (e.g., using poly-l-lysine coated) coverslips or chamber wells. In the example experiment, we used the BT-474 cell line (ATCC catalog number HTB-20), grown on chambered cover glass.

2.2. Fluorophores

1. The donor and acceptor fluorophores, similarly to flow-cytometric FRET, must be appropriate for the lasers available (22), and it is advisable to use fluorophores that emit in the red and far red spectral range, where cellular autofluorescence becomes progressively weaker (23).
2. The most straightforward way of labeling cells is using fluorophore conjugated antibodies, that are usually available with the most commonly used dyes (see Table 1 and Notes 1 and 2).
3. Labeling the proteins of interest with fluorescent proteins is another feasible option. The most useful fluorescent protein pairs are also listed in Table 1.
4. In the following sections, we describe the use of fluorescently labeled antibodies; however, these protocols are applicable to fluorescent fusion proteins as well.

2.3. Labeling of Proteins Using Fluorescent Antibodies

1. PBS: 137 mM NaCl, 3 mM KCl, 10 mM Na_2HPO_4, 1.8 mM KH_2PO_4, pH 7.4 (see Note 3).
2. PBS + 0.1% BSA (for extracellular labeling).
3. PBS + 0.1% BSA + 0.1% (v/v) Triton X-100 (for intracellular labeling).

4. PBS + 1% formaldehyde (for fixation).
5. Fluorescently labeled antibodies, which are conjugated with donor and acceptor dyes, respectively.

2.4. Evaluation Software

Acceptor photobleaching FRET measurements can be efficiently analyzed using the AccPbFRET ImageJ plugin (24). The newest version of the program can be freely downloaded from its homepage (25).

3. Methods

3.1. Sample Preparation and Labeling

3.1.1. Sample Preparation

1. In general, the following samples are necessary:
   (a) Sample 1. Unlabeled cells.
   (b) Sample 2. Cells with protein A labeled with the donor-conjugated antibody.
   (c) Sample 3. Cells with protein B labeled with the acceptor-conjugated antibody.
   (d) Sample 4. Cells with protein A labeled with the donor-conjugated and protein B labeled with the acceptor-conjugated antibody. Distinct samples of such labeling should be prepared for treatments with the drug to be tested, and for control (see Notes 4 and 5).
2. It is advisable to have a negative and a positive control with known molecular distances or FRET efficiencies obtained earlier using the same method. For example, in the case of pertuzumab, two epitopes on ErbB2 can be stained with pertuzumab and trastuzumab, respectively. This provides a good positive control as these epitopes are very close to each other on the same molecule.

3.1.2. Labeling Extracellular Epitopes

1. Wash cells twice with ice-cold PBS + 0.1% BSA.
2. Label cells with (usually) 5–50 μg/mL final concentration of antibodies (should be above saturating concentration that has to be determined previously) in PBS + 0.1% BSA for 10–30 min on ice (see Note 6).
3. Wash cells twice with ice-cold PBS.
4. Fix cells with PBS + 1% formaldehyde for 10 min on ice and then 10 min at room temperature.
5. Store samples at 4°C until measurement. Samples can be stored at least for a week.

3.1.3. Labeling Intracellular Epitopes

1. Wash cells with ice-cold PBS + 0.1% BSA + 0.1% (v/v) Triton X-100.
2. Repeat washing.

3. Label cells with (usually) 5–50 μg/mL final concentration of antibodies (should be above saturating concentration that has to be determined previously) in PBS + 0.1% BSA + 0.1% (v/v) Triton X-100 for 30–60 min on ice.
4. Wash cells twice with ice-cold PBS + 0.1% BSA + 0.1% (v/v) Triton X-100.
5. Fix cells with PBS + 1% formaldehyde for 10 min on ice and then 10 min at room temperature.
6. Store samples at 4°C until measurement. Samples can be stored at least for a week.

3.2. Measurement Using a Confocal Microscope

1. Before the measurement, set up the confocal microscope including the objective (at least 40× for single-cell imaging), lasers and detection channels according to the fluorophores used. The donors AlexaFluor546, AlexaFluor555 and Cy3 can be excited at 543 nm and detected through a 560–615 nm emission filter. Acceptors Cy5, AlexaFluor633, and AlexaFluor647 can be excited at 633 nm and detected through a 650 nm longpass filter.
2. Images have to be taken before and after photobleaching the acceptor in two channels:
   - Donor excitation/donor emission, and
   - Acceptor excitation/acceptor emission.
3. Do the measurement of (a) unlabeled, (b) donor-only, (c) acceptor-only (the latter two are needed for calculating correction factors) and (d) donor + acceptor samples of untreated and treated cells. Take at least ten images of each sample (see Note 7).

3.3. Evaluation of FRET Images

Analysis of confocal images and calculation of energy transfer efficiencies can be done using the AccPbFRET ImageJ plugin which uses the following calculations (24). When we include all necessary corrections in Eq. 2, FRET efficiency $E_{(i,j)}$ is obtained pixel-by-pixel according to the following equation:

$$E_{(i,j)} = 1 - \frac{(1-\alpha)(F_{\mathrm{D1}(i,j)} - \delta F_{\mathrm{A1}(i,j)})}{\gamma\left(F_{\mathrm{D2}(i,j)} - (\alpha\delta + (1-\alpha)\varepsilon)F_{\mathrm{A1}(i,j)}\right) - \alpha\left(F_{\mathrm{D1}(i,j)} - \delta F_{\mathrm{A1}(i,j)}\right)} \quad (3)$$

where $F_{\mathrm{D1}(i,j)}$ and $F_{\mathrm{D2}(i,j)}$ are the donor fluorescence values of the pixel (i,j) before (1) and after (2) photobleaching the acceptor, and $F_{\mathrm{A1}(i,j)}$ the acceptor fluorescence for the same pixel before photobleaching. All F values are background corrected throughout. α, γ, δ, and ε are correction factors that are described below.

In some cases, photobleaching of the acceptor is not complete. We can correct for this phenomenon assuming that photobleaching occurs indiscriminately to all acceptor molecules, and there is not more than one acceptor per donor molecule present. To correct for incomplete acceptor bleaching, the correction factor α is calculated as

$$\alpha = \left\langle F_{A2(i,j)} / F_{A1(i,j)} \right\rangle \quad (4)$$

where $F_{A2(i,j)}$ and $F_{A1(i,j)}$ are intensities in the acceptor channel in pixels above threshold of the donor and acceptor labeled sample, before (1) and after (2) photobleaching. The <> signs denote mean value.

Although it is usually unavoidable, caution needs to be taken to minimize photobleaching of the donor during the measurement. The correction factor γ for unwanted photobleaching of the donor during the image acquisition procedure can be calculated either as

$$\gamma = \left\langle F_{Dd1(i,j)} \right\rangle / \left\langle F_{Dd2(i,j)} \right\rangle \quad \text{or} \quad \gamma = \left\langle F_{Dd1(i,j)} / F_{Dd2(i,j)} \right\rangle \quad (5)$$

where $F_{Dd1(i,j)}$ and $F_{Dd2(i,j)}$ are donor fluorescence intensities of donor only samples (D_d) in pixels above threshold before (1) and after (2) photobleaching the acceptor.

The program offers (similarly as for the correction factors δ and ε), the possibility to calculate the factor on a pixel-by-pixel basis and then average pixels above threshold for raw data, or, alternatively, to average the raw data for these pixels and then calculate an average correction factor.

In the case that the acceptor dye also fluoresces in the donor channel, FRET would be underestimated without correcting for this cross talk (9). The appropriate correction factor δ is calculated as

$$\delta = \left\langle F_{Dd1(i,j)} \right\rangle / \left\langle F_{Aa1(i,j)} \right\rangle \quad \text{or} \quad \delta = \left\langle F_{Dd1(i,j)} / F_{Aa1(i,j)} \right\rangle \quad (6)$$

where $F_{Da1(i,j)}$ and $F_{Aa1(i,j)}$ are signals in the donor and acceptor channels in pixels above threshold of an "acceptor only" labeled sample, before (1) photobleaching the acceptor.

In some cases, photobleaching the acceptor can yield a photoproduct with distinct absorption and emission properties, which can contribute to the post-bleach donor signal, resulting in the overestimation of FRET efficiency (10). The correction factor ε for such acceptor-photoproduct is calculated as

$$\varepsilon = \left\langle F_{Da2(i,j)} \right\rangle / \left\langle F_{Aa1(i,j)} \right\rangle \quad \text{or} \quad \varepsilon = \left\langle F_{Da2(i,j)} / F_{Aa1(i,j)} \right\rangle \quad (7)$$

where $F_{Da2(i,j)}$ and $F_{Aa1(i,j)}$ are intensities in the donor and acceptor channels in pixels above threshold of an "acceptor only" labeled sample, before (1) and after (2) photobleaching.

The necessity for using correction factors (apart from that for unwanted donor photobleaching and incomplete acceptor photobleaching) can be avoided by selecting suitable fluorophores. In the case of the AlexaFluor 546–AlexaFluor 647 donor–acceptor pair, it is usually sufficient to correct for unwanted donor photobleaching, and therefore to analyze only the donor images.

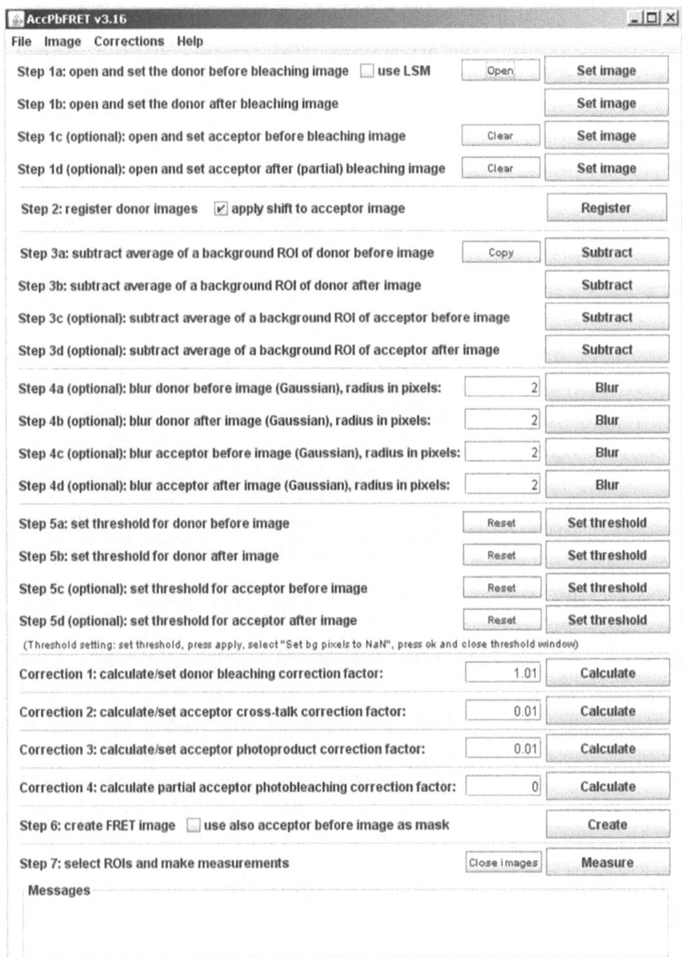

Fig. 2. Analysis of images using the AccPbFRET program. This figure shows the main window of the ImageJ plugin AccPbFRET which is used to analyze acceptor photobleaching FRET images. The analysis consists of calculation of correction factors which is performed once at the beginning, followed by seven steps. Those have to be done with all FRET images of the donor and acceptor-labeled samples.

The AccPbFRET software (24) leads the user through the FRET analysis process of seven steps, which include opening and registering the images, background subtraction, optional Gaussian blur, thresholding, making corrections, creating the FRET image, and making measurements. These steps can be seen in Fig. 2, which shows the main window of the AccPbFRET plugin and the corresponding parameter settings. The single steps were performed as follows:

In step 1, images can be opened using the “Open” button. After opening, these images can be set as donor before/after, or acceptor before/after photobleaching using the “Set image” buttons. Multichannel LSM images (either having before and after images, or donor and acceptor channel images, or both) can be opened in one step when the "use LSM" checkbox is selected. Otherwise, the opened multichannel image has to be split (item available from the "Image" menu) before setting it. To register (correct small *xy* shifts) the donor images in step 2, press the "Register" button. If the "apply shift to acceptor image" checkbox is checked, acceptor image pairs will automatically be registered using the same shift. To subtract background (the average of pixels in a preselected ROI) in step 3, the "Subtract" button has to be pressed for each image. The "Copy" button will copy the ROI of the first image to the others before subtraction. Step 4 is an optional Gaussian blur. In step 5, thresholds can be set for all images. This always has to be done to exclude background pixels from the analysis. Steps 4 and **5** can be reverted using the "Reset" buttons. Correction values are calculated by the program based on images of the single-labeled samples. These values are used in step 6, but these calculations have to be done at the beginning of the analysis of FRET images (i.e., before starting with step 1). When the "Calculate" buttons are pressed, small windows pop up explaining which images have to be opened and set for calculation of the corresponding correction factors. Donor bleaching correction is always necessary, but the acceptor and acceptor photoproduct cross talk corrections are usually not needed when using optimal fluorophores. Pressing "Create" in step 6 will create the FRET image. ROIs can be selected on this and measurements can be made in step 7 using the "Measure" button.

3.4. Interpretation of FRET Efficiencies

The well-established acceptor photobleaching FRET method described in this chapter proved its applicability in the last years in solving many questions of biology (26–29). By using this method we can obtain reliable results without the need for very expensive instrumentation and software. However, it is always important to use positive and negative controls to prove the reliability of the results, and negative FRET data need to be interpreted with caution. However, when evaluating pharmacological disruption of protein–protein interactions, decrease in FRET efficiency after treatment with a drug can be a good proof of prevention of molecular associations.

Figure 3 shows the results of the analysis of disruption of ErbB1–ErbB2 interaction by the antibody pertuzumab and by pertuzumab F(ab)2 fragments. ErbB1 and ErbB2 were labeled on BT-474 cells with anti-EGFR mAb 528 and anti-ErbB2 mAb trastuzumab. The average FRET efficiency is about 0.18 in the case of untreated cells. This value decreases to 0.12 and 0.08 after pertuzumab and pertuzumab F(ab)2 treatment, respectively. These results clearly show that this antibody disrupts ErbB1–ErB2 interaction, which is expected to result in slower tumor growth. Interestingly, the F(ab)2 fragment is even more effective than the whole antibody. The positive control in this system is FRET between

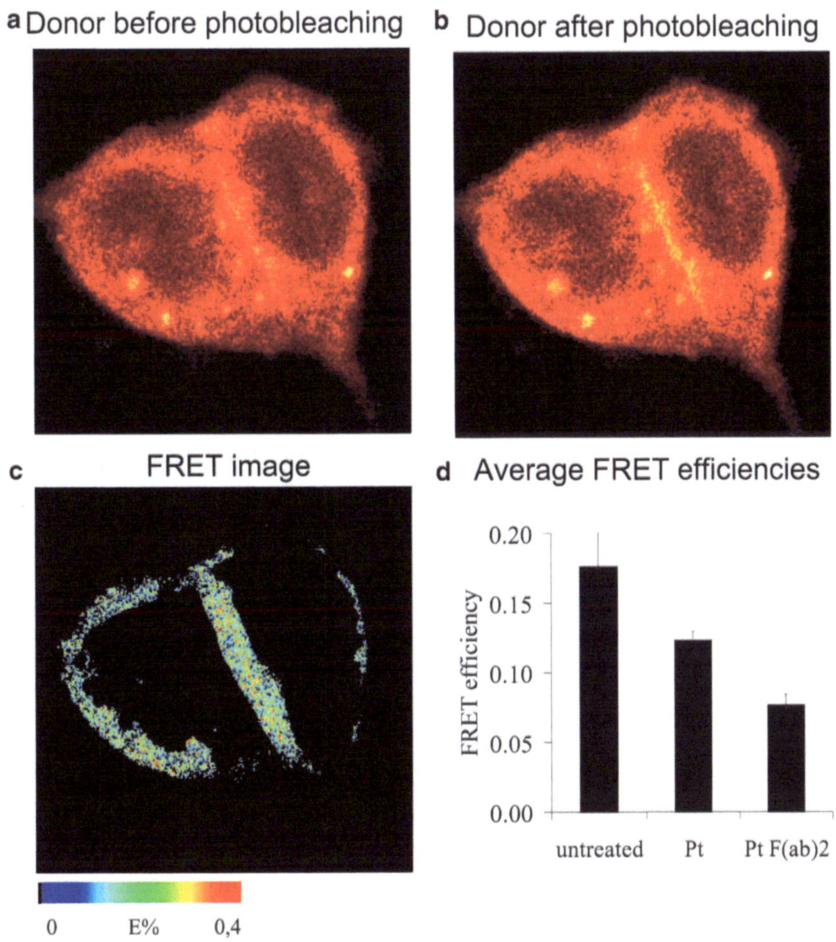

Fig. 3. Analysis of distruption of ErbB1-ErbB2 interaction by Pertuzumab. In this figure, we demonstrate how the ErbB2 dimerization inhibitor Pertuzumab can be validated using the acceptor photobleaching FRET technique. (**a**, **b**) Show the donor (ErbB1), labeled with monoclonal Ab 528 directly conjugated to AlexaFluor546, before and after photobleaching the acceptor, which is the directly conjugated AlexaFluor647 dye on the anti-ErbB2 antibody trastuzumab. The antibodies bind to their respective targets ErbB1 and ErbB2 on BT-474 cells. The images are shown in indexed color using a red-orange lookup table. Note the increase of the donor signal after acceptor photobleaching. (**c**) FRET image created by the AccPbFRET software, shown in indexed color using a rainbow lookup table, representing FRET efficiencies (E%) from 0 to 40 as indicated on the color scale bar. (**d**) The average FRET efficiency (±SEM) of cells was calculated for untreated cells, and cells treated for 30 min before the measurement either with 10 mg/l Pertuzumab whole antibody (Pt) or 6 mg/l Pertuzumab F(ab)2 fragment (Pt F(ab)2).

donor-labeled pertuzumab and acceptor-labeled trastuzumab, where FRET efficiencies averaging to 0.32 are measured, coherent with both antibodies binding to the same target molecule on distinct, but not distant epitopes. Obviously, the effect of pertuzumab cannot be tested and would be irrelevant in this case.

Since FRET has very sharp distance dependence, it is difficult to measure relatively long distances because FRET efficiency gets very low. Therefore, the absence of FRET is not a direct proof of the absence molecular proximity. Also, absence of FRET can be caused by steric hindrance even for neighboring protein molecules. On the other hand, presence of FRET to any appreciable extent above the experimental error of measurement (about 0.05) is strong evidence of molecular interactions.

4. Notes

1. Excitation and emission spectra of molecules for donor–acceptor pairs can be found at one of the following Web sites: Becton-Dickinson Fluorescence Spectrum Viewer (http://www.bdbiosciences.com/spectra), Invitrogen-Molecular Probes Fluorescence Spectra Viewer (http://www.probes.invitrogen.com/servlets/spectraviewer).
2. Direct labeling with conjugated primary antibodies should be used when possible. Indirect labeling usually diminishes FRET owed to the greater separation of fluorophores, and it is also possible that FRET occurs between otherwise more distant targets if a large molecular labeling complex is built up (16). The polyclonal nature of most secondary antibodies can also cause difficulties in interpreting the results.
3. HEPES buffer: 125 mM NaCl, 5 mM KCl, 20 mM HEPES, 1 mM $CaCl_2$, 1.5 mM $MgSO_4$, pH 7.2; can also be used instead of PBS, both for extra- and intracellular labeling.
4. It is advisable to test the possibility of reaching higher FRET efficiency with the untreated sample from epitope A to B, or from epitope B to A. If one epitope is expressed in great molar excess of the other, labeling that one with the acceptor usually yields a more efficient FRET, since all donors can have an acceptor pair.
5. Although low expression levels limit the application of FRET methods, acceptor photobleaching FRET can be successfully used with proteins expressed at as low as 100–150 thousand per cell.
6. Time of labeling with fluorescent antibodies should be optimized previously. Usually, incubation beyond 10 min using saturating concentration does not increase labeling intensity by more than 10%.
7. The selection of dyes is based not only on spectral characteristics, but it is also important to have an easily photobleachable acceptor, in order to avoid shifting of the sample during the time required for photobleaching. During photobleaching of the acceptor, the sample might move if conditions are not optimal. Shifts are usually caused by thermal difference of the sample and the environment and air currents from air conditioning.

Using the AccPbFRET program, small *xy*-shifts can be corrected. In case of a *z*-shift, it is not possible to make corrections and these images cannot be used for analysis.

Acknowledgments

The authors were supported by the following grants: Hungarian National Research Fund K75752 and NK101337; Hungarian National Development Agency TAMOP-4.2.1/B-09/1/KONV-2010-0007 and 4.2.2/A-11/1/KONV-2012-0025; Baross Gabor programme (REG-EA-09-1-2009-0010) and ETT 362-01/2009 grants.

References

1. Förster T (1946) Energiewanderung und fluoreszenz. Naturwissenschaften 6:166–175
2. Clegg RM (2002) FRET tells us about proximities, distances, orientations and dynamic properties. J Biotechnol 82:177–179
3. Jares-Erijman EA, Jovin TM (2006) Imaging molecular interactions in living cells by FRET microscopy. Curr Opin Chem Biol 10:409–416
4. Vogel SS, Thaler C, Koushik SV (2006) Fanciful FRET. Sci STKE. 331:re2.
5. Vereb G, Matkó J, Szöllősi J (2009) Cytometry of fluorescence resonance energy transfer. In: Darzynkiewicz Z, Robinson JP, Roederer M (eds) Essential cytometry methods. Elsevier, Oxford, pp 55–108
6. Stryer L, Haugland RP (1967) Energy transfer: a spectroscopic ruler. Proc Nat Acad Sci USA 58:719–726
7. Vereb G, Nagy P, Szöllősi J (2011) Flow cytometric FRET analysis of protein interaction. Methods Mol Biol 699:371–392
8. Jares-Erijman EA, Jovin TM (2003) FRET imaging. Nat Biotechnol 21:1387–1395
9. Fazekas Z, Petrás M, Fábián A et al (2008) Two-sided fluorescence resonance energy transfer for assessing molecular interactions of up to three distinct species in confocal microscopy. Cytometry A 73:209–219
10. Bastiaens PIH, Jovin TM (1998) Fluorescence resonance energy transfer microscopy. In: Celis JE (ed) Cell biology: a laboratory handbook, 2nd edn. Academic Press, New York, pp 136–146
11. Szentesi G, Vereb G, Horváth G et al (2005) Computer program for analyzing donor photobleaching FRET image series. Cytometry A 67A:119–128
12. Sun Y, Wallrabe H, Seo SA et al (2010) FRET microscopy in 2010: the legacy of Theodor Forster on the 100th anniversary of his birth. Chemphyschem 12:462–474
13. Zal T (2008) Visualization of protein interactions in living cells. Adv Exp Med Biol 640: 183–197
14. Buranachai C, Kamiyama D, Chiba A et al (2008) Rapid frequency-domain FLIM spinning disk confocal microscope: lifetime resolution, image improvement and wavelet analysis. J Fluoresc 18:929–942
15. Vereb G, Matkó J, Szöllősi J (2004) Cytometry of fluorescence resonance energy transfer. Methods Cell Biol 75:105–152
16. Mocanu MM, Fazekas Z, Petrás M et al (2005) Associations of ErbB2, beta1-integrin and lipid rafts on Herceptin (Trastuzumab) resistant and sensitive tumor cell lines. Cancer Lett 227: 201–212
17. Friedlander E, Barok M, Szöllősi J et al (2008) ErbB-directed immunotherapy: antibodies in current practice and promising new agents. Immunol Lett 116:126–140
18. Citri A, Gan J, Mosesson Y et al (2004) Hsp90 restrains ErbB-2/HER2 signalling by limiting heterodimer formation. EMBO Rep 5:1165–1170
19. Zsebik B, Citri A, Isola J et al (2006) Hsp90 inhibitor 17-AAG reduces ErbB2 levels and inhibits proliferation of the trastuzumab resistant breast tumor cell line JIMT-1. Immunol Lett 104:146–155
20. Franklin MC, Carey KD, Vajdos FF et al (2004) Insights into ErbB signaling from the structure of the ErbB2-pertuzumab complex. Cancer Cell 5:317–328

21. Agus DB, Gordon MS, Taylor C et al (2005) Phase I clinical study of pertuzumab, a novel HER dimerization inhibitor, in patients with advanced cancer. J Clin Oncol 23:2534–2543
22. Horváth G, Petrás M, Szentesi G et al (2005) Selecting the right fluorophores and flow cytometer for fluorescence resonance energy transfer measurements. Cytometry A 65: 148–157
23. Sebestyén Z, Nagy P, Horváth G et al (2002) Long wavelength fluorophores and cell-by-cell correction for autofluorescence significantly improves the accuracy of flow cytometric energy transfer measurements on a dual-laser benchtop flow cytometer. Cytometry 48:124–135
24. Roszik J, Szöllősi J, Vereb G (2008) AccPbFRET: an ImageJ plugin for semi-automatic, fully corrected analysis of acceptor photobleaching FRET images. BMC Bioinformatics 9:346
25. AccPbFRET plugin, http://www.biophys.dote.hu/accpbfret/.
26. Alvarez-Curto E, Pediani JD, Milligan G (2010) Applications of fluorescence and bioluminescence resonance energy transfer to drug discovery at G protein coupled receptors. Anal Bioanal Chem 398:167–180
27. Cha B, Zhu XC, Chen W et al (2010) NHE3 mobility in brush borders increases upon NHERF2-dependent stimulation by lyophosphatidic acid. J Cell Sci 123:2434–2443
28. Goh WI, Lim KB, Sudhaharan T et al (2011) mDia1 and WAVE2 interact directly with IRSp53 in filopodia and are involved in filopodium formation. J Biol Chem 287(7):4702–4714
29. Hernandez FP, Sandri-Goldin RM (2010) Herpes simplex virus 1 regulatory protein ICP27 undergoes a head-to-tail intramolecular interaction. J Virol 84:4124–4135

Chapter 12

Systematic Analysis of Complex Signal Transduction Pathways Using Protein Fragment Complementation Assays

Thomas I. Koblizek, Ann Siehoff, and Anthony Pitt

Abstract

The state of a signal transduction pathway can be assessed by monitoring a given point, or a signaling node, of interest within that pathway by high content analysis. The activity at these nodes may be correlated with the general effect on cell number and morphology at the same time. Here we describe a method to analyze protein-protein interactions by protein fragment complementation assays. Complex signal transduction pathways become accessible by looking at them in their native cellular context, with all competitive and feedback mechanisms in place. Analyzing protein-protein interactions directly makes this method widely usable for many protein families and is independent of an intrinsic enzymatic activity.

Key words: High content analysis, Protein fragment complementation, Protein-protein interactions, Cell based assay, Live cell imaging

1. Introduction

Fundamentally all biochemical processes within a cell are mediated via a series of protein–protein interactions (1). Many of these interactions are highly dynamic and reflect the current status of the biochemical process to which they are connected. Often these protein–protein interactions show no gross change in the degree of interactions taking place but do exhibit significant translocation events within the cell, therefore, making high content analysis crucial. To fully understand the biology and in order to study compound pharmacology, we require technologies that allow us to visualize the true dynamics of these interactions in real time, in living cells, and therefore to determine their temporal and spatial localization.

A major challenge for drug discovery therefore remains the need for assays that capture diverse targets and pathways in their native

Jürgen Moll and Riccardo Colombo (eds.), *Target Identification and Validation in Drug Discovery: Methods and Protocols*, Methods in Molecular Biology, vol. 986, DOI 10.1007/978-1-62703-311-4_12,

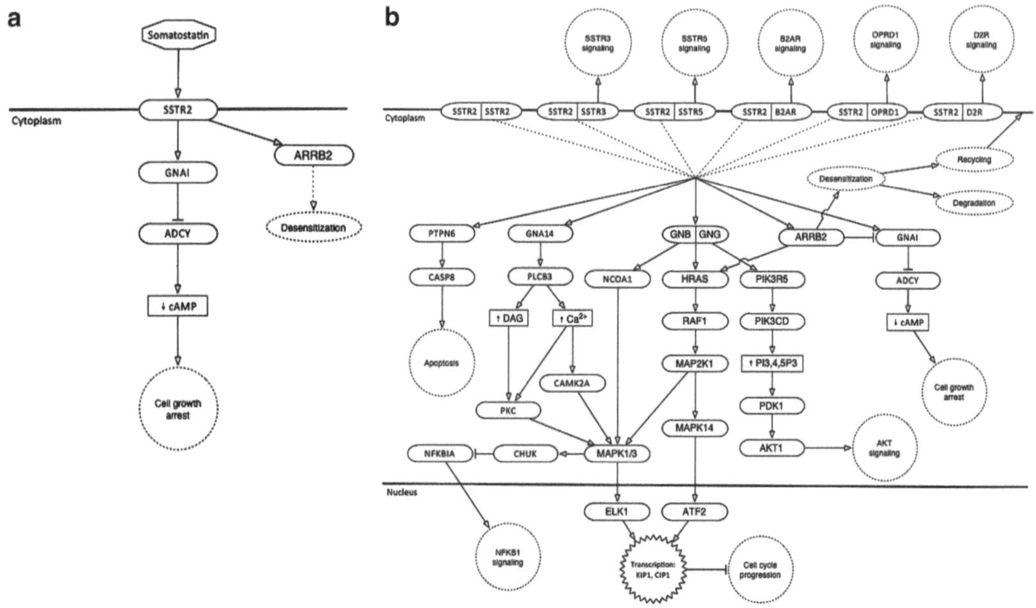

Fig. 1. (**a**) Linear depiction of the somatostatin pathway. Classic linear model of a GPCR (SSTR2) related signal transduction pathway. The GPCR is localized in isolation on the cell membrane and, after binding an external agonist, promotes an intracellular response via interaction with G-proteins which triggers the modulation of a second messenger molecule. GNAI = G-alpha-i, ADCY = adenylate cyclase, ARRB = beta-arrestin 2, SSTR2 = somatostatin receptor 2. (**b**) Somatostatin signaling and its connections to other key pathways. An updated view of the role of SSTR2. Rather than a simple linear pathway SSTR2 is connected to a complex network of interacting pathways influencing multiple biochemical processes. SSTR2 is also found to exist in the form of GPCR homo- and heterodimers on the cell membrane where the dimer partners modulate the activities of each other.

state. Protein fragment complementation assays (PCAs) represents a cell-based assay platform that facilitates the analysis of a broad range of signaling pathways at any time point of interest by analyzing protein–protein interactions. Each protein partner of interest is fused to one fragment of a rationally dissected reporter protein. If the fusion proteins are co-expressed in a living cell, the fragments of the reporter protein can interact, refold, and generate a detectable signal only when the two test proteins form a complex (2). Using inherently fluorescent or bioluminescent proteins as reporters, PCA can be combined with high-throughput automated microscopy and image analysis. This strategy captures the dynamics of protein complexes and their subcellular location, as a result of cellular responses to drugs or genetic interference (3).

PCAs facilitate analysis of formerly “unassayable” processes like ubiquitinylation and histone (de-)acetylation in a live-cell context (4). It also significantly broadens the spectrum of assays around well-known targets such as GPCRs enabling analysis of GPCR homo- and heterodimerization and biased signaling (5).

A prototype of a simple linear pathway in cells is illustrated by the G-protein-coupled somatostatin receptor SSTR2 (Fig. 1a). In this model the GPCR resides in isolation on the cell membrane and, after binding an agonist, elicits an intracellular response via G-proteins, which results in the modulation of a second messenger molecule leading ultimately to cell growth arrest. The GPCR is then subsequently desensitized by interaction with beta-arrestin 2 (6). However, the real picture of SSTR2 signaling (Fig. 1b) is far more complex where the simple linear model is replaced by a network of interacting pathways typified by protein–protein interactions that result in the modification of several biochemical processes within the cell. It has also been observed that SSTR2 does not exist in isolation on the cell membrane but typically forms heterodimers which modulate its response, for example, by upregulating or downregulating its signaling activity or the rate of desensitization. In addition SSTR2 exerts an effect on signaling pathways of its dimerization partner and its associated biochemical networks (7).

2. Materials

1. Cell culture medium: McCoy's 5A Medium with L-glutamine, 10% FCS.
2. Trypsin–EDTA: 0.5 mg/ml trypsin, 0.2 mg/ml EDTA, in PBS.
3. Freezing medium: McCoy's 5A Medium with L-glutamine, 10% FCS, 5% DMSO.
4. Fixation medium: Hank's BSS containing 4% paraformaldehyde plus 8–10 μM DRAQ5™.
5. U2-OS cells stable transfected with SSTR2/β-arrestin2 YFP (see Note 1).

3. Methods

3.1. Thawing of Cells/ Initiation of Culture Process

Perform the following steps in a sterile atmosphere:

1. Prepare a T75 flask with 15 ml of culture medium and preincubate in a 37°C, 5% CO_2, humidified incubator.
2. Quickly thaw the vial of U2-OS cells in a 37°C water bath.
3. Transfer cell suspension into the T75 flask containing the prewarmed medium.
4. Place the T75 flask in a 37°C, 5% CO_2, humidified incubator.
5. Exchange medium after the majority of cells are attached, usually 6–18 h after thawing.

When cells have reached 80–90% confluency (after 3–5 days), passage cells as described below.

3.2. Maintenance of Cells

1. Aspirate medium from the T75 flask.
2. Add 10 ml of sterile PBS, carefully rinse cells, and aspirate completely.
3. Add 1 ml of prewarmed trypsin–EDTA to the T75 flask to detach cells.
4. Incubate for 5–10 min at 37°C.
5. Add 9 ml of prewarmed medium and pipette gently up and down until a single-cell suspension is reached.
6. Count cells.
7. Seed $1–2 \times 10^6$ cells per T75 flask.
8. Do not allow cells to become more than 80–90% confluent.
9. Keep cells in culture for at least 10 days before assaying them (see Note 2).

3.3. Cryopreservation of Cells

1. Harvest cells as described in steps 1–6 of Subheading 3.2.
2. Centrifuge the cell suspension at $150 \times g$ for 10 min.
3. Aspirate the supernatant and resuspend the cell pellet in freezing medium (McCoy's 5A Medium with L-glutamine, 10% FCS, 5% DMSO) to a concentration of 1×10^6 cells/ml.
4. Transfer 1 ml of the cell suspension to a cryovial.
5. Place vials in an insulated container and store at −80 °C overnight (see Note 3).
6. For long-term storage, transfer vials to liquid nitrogen (see Note 4).

3.4. Preparation of Assay Plates and Stimulation of Cells

1. Harvest cells from culture vessels as described in steps 1–6 of Subheading 3.2.
2. Dilute cells in culture medium (without selection antibiotics) to 3.25×10^4 cells/ml.
3. Seed 40 μl in an appropriate 384-well plate and incubate in a 37°C, 5% CO_2, humidified incubator.
4. 24 h after seeding, remove culture medium by aspiration and add 40 μl of prewarmed assay medium (McCoy's 5A, 0.2% FCS, no selection antibiotics).
5. After 30 h, add 5 μl of SST-14 from a 9× stock solution in assay medium (McCoy's 5A, 0.2% FCS, no selection antibiotics).
6. Incubate in a 37°C, 5% CO_2, humidified incubator for 30 min.
7. Remove 40 μl of culture medium and add 30 μl of Hank's BSS containing 4% paraformaldehyde plus 8–10 μM DRAQ5™ (see Note 5).
8. Incubate at room temperature for 30 min.

Table 1
Imaging parameters

YFP detection	Excitation 488 nm Emission 535 nm
DRAQ5 detection	Excitation 635 nm Emission 690 nm
Laser power	488 nm: 3500 μW 635 nm: 1200 μW
Objective	20× water immersion (0.7 NA, WD = 0.6 mm)
Acquisition time	200 ms
Images per well	8–10

9. Remove fixative solution by aspiration and wash twice with 100 μl of Hank's BSS per well.
10. After the second wash, overlay cells with 100 μl of Hank's BSS and seal plate with a foil lid; proceed to imaging or store plates at 4°C overnight.

3.5. Imaging

Imaging is performed using a sensitive, high-resolution confocal system like the Opera® instrument (Perkin Elmer). Table 1 summarizes the relevant imaging parameters used (see Note 5).

4. Notes

1. U2-OS sells are often used for high content assays since they are well suited because of their consistent, flat morphology and their large size (Fig. 2).
2. Cell-based assay variability is to some extent caused by the condition and viability of the cells. To reduce inconsistency, cell number and division cycles need to be tracked, seeding densities need to be kept constant, and cells kept from getting too dense. When passage numbers are getting too high or deviations from usual growth pattern are being observed, cultures should be discarded and restarted with a lower passage number. Also, cell signaling pathways are linked to cell density, so culture density at the time of the stimulation is critical and should be titrated carefully.
3. When freezing down the cells, temperature decrease should not be too rapid, to prevent the formation of ice crystals.
4. In general, cells should not be kept at −80°C for too long since viability decreases over time.

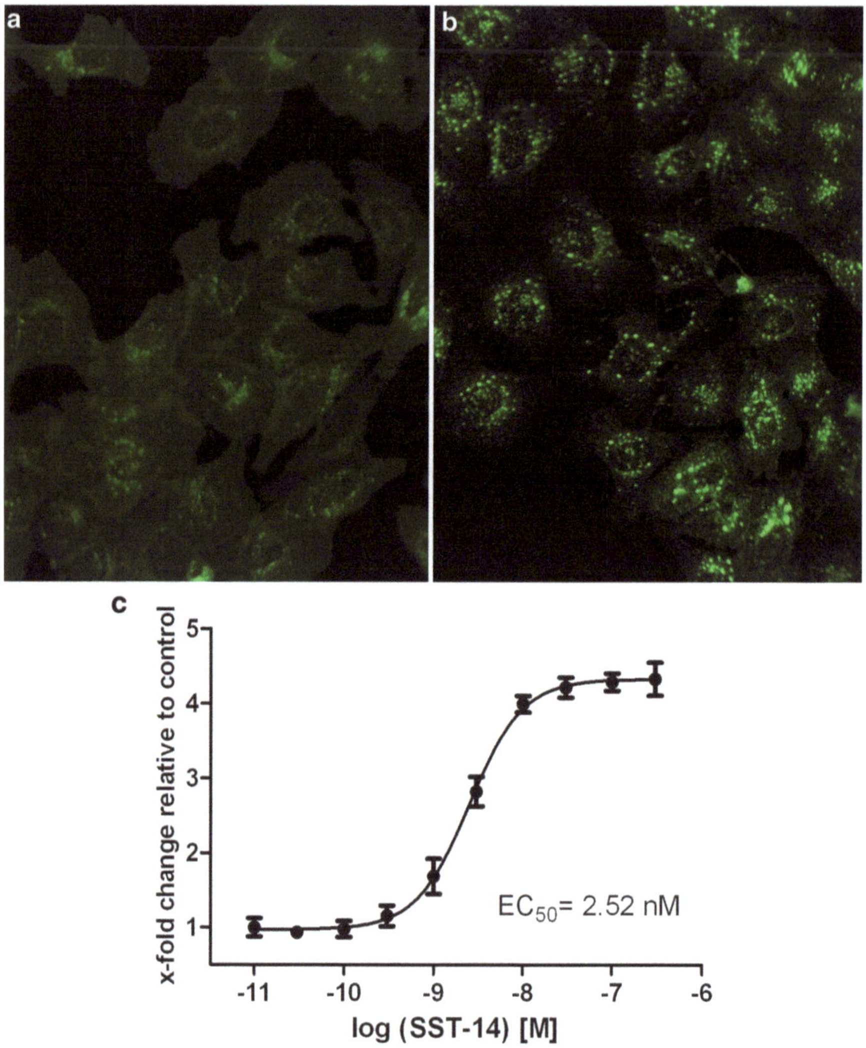

Fig. 2. SSTR2/β-arrestin-2 YFP U-2 OS cells. (**a**) In unstimulated SSTR2/β-arrestin-2 YFP U-2 OS cells, the complex is already formed and localized predominantly at the plasma membrane. (**b**) Stimulation with somatostatin-14 (SST-14) for 30 min leads to a translocation of the YFP signal from the plasma membrane to intracellular vesicular structures. (**c**) Dose response to SST-14. Cells were treated with increasing concentrations of SST-14 for 30 min. A spot detection algorithm was used to measure activation and internalization of the SSTR2/β-arrestin-2 complex. Each data point represents the average of 24 images acquired in three wells, normalized to the average of eight vehicle control wells (64 images).

5. There are a number of stains to label the nucleus for the purpose of detection in high content analysis. We use DRAQ5™ to label not only the nuclei but also the cytosol to a lesser extent. With thresholds properly adjusted, imaging software can detect the nuclei and cytosol in the same channel in two consecutive steps with just one staining.
6. For image analysis, algorithms will depend on the imaging solution used respectively. Various strategies have been employed to obtain consistent results, provided the actual number of cells analyzed per sample is sufficient (>3,000 cells).

References

1. Tyagi M, Hashimoto K, Shoemaker BA et al (2012) Large-scale mapping of human protein interactome using structural complexes. EMBO Rep 13:266–271
2. Remy I, Michnick SW (2001) Visualization of biochemical networks in living cells. Proc Natl Acad Sci USA 98:7678–7683
3. Michnick SW, Ear PH, Manderson EN et al (2007) Universal strategies in research and drug discovery based on protein-fragment complementation assays. Nat Rev Drug Discov 6:569–582
4. MacDonald ML, Lamerdin J, Owens S et al (2006) Identifying off-target effects and hidden phenotypes of drugs in human cells. Nat Chem Biol 2:329–337
5. Prinster SC, Hague C, Hall RA (2005) Heterodimerization of G protein-coupled receptors: specificity and functional significance. Pharmacol Rev 57:289–298
6. Luttrell LM, Gesty-Palmer D (2010) Beyond desensitization: physiological relevance of arrestin-dependent signaling. Pharmacol Rev 62:305–330
7. Teijeiro R, Rios R, Costoya JA et al (2002) Activation of human somatostatin receptor 2 promotes apoptosis through a mechanism that is independent from induction of p53. Cell Physiol Biochem 12:31–38

Chapter 13

Reverse Phase Protein Microarrays and Their Utility in Drug Development

Maria Isabella Sereni, Mariaelena Pierobon, Roberto Angioli, Emanuel F. Petricoin III, and Mitchell J. Frederick

Abstract

The majority of human diseases, including cancer, are characterized by abnormal protein function. Proteins regulate virtually every cellular process and exhibit multiple kinds of post-translational modification that modulate expression levels and activation states, such as phosphorylation by protein kinases. Additionally proteins interact with each other in complex regulatory networks and signal transduction pathways modulated by feedback mechanisms. These pathways are disrupted in disease and altered by therapeutic drugs. Reverse phase protein microarray (RPMA) technology allows simultaneous measurement of numerous phosphorylated, glycosylated, cleaved, or total cellular proteins from complex mixtures in many samples at once. Therefore, RPMAs can provide a portrait of a cell's signaling pathways in diseased states, before and after treatment with drugs, and allows comparison of changes in drug-resistant and sensitive cells. Furthermore, the technology offers a means of connecting genomic abnormalities in cancer to targetable alterations in protein signaling pathways, even for genetic events that seem otherwise undruggable. Consequently, the RPMA platform has great utility in many steps of drug development including target identification, validation of a pharmaceutical agent's efficacy, understanding mechanisms of action, and discovery of biomarkers that predict or guide therapeutic response. RPMAs have become a powerful tool for drug development and are now being integrated into human clinical cancer trials, where they are being used to personalize therapy.

Key words: Reverse phase protein microarrays, RPMA, Drug development, Drug discovery, Biomarkers, RPMA in Drug Development

1. Introduction

Although DNA is highly regarded as the genetic "blueprint" of an organism, *proteins* encoded by genes within the DNA are the true effector molecules of the cell. The human genome contains anywhere from 20,000 to 25,000 genes encoding proteins, which control cell structure, metabolism, growth, differentiation, gene expres-

Jürgen Moll and Riccardo Colombo (eds.), *Target Identification and Validation in Drug Discovery: Methods and Protocols*, Methods in Molecular Biology, vol. 986, DOI 10.1007/978-1-62703-311-4_13,

sion, homeostasis, and many other processes including cell death. In fact, proteins directly or indirectly regulate nearly every cellular function (1, 2). It is not surprising, therefore, that the majority of human diseases—both genetic and nongenetic in nature—can be traced back to abnormal protein function. It is precisely for this reason that the vast majority of pharmaceutical agents currently available to treat diseases target proteins (3).

Despite the fundamental importance of proteins in biology, scientists presently lack exquisitely high-throughput techniques capable of examining the entire proteome, analogous to the way microarray technology and next-generation sequencing platforms permit the facile and comprehensive profiling of RNA and DNA at the level of the transcriptome and genome, respectively. This has been a key barrier in elucidating the complex behaviors of proteins and their alterations in cellular processes and disease. Understanding abnormal protein function is further complicated by the numerous kinds of modifications that normally occur to regulate protein activity. These posttranslational modifications which largely include phosphorylation, cleavage or processing, conformational changes, acetylation, glycosylation, and ubiquitination, control protein stability, activation status, subcellular localization, as well as interactions with other proteins. In fact, it has been estimated that more than one million different protein species arise from combinations of modifications that occur on proteins encoded by the human genome (4). Consequently, it is difficult to fully predict the behavior of a cell simply based on DNA and RNA profiling alone, which does not always correlate with protein levels or take into account posttranslational modifications. Information about the status of proteins and their modifications is paramount to unraveling mechanisms of disease and other cellular processes.

In addition to the myriad of posttranslational modifications, there is a further layer of complexity to studying proteins. Proteins do not operate as single molecules, but rather as an elaborate system in concert with other proteins based on reciprocal interaction in various cellular functions. This protein network is driven by complex intracellular as well as extracellular (i.e., microenvironment) protein–protein interactions and signals. Consequently, disruption of a protein in disease or modulation by therapeutic drugs often triggers a cascade of events or changes in other proteins due to feedback mechanisms and activation of compensatory pathways. Therefore, it is crucial when elucidating the cause of a disease, identifying candidate pharmaceutical targets, or validating mechanisms of drug response or failure to simultaneously consider the status and impact of multiple proteins and the signaling networks or pathways they make up. This highlights the great need and utility for techniques that permit the analysis of many proteins at once, which are also capable of detecting posttranslational modifications.

Historically, two-dimensional (2-D) gel electrophoresis and mass spectrometry have been used as principal techniques for protein investigations. However, both of these techniques have serious drawbacks and limitations. With 2-D gels, there is an inherent difficulty in determining the identity of proteins and their post-translational modifications, the procedures are technically challenging and difficult to perform in a reproducible fashion, and electrophoresis does not easily permit multiplexing many samples, making it a relatively low-throughput platform. Furthermore, it is difficult to resolve the many thousands of protein species normally present in biological samples on gels, along with their modified forms which exist in vastly different amounts, spanning a very large dynamic range. Similarly, mass spectrometry does not work well for analyzing highly complex mixtures containing thousands of proteins present over a large range of concentrations and invariably identifies the most abundant proteins, rather than the most important. Mass spectrometry is also not considered a high-throughput technique in the sense that many samples cannot be analyzed at once.

With reverse phase protein microarray(s) (RPMAs) on the other hand, numerous phosphorylated, glycosylated, cleaved, or total cellular proteins from complex mixtures can be analyzed in many samples simultaneously. Therefore, RPMAs are well suited for studying changes in cellular protein expression levels and signaling that occur as both the cause and consequence of disease and in response to therapeutic drugs. There are four essential and somewhat overlapping areas of drug development in which RPMA technology is useful: (1) target discovery, (2) target validation, (3) understanding mechanisms of sensitivity and resistance, and (4) identification of biomarkers that predict therapeutic response. This chapter addresses the past, present, and future utility of RPMA in drug development, with an emphasis on cancer as a disease model.

2. RPMA Technology

The term "reverse phase protein microarray" was first coined in 2001 by Paweletz et al. (5) in the landmark paper which introduced the platform and demonstrated its utility for analyzing proteins derived from laser capture microdissected (LCM) prostate cancer. RPMAs have also been referred to by other names in the literature including "lysate arrays," "reverse phase lysate arrays," "protein microarrays," and "reverse phase protein arrays" (6). A very nice historical perspective on the development of RPMA technology is presented in a review by Mueller et al. (6). A detailed description of the methods for performing RPMA is beyond the scope of this chapter and would be redundant because there are multiple outstanding publications that already describe the specific step-by-step laboratory protocols (7–11). A list of these excellent sources of reference along with other previous *reviews* addressing RPMAs and their pleiotropic role in research and potential utility in personalized medicine appears in Table 1.

The basic principles of the RPMA platform will be summarized here and compared to other existing techniques so that the application to drug development can be understood and appreciated. RPMAs are a highly robust platform utilizing a bait and capture principle. They resemble miniaturized dot blots because lysates are first immobilized on a medium and then detected with specific antibodies. A schematic diagram illustrating the basic workflow and principals of the RPMA platform appears in Fig. 1. Essentially, protein lysates prepared from samples of interest are subjected to denaturing conditions and then printed as tiny spots onto a solid substrate (e.g., nitrocellulose-coated glass slides) using advanced robotics to produce many parallel arrays at once with high reproducibility and precision. Each array, which can contain hundreds of spots or samples, is then pre-blocked to prevent nonspecific reactivity, individually incubated with a highly specific antibody to a target of interest, hybridized with a secondary detection antibody and signal amplification reagent, and finally visualized after adding a colorimetric or fluorogenic detection agent. The various signal amplification and detection systems have been described in detail elsewhere (7, 11–13). As RPMAs can easily accommodate hundreds of spots, it is an ideal platform for making comparisons between diseased and non-diseased tissue, before and after drug treatment samples, or for kinetic studies.

Owing to the signal amplification methods utilized, RPMAs are extremely sensitive and can detect target proteins in the femtomole range (14). This can translate to detecting a target protein from less than one cell equivalent of protein if the analyte is abundant enough and the antibody affinity high enough. For example, VanMeter et al. (15) demonstrated that as few as 3,000 molecules of phosphorylated vascular endothelial growth factor receptor (VEGFR), which is almost exactly one femtomole, could be detected by RPMA (15). As those experiments were conducted with A431 cells estimated to have approximately 100,000 receptors per cell, this means that phosphorylated VEGFR was detectable from less than one cell equivalent of protein. In practice, however, lysates from a higher number of cell equivalents are usually spotted on arrays because of the wide dynamic range in abundance for various target protein species as well as variations in antibody binding affinities. Typically, 120 arrays are printed from lysates derived from 20,000 LCM cells, often with each sample present in triplicate. Therefore, this amounts to roughly 50 cell equivalents or approximately 10 ng total protein lysate per spot, assuming that there is roughly 200 μg protein per one million cells.

Table 1
Recent reviews published about RPMAs

First author	Title	Comments	Reference
Espina, V.	Reverse phase protein microarrays for monitoring biological responses	Step-by-step methods for RPMA	(10)
Espina, V.	Reverse phase protein microarrays for theranostics and patient-tailored therapy	Step-by-step methods for RPMA	(8)
Pierobon, M.	Reverse-phase protein microarrays	Step-by-step methods for RPMA	(7)
Gallagher, R.I.	Reverse phase protein microarrays: fluorometric and colorimetric detection	Compares fluorometric and colorimetric detection for RPMA, with protocols given	(11)
Berg, D.	Molecular profiling of signaling pathways in formalin-fixed and paraffin-embedded cancer tissues	Discussed how to extract protein from FFPE specimens and use in RPMA platform	(60)
Pierobon, M.	Reverse phase protein microarrays for clinical applications	General review on RPMA platform and use in cancer research	(34)
Wilson, B.	Monitoring proteins and protein networks using reverse phase protein arrays	General review on RPMA platform and use in research and clinical application	(9)
Espina, V.	Phosphoprotein stability in clinical tissue and its relevance for reverse phase protein microarray technology	Reviews issue related to preservation of phosphoproteins prior to RPMA; step-by-step methods included	(61)
Malinowsky, K.	Targeted therapies in cancer—challenges and chances offered by newly developed techniques for protein analysis in clinical tissues	General review on utility of RPMA in analyzing clinical specimens, with emphasis on using FFPE samples	(62)
Espina, V.	Reduction of preanalytical variability in specimen procurement for molecular profiling	Review of factors effecting phosphoprotein stability, and guidelines for handling tissue	(43)
Mueller	Reverse phase protein microarrays advance to use in clinical trials	Historical review of RPMA as well as review of current use in human clinical cancer trials	(6)

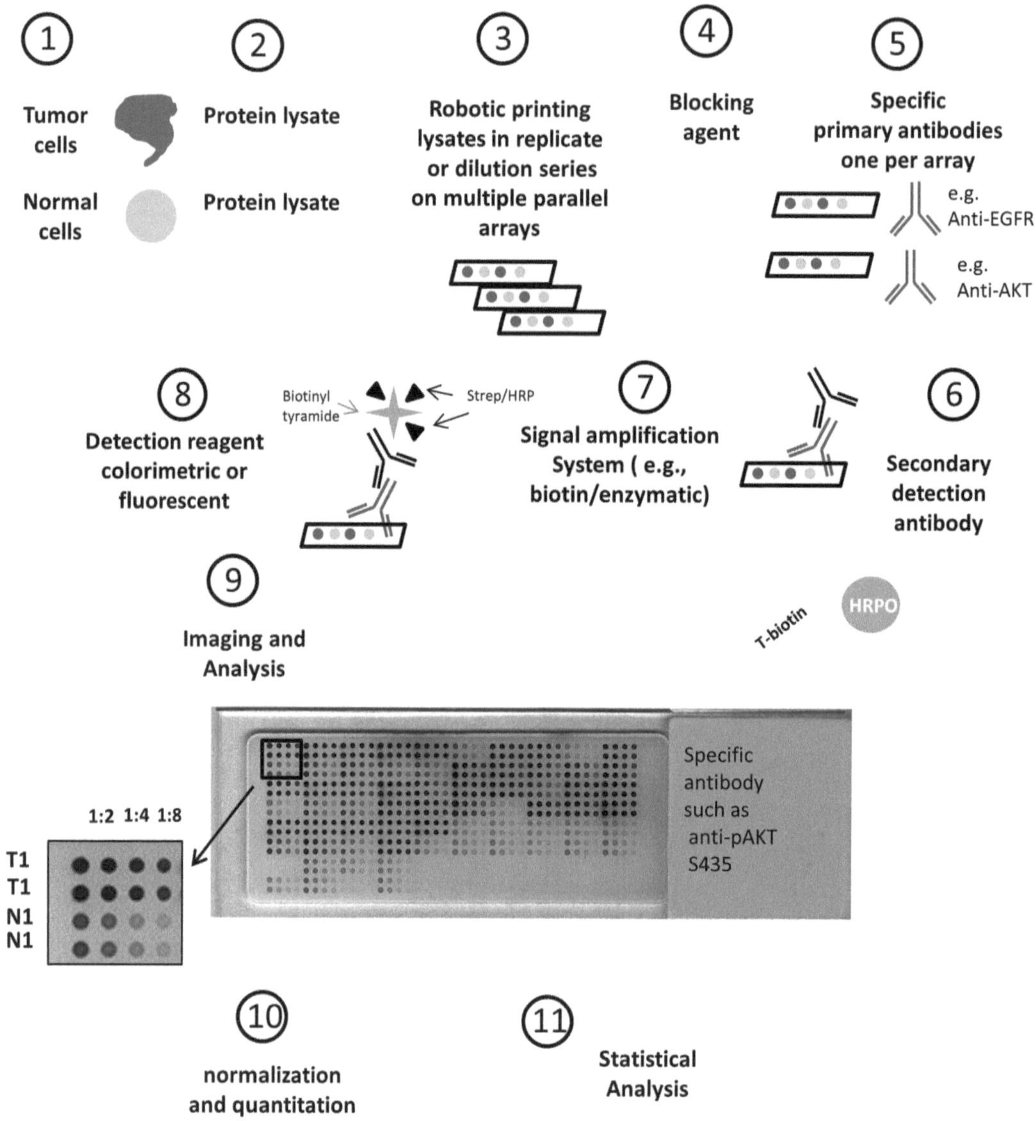

Fig. 1. Schematic diagram illustrating the principles of RPMA platforms. (1) Samples are isolated (e.g., tumor or normal tissue); (2) protein lysates are prepared from samples; (3) lysates are printed onto arrays' surface with advanced robotics; (4) after drying, arrays are treated with a blocking agent to prevent nonspecific binding of antibodies; (5) each array is then incubated individually with a highly specific primary antibody; (6) biotinylated secondary antibody is added to react with primary antibody; (7 and 8) some type of signal amplification chemistry is used usually involving biotin. For example, addition of a streptavidin–biotin–peroxidase complex can catalyze covalent addition of biotinylated tyramide, which has free biotin sites for binding new streptavidin–biotin–peroxidase that can catalyze addition of a colorimetric substrate in a second enzymatic reaction. Alternatively, the free biotin sites can bind streptavidin conjugated fluorescent dye; (9) colorimetric substrate or fluorescent dye is imaged with a scanner and image analysis performed; in the example shown, replicates of lysate from a tumor specimen "T1" and the patient-matched normal tissue "N1" are shown in the magnified region of the RPMA slide. Typically, lysates are spotted in serial dilution beginning with the undiluted sample often referred to as the "neat," followed by serial twofold dilutions (e.g., 1:2, 1:4, 1:8); (10) some method of protein normalization is employed, usually based upon total protein quantification from parallel arrays stained with SYPRO Ruby Red; (11) normalized data is analyzed by statistical means or unsupervised hierarchical clustering.

If the cells are taken from tissue, the total protein levels per spot could be even higher, given that there are many extracellular proteins present in tissue as well. Regardless, 50 cell equivalents or 10 ng protein permits investigators to directly measure the amount of protein present in each spot by staining a few parallel arrays with the fluorescent dye SYPRO Ruby Red, which has a linear range of sensitivity for detecting protein between 1 ng and 1 μg (7). Controlling for the actual amount of total protein present in samples facilitates proper normalization for downstream quantitative analysis, and it is especially important when the protein concentration of the original sample is not easily determined, as in the case of LCM cells or fine needle aspirates.

RPMAs are capable of profiling the expression levels and activation states of multiple proteins from complex mixtures and very small inputs of starting material, making them ideal for analyzing biopsies or clinical specimens, where sample amounts are often limiting. Indeed, RPMAs can be used to analyze many types of samples including surgically excised tissue biopsies, fine needle aspirates, serum and other bodily fluids, as well as cultured cells. The technique is quantitative over a very large dynamic range and is highly reproducible with a low coefficient of variation (i.e., < 15%) observed within and between arrays owing to the use of robotics for printing and automated stainers for detection. Additionally, RPMAs are an intermediate throughput platform which allows hundreds of protein endpoints to be simultaneously analyzed in hundreds of samples within one experiment, generating tens of thousands of data points for analysis.

Reverse phase arrays differ substantially from forward phase capture arrays (otherwise known as antibody arrays) because protein lysate itself is printed on the array as bait used to capture the specific antibody, rather than printing the antibody first to capture the protein from solution. Therefore, many hundreds of samples can be printed on the same RPMA, and hundreds of parallel or identical arrays can be printed at once using advanced robotics. Each RPMA containing the multiple samples, identifiable by their geographic location, is incubated with a unique and specific detection antibody.

There are several important advantages RPMAs have over antibody arrays and other proteomic techniques such as immunohistochemistry or tissue arrays. Antibody arrays usually require a second specific antibody, made in a different species, for each captured protein to be visualized in a manner analogous to enzyme-linked immunosorbent assays (ELISA). Therefore, it becomes difficult to simultaneously optimize the antibody–antigen hybridization conditions for so many antibodies at once present on antibody arrays while minimizing nonspecific cross-reactivity and ensuring that proteins over a wide range of concentrations can be quantitated in a linear fashion (14). Antibody arrays also consume or require much higher inputs of protein than reverse phase arrays. With antibody arrays, there is generally only one sample applied to each array, so comparisons of the same analyte between multiple samples take place following slightly different incubation or antibody hybridization reactions. This would be analogous to quantitatively comparing the signal for two samples run on completely different gels by western blotting, which is generally unacceptable. There is a similar caveat when performing classic immunohistochemistry because each tissue section to be compared is incubated independently with reagents. This problem is less of an issue with tissue arrays. However, immunohistochemistry performed on tissue arrays is difficult to objectively quantitate, and comparisons can be difficult to interpret owing to the very poor dynamic range of signals.

Generally, forward phase antibody arrays are not optimal for cross-sample comparisons or quantitating the same analyte in many samples at once. However, it is possible to incubate a mixture of several samples on one antibody array, but that requires post-lysate labeling the proteins from different samples with fluorescent tags, which can introduce unwelcome experimental variability. In the absence of post-lysate labeling of proteins, forward phase arrays require two sets of antibodies capable of recognizing epitopes in a more native state (i.e., in the absence of denaturing detergents) which limits the number of useful antibodies available and hence the repertoire of recognizable analytes. Likewise, tissue arrays rely on antibodies that work in immunohistochemistry, which is a minority of antibodies. With RPMAs, on the other hand, only one specific antibody per analyte is required, and quantitative comparisons of the same analyte among multiple samples are performed following a single incubation or hybridization reaction, and no labeling of proteins is required. As the protein epitopes recognized on RPMA are in a denatured state, many antibodies that have been developed for traditional western blotting applications are capable of working for RPMA as well, provided the antibodies show no significant cross-reactivity with other proteins. Therefore, some validation is required to show that the antibody detects predominately the target of interest at the correct molecular weight on western blots, as judged by the absence of nonspecific bands. Further validation is often performed to show that an antibody can detect quantitative differences between samples on the RPMA platform when levels of the target protein are purposely manipulated by a treatment or through serial dilution.

3. Use of RPMA Technology in Development of Anticancer Drugs

RPMA platforms were originally devised at laboratories within the National Cancer Institute (5, 6), where the impetus to develop the technology came partly from the need to better understand the biology of cancer as well as discover novel candidate drug targets for this disease. It is now firmly established that cancer is caused by accumulation of genetic and epigenetic abnormalities, resulting in the inappropriate activation of oncogenes and inactivation of tumor suppressor genes. The genomic alterations that give rise to cancer manifest themselves by altering the expression levels, posttranslational modification states, and activation status of numerous proteins intimately involved in normal cellular homeostasis. Proteins interact with each other in complex regulatory networks, with feedback loops and input signals from the tumor microenvironment, which can dramatically affect the function of the abnormal proteins encoded by cancer genes. Disruptions in cellular homeostasis and their regulatory signaling networks lead to the eight hallmarks of cancer (16), including sustained proliferative signaling, evading growth suppressors, resisting cell death, immortality, inducing angiogenesis, activating invasion and metastasis, reprogramming energy metabolism, and evading immune destruction.

Each of the hallmarks of cancer, similar to all normal cellular processes, is regulated by protein phosphorylation. The ubiquitous role of protein phosphorylation in cancer has lead to a virtual explosion in the number of small molecule inhibitors and pipeline drugs being developed by pharmaceutical companies to target protein kinases. Protein profiles of a tumor provide a readout for the net consequences of genomic alterations in cancer. RPMAs are particularly well suited for profiling total protein expression and levels of phosphorylation that depict the activation status of kinase pathways. Consequently, the RPMA platform is valuable for nearly every stage of anticancer drug development.

4. RPMAs and the Drug Development Cycle

Drug development frequently begins with discovery of important candidate targets, often through studying the biology of a disease. RPMAs are useful in the multiple steps of drug development, which is not always a linear process but proceeds in an iterative and dynamic fashion as illustrated in Fig. 2. A common way to look for targets in cancer therapy is to generate RPMA profiles of tumors and make comparisons between the tumors or to corresponding nonmalignant tissue. Unsupervised two-way clustering of analytes and samples can reveal subtypes of tumor with differential activation of pathways that should be targeted differently, thus defining initial drug target candidates. Next, RPMAs can be used to validate whether pharmaceutical agents can modulate intended pathways both in vitro and in vivo. During the process of validation, drug treatment may be found by RPMAs to modulate other proteins either downstream of the initial target or previously unknown to

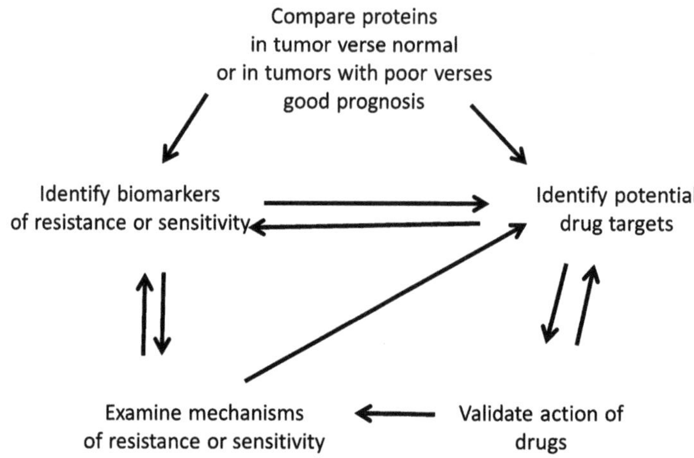

Fig. 2. Dynamic steps in the drug development cycle. As explained in the text, multiple strategies can be used for initial drug target discovery, and at various steps in validating the drug target, mechanisms of action, or defining biomarkers of response, new drug targets can emerge.

be connected to a pathway. These additionally modulated proteins could then become new candidate drug targets illustrating that validation of drugs in itself can lead to identification of further drug targets (Fig. 2).

RPMAs can also aid in verifying mechanisms leading to drug resistance or sensitivity. When the pathways mediating resistance are identified by RPMA, they can then become additional drug targets (Fig. 2). Even when proteins are not directly involved with the mechanism of a drug phenotype, they can still be suitable biomarkers for drug efficacy if they are co-regulated along with the molecules mediating the response. Alternatively, a broad biomarker screen of drug-resistant and drug-sensitive cells can often shed light on the mechanism of a drug or directly suggest alternative drug targets. Sometimes, the initial drug discovery may be driven by using RPMAs to look for biomarkers of poor prognosis in cancer, and those biomarkers and their associated pathways can become new candidate drug targets (Fig. 2). In the sections that follow, specific examples of how RPMAs have aided and continue to facilitate discovery and testing of anticancer agents will be reviewed. For convenience, the discussion is organized into the various stages of drug development, but many studies are illustrative of multiple steps and the overlap will be apparent.

5. Drug Target Discovery Using the RPMA Platform

Metastatic disease is often the cause of mortality for most types of cancer, highlighting the need to understand differences in the drug sensitivities and activation pathways that exist between primary and metastatic tumors. Unfortunately, tissue biopsies from metastatic cancer are often difficult to obtain for research because metastatic lesions are not frequently treated by surgical excision. Nevertheless, there have been several publications comparing the signaling pathway differences in metastases using RPMA platforms. Phosphorylation of c-Kit was found to be elevated in 5/6 metastatic ovarian carcinomas compared to primary tumors (17) using RPMAs. As c-Kit is a known oncogene that drives proliferation and survival of other tumor types, the RPMA results suggested that the c-Kit pathway may be an interesting drug target for patients with advanced stage ovarian carcinoma, who usually present with metastatic disease. A phase II clinical trial (NCT00510653) examining the efficacy of Gleevec, an inhibitor of c-Kit, in patients with metastatic ovarian cancer was initiated at the M.D. Anderson Cancer Center in 2007 and is still ongoing at the time of this publication. More recently, increased activation of the c-Kit, PDGFR, and c-Abl pathways in liver metastasis from patients with stage IV colon cancer was found by RPMA profiling (ASCO 2010, abstract number TPS194). Identification of these targets led directly to a phase II/III trial examining the efficacy of Gleevec (i.e., Imatinib) in treating patients with metastatic colon cancer to the liver (NCT00867334).

In another study, the RPMA profiles of brain metastases derived from breast or lung cancer were compared (18). Metastatic breast cancers had higher phosphorylation of ErbB2/Her2 and insulin-like growth factor 1 receptor (IGF-1R), whereas subsets of metastatic lung tumors showed higher activation of the EGFR, ERK, and mTOR signaling pathways. These findings have obvious implications for treatment, as there are FDA-approved drugs and experimental agents in clinical trials for inhibiting all these pathways. Indeed, Lapatinib (a dual inhibitor of both ErbB2 and EGFR) has shown antitumor activity in preclinical models of brain metastasis from breast cancer and has demonstrated modest efficacy alone and in combination with chemotherapy in patients with metastatic breast cancer to the brain (18). The metastatic lung cancers demonstrated a high degree of heterogeneity and could be subdivided into five groups on the basis of pathways activated, underscoring the need to personalize treatment as will be discussed in the biomarker section below.

RPMAs can also be used to make links between genomic abnormalities in cancer and the signaling pathways they disrupt. Although EGFR protein is known to be overexpressed in the majority of head and neck cancers, mutations of the EGFR are rare in this disease, and amplifications of the gene only occur in roughly 10% of patients (19). In a study utilizing RPMAs to compare the phosphoproteomic differences between primary head and neck tumors and patient-matched nonmalignant control tissue, Frederick et al. (20) detected highly elevated phosphorylation of the EGFR on multiple amino acid sites in just a subset of patient tumors which all turned out to contain large focal amplifications of the EGFR locus. The same tumors also had very high levels of phosphorylated Her2 receptor and the Shc signal adaptor protein, linking these signaling pathways to the EGFR amplifications.

In some cases, the link between genomic alterations and the disrupted proteomic pathways in cancer can lead to discovery of potential drug targets, even for genetic alterations that may otherwise be difficult to inhibit or may currently be considered "undruggable." Translocations involving the mixed lineage leukemia (MLL) gene are associated with a variety of hematological malignancies, including acute myelogenous leukemia and acute lymphoblastic leukemia (ALL) of both T-cell and B-cell lineage. The MLL oncogene encodes for a histone-lysine *N*-methyltransferase that is a positive global regulator of gene transcription. Accordi et al (21). used RPMA to compare B-cell precursor (BCP)-ALL tumors from patients with or without an MLL translocation and found elevated levels of total cyclin E and elevated phosphorylation of AMPKα, AMPKβ, LKB1, BCL-2, and eNOS/NOS III to be associated with presence of the translocation. Their results were also confirmed independently by western blotting experiments using lysates from patient samples. As LKB1 is upstream of AMPK and eNOS and BCL-2 are downstream, the findings strongly suggest that BCP-ALL patients with MLL translocations have aberrant activation of the AMPK pathway. Moreover, inhibtion of AMPK in two established BCP-ALL tumor lines possessing MLL translocations induced apoptosis with an associated decrease in phosphorylation of BCL-2 and eNOS, confirming that they are downstream targets. Consequently, RPMAs were used to identify the AMPK pathway as a new potential therapeutic target for a subset of BCP-ALL patients harboring MLL rearrangements.

The ability to connect genomic abnormalities to signal pathway alterations using RPMA can be exploited in direct experimental approaches to hunt for better ways to target the seemingly undruggable alterations found in cancer. If siRNA or shRNA is used to knockdown an oncogene in cell lines, pathways that exhibit a concomitant decrease in activation status, as assessed by RPMA, may be involved in mediating the oncogene function and could be viable drug candidates. A similar strategy could be applied for discovering ways to target loss of tumor suppressor function in cancer as well. RPMAs can be used to determine pathways that are elevated and potentially targetable owing to loss of tumor suppressor function by identifying which pathways are inhibited following forced re-expression of the tumor suppressor gene. These approaches are not limited to finding events downstream of observable genetic alterations. For example, RPMA could be applied to profile changes in total or phosphoprotein levels in cells treated with an inhibitor targeting a specific kinase pathway or following siRNA knockdown of the gene encoded by a key protein in the pathway in order to discover other druggable targets downstream that may be mediating a specific function.

Combining a systems biology approach with RPMA technology can be a powerful way to identify new drug targets, map complex signaling pathways, and design better combination therapy. An shRNA or siRNA library screen can be combined with the RPMA platform to map complex signaling networks and identify potential targets upstream of an interesting pathway. If knocking down a kinase decreases the activation status or levels of a protein, then the kinase is most likely a positive regulator (possibly upstream), whereas knocking down negative regulators of a pathway will result in the pathway being increased. Lu et al. (22) monitored changes in total and phosphoprotein levels in a breast cancer cell line following siRNA-mediated knockdown of 541 different kinases. Unexpectedly, they identified the AKT downstream target GSK3A as being able to positively regulate phosphorylation of AKT, predicting that GSK3α may act both upstream and downstream of AKT activation. On the other hand, knockdown of certain MAPK members led to elevated pATK suggesting that the MAPK pathway negatively regulates the AKT pathway through a negative feedback loop (22). These findings may have direct implications for designing better combinations of molecular-targeted therapies. For example, drugs that inhibit certain members of the MAPK pathway in breast cancer may have the undesirable consequences of elevating AKT phosphorylation, which suggests that they should be combined with inhibitors for the AKT pathway.

Many other studies have used the RPMA platform to discover potential drug targets in cancer. Activation of a Cox-2–EGFR pathway in colon cancer tumors from patients who present with synchronous liver metastasis (23), and elevated markers of the autophagy pathway in breast ductal carcinoma in situ (24) have been detected with RPMAs. In their phosphoproteomic analysis of head and neck cancer utilizing RPMAs, Frederick et al. (20) identified multiple potential drug targets in addition to the EGFR. Despite the fact that the head and neck tumors were chemo- and radiation-treatment naïve, there was considerable activation of both checkpoint kinases I and II evident by RPMA and extensively validated by IHC. Based upon these observations, preclinical studies examining the efficacy of checkpoint kinase inhibitors combined with standard chemotherapy have been initiated with head and neck cancer cell lines (manuscript in preparation). Highly elevated phosphorylation of atypical protein kinase C (PKCi) on an activating residue was also frequently found in primary head and neck tumors, compared to matched nonmalignant controls. Frequent activation of PKCi in the primary tumors, together with the observation that the chromosomal locus containing the gene encoding

PKCi was amplified in established head and neck cancer cell lines, suggested that PKCi may function as an oncogene in head and neck cancer, similar to its role in non-small cell lung carcinoma.

6. Validation of Drug Targets Using RPMAs

Kinases have been the subject of intense drug discovery efforts in cancer therapeutics because phosphor-proteomic networks regulate the malignant behavior of tumor cells (25). A number of pharmaceuticals targeting either protein kinases or growth factor receptors associated with tyrosine kinases have already gained approval by the Food and Drug Administration (FDA) for treating cancer. In addition, experimental drugs targeting more than 30 of the known 515 human kinases have made their way into human clinical cancer trials (25). Many more kinase inhibitors are currently in pharmaceutical developmental pipelines. RPMAs are an ideal platform for validating the efficacy with which these agents, and other types of drugs, inhibit their targets and modulate intracellular signaling pathways.

Although in vitro screens for compounds capable of inhibiting kinases frequently employ cell-free systems using recombinant proteins and artificial substrates, RPMA technology can be used to demonstrate whether a drug can block activity of the native kinase using endogenous substrates within cells and tissues. Quantitative analysis of drug-induced changes by RPMA can establish the potency and specificity of inhibitors. RPMA profiles can be compared following genetic knockdown of targets (i.e., siRNA) to confirm the specificity of downstream targets. Studies have demonstrated the utility of RPMA technology for assessing the potency and specificity of inhibitors targeting the EGFR, MEK, IGF-1R, JAK, and PI3K pathways in cell lines (25, 26). As will be discussed in the biomarker section below, many human drug trials are incorporating RPMA platforms into their design because the technology allows quantitative assessment of signaling pathways, in cancerous and normal tissue, before and after drug exposure (6).

7. Understanding Drug Mechanisms Through RPMA Technology

Signal transduction pathways and protein regulatory networks are not linear (26) but contain branches, negative feedback loops, and common points of connection between seemingly parallel pathways known as nodes. Furthermore, these pathways and networks are tightly controlled within cells so that disruption of any point has a rippling effect often activating compensatory mechanisms through which cells try to reestablish homeostasis. Owing to this high degree of dynamic interconnectivity, it is frequently difficult to distinguish which molecules are critical for the phenotypes associated with drug response, even when the initial target is obvious. RPMAs are a valuable tool for examining drug mechanisms because the level and activation status of many proteins at once can be simultaneously examined in both sensitive and resistant cells, before and after treatment.

Aberrant growth factor signaling or BRAF mutations leading to downstream activation of the ERK1/2 pathway in cancers have prompted testing of experimental drugs targeting MEK, the intermediate druggable target in these pathways, directly upstream of ERK activation. Treatment of breast cancer cell lines with a MEK inhibitor was shown by RPMA to result in activation of the AKT pathway (25). As the AKT pathway promotes tumor cell survival, unintended activation of this pathway in response to MEK inhibitors could be a potential escape mechanism. Indeed, Gopal et al. (27) used RPMA to examine mechanisms of sensitivity and resistance to a MEK inhibitor in melanoma cell lines containing activating BRAF mutations. Sensitivity to the MEK inhibitor in the melanoma cell lines did not correlate with the ability of the drug to inhibit the MEK pathway, which occurred in both sensitive and resistant cells. However, melanoma cell lines that were resistant either had high basal activation of the AKT pathway or had strong compensatory induction of the AKT pathway following treatment with the MEK inhibitor, which seemed to be mediated through the IGF-1R. Targeting the AKT or IGF-1R pathway was able to sensitize resistant melanoma cells to the MEK inhibitor (27). These experiments, as well as previously discussed studies such as the one linking resistance to hormone therapy in breast cancer to elevated activation of IGF-1R and PI3K/mTOR (28), illustrate an important point: pathways associated with drug resistance discovered by RPMA may suggest new drug targets (Fig. 2) or better combinations of therapy to overcome resistance.

RPMA platforms have also been used to identify potential mechanisms of sensitivity to drugs. NSC-741909 is an anticancer compound which kills a number of different tumor types from the NCI-60 panel and is the derivative of another molecule originally identified through its ability to selectively kill cells with a K-RAS mutation. Analysis of 77 protein endpoints by RPMAs, following NSC-741909 treatment of a sensitive lung cancer line, revealed drug-induced elevation of JNK phosphorylation as an early and persistent event (29). Follow-up experiments demonstrated that the drug worked by decreasing levels of a JNK phosphatase. Inhibitors of JNK blocked NSC-741909-induced cell death, confirming that JNK kinase activation, as detected by RPMA, was a major pathway mediating the effects of the drug (29).

8. RPMAs in Personalized Medicine and Biomarker Discovery

Although many tumors of common histological type look alike under the microscope, there is considerable variation in their genomes. This is one of the primary reasons why patients with the same kind of cancer can have completely different clinical courses of disease or responses to therapy. The consequences of DNA alterations in cancer are perhaps best understood in the context of associated changes in the levels and activation status of proteins. Recent advances in technology now permit more comprehensive, rapid, and cost-effective profiling of tumors at the molecular level and have ushered in a new era of "personalized medicine." The goal of personalized medicine is to use information about the specific molecular alterations in a tumor to pick the most efficacious therapy for that patient. Molecular alterations in tumor cells that can be used to predict prognosis, sensitivity, or resistance to therapy are known as "biomarkers." When the biomarkers themselves are therapeutic targets, they are called "theranostic" (30) markers.

Traditional cytotoxic chemotherapeutics target pathways that are likely to be present in all rapidly growing cells, and most of these drugs have serious toxic side effects on normal cells in the body. Although these drugs were designed to target cancer, generally only a minority of patients with advanced stage disease have tumors that respond well to standard chemotherapeutics. A new class of pharmaceutical agents, known as "molecular-targeted therapies," are being incorporated into cancer treatment. These molecular-targeted therapies work against very specific genes and proteins in a tumor, and it is expected that only a subset of tumors with alterations in those specific genes and pathways will respond well to the treatment. In many cases, specific genomic alterations have been found to predict dramatic clinical responses to targeted therapeutics. Clinicians and scientists are now appreciating that molecular-targeted therapies work best when there is a genetic alteration in a patient's tumor that is linked to the specific pathways targeted by a drug (31–33). As examples, lung cancer tumors with activating mutations in the EGFR respond best to EGFR inhibitors, melanoma patients with BRAF mutations respond well to MEK and Raf inhibitors, and breast cancers with Her2 amplifications respond strongly to Her2-targeting agents. Finally, leukemias with BCR–ABL translocations and gastrointestinal stromal tumors with mutations in c-Kit show marked response to Gleevec, which targets both of these kinases (31). Consequently, there is a great need to discover appropriate biomarkers in a patient's tumor that will predict response to molecular-targeted therapies, standard chemotherapies, and their combinations.

Whether genes, proteins, noncoding RNAs, or a combination will make the best biomarkers is presently unknown, and the answer may vary widely depending upon the type of cancer and therapy or outcome being considered. As already emphasized, proteins are the true effector molecules of the cell. Genes altered in cancer must be copied into messenger RNA (mRNA) molecules that are then translated into proteins, and not all genes are expressed in every cell. Even noncoding RNAs are thought to exert their phenotypes by regulating protein expression. Extracellular messages from the tumor microenvironment also influence intracellular protein levels and their activation status. Therefore, it can be challenging to predict how a mutation in a gene will affect the activation status or levels of the protein it encodes and how that event will ultimately disrupt complex protein signaling networks.

RPMAs provide a snapshot of the activation status and protein profiles present within a cell (34) and are therefore a good readout for the net effect produced by genomic alterations in cancer. Given the central and causative role aberrant cell signaling plays in tumors, phosphoproteins are a very important class of biomarkers in cancer. Exceptional and potentially innovative opportunities exist for protein microarray technology. Because most new drug targets for cancer and many other diseases are signaling-related, a proteomic approach now makes it possible to create a diagnostic portrait based on the activity of the drug targets, stratifying patients who will respond to a particular therapy and who will not. Thus, offering to clinicians the knowledge of which pathways are active in a patient's tumor will enable them to specifically choose targeted therapy, leading to dramatic and significant improvements in therapeutic efficacy and patient survival (35–37). The possibility to generate a portrait based on the activation states of a drug target represents the power of theranostic medicine which is based on a therapeutic drug target that is also a diagnostic biomarker, such as ErbB2 for breast cancer. Comprehensive proteomic profiling and trial-focused endpoint profiling will be critical for selection of patient-tailored molecular therapy. RPMA technology can help to select patients who may be more likely to benefit from specific targeted therapies and will bring to reality the clinical adoption of molecular stratification. Not surprisingly, RPMA platforms have demonstrated great utility in defining biomarkers for predicting cancer prognosis and response to therapy.

An early study employing RPMA demonstrated elevated activation of the AKT/mTOR pathway in pediatric rhabdomyosarcoma patients with poor survival (38). RPMA analysis of archival formalin-fixed paraffin-embedded (FFPE) esophageal cancers revealed that patients with a poorer prognosis had elevated markers of the Her2 pathway (39), suggesting that drugs targeting Her2 could be useful in this population. Using an experimental in vitro model, Miller et al. (28) demonstrated by RPMA that breast cancer cell lines which lose estrogen dependency have higher activation of the PI3K/mTOR pathway (possibly mediated through the

IGF-1R receptor) and that blocking the PI3K pathway with kinase inhibitors or knocking down IGF-1R could prevent emergence of hormone refractory breast cancer cells. The same pathways may be useful targets in patients, as they also found elevated levels of phosphorylated IGF-1R and evidence for PI3K activation using RPMAs in primary breast cancers exhibiting poor outcome following hormone-based therapy (28). In a similar fashion, RPMA profiles were used to compare tumors from patients with BCP-ALL that respond poorly to prednisone (21), which is one of the standard treatments. Patients with shorter relapse time and poor response to prednisone were found to have elevated cyclin E and decreased levels of the inactivated form of LCK kinase, suggesting their tumors could possibly have elevated LCK kinase activity. Future studies are needed to determine whether activation of LCK kinase contributes to prednisone resistance in BCP-ALL, and a siRNA library screening approach could be used to identify possible drug targets capable of inhibiting cyclin E in this disease. These studies clearly illustrate that the biomarkers of poor prognosis or treatment failure identified by RPMAs can suggest additional or novel therapeutic targets for subsets of patients with bad outcome. Moreover, the biomarkers could be used in up-front screening strategies to stratify patients for alternative treatments.

9. Use of RPMA Technology in Clinical Cancer Trials

In many cases, only a small subset of patients responds well to molecular-targeted therapies in clinical cancer trials (31–33). The need to discover better biomarkers for predicting response to therapies and for guiding treatment selection in cancer patients has sparked an increase in the number of human cancer trials that are now incorporating RPMA platforms into their design. Currently, there are 12 clinical trials involving RPMA platforms enlisted in the US National Library of Medicine (http://clinicaltrials.gov). A thorough review of past and ongoing clinical trials utilizing RPMAs was recently published by Mueller et al. (6). As there are presently no FDA-approved RPMA-based diagnostic tests for cancer, most clinical trials have utilized RPMAs as a biomarker discovery tool, rather than to guide treatment.

RPMAs are one of the platforms being used to look for biomarkers of response in a phase II trial (NCT00798655) evaluating progression-free survival in patients with locoregionally advanced head and neck cancer who will receive the EGFR antagonizing antibody, panitumumab, in conjunction with postoperative chemoradiotherapy. In the trial, baseline biomarkers related in the EGFR, Src, MAPK, AKT, STAT1, STAT3, STAT5, p21, p27, PARP, E-cadherin, ErbB3, VEGF, and IL-8 pathways will be measured by RPMA from archival FFPE specimens and correlated with progression-free survival. In the I-SPY 2 trial for patients with high-risk breast cancer (NCT01042379), data from RPMA are being used in conjunction with other platforms to identify biomarkers that would predict efficacy of novel targeted therapies in combination with standard chemotherapies. In an ongoing clinical trial utilizing chloroquine to treat patients with breast ductal carcinoma in situ (NCT01023477), pre- and posttreatment biopsies will be analyzed by RPMA to look for predictive biomarkers as well as validate whether the drug is targeting the autophagy pathway. One of the reasons for initiating this clinical trial was because of RPMA data from the previously mentioned study, which demonstrated increased activation of the autophagy pathway in breast ductal carcinoma in situ samples (24).

Although there are no FDA-approved RPMA-based tests yet, it is still possible to make treatment decisions in clinical trials using RPMA data under certain circumstances. Such circumstances exist for patients with advanced stage disease who have run out of options and who are not being denied standard care. In this situation, an RPMA test which is not FDA-approved can be used to select drugs which are themselves FDA-approved (i.e., nonexperimental/investigational agents) for on- or off-label indications and to treat patients in a clinical trial provided there is no extra risk to the patient, and the RPMA tests are performed in a Clinical Laboratory Improvement Amendment (CLIA) compliant manner. Two examples of such trials are the "Side-Out" phase II/III clinical trial for metastatic breast cancer (NCT01074814) and the NITMEC phase II/III trial for metastatic colorectal cancer (NCT00867334). RPMA profiles obtained from biopsies of metastatic breast cancer lesions are part of a set of data obtained from multiple platforms that will be used to select therapy using FDA-approved drugs in the Side-Out trial. Molecular alterations present in metastatic disease are analyzed using immunohistochemistry, DNA microarrays, and RPMA, and the combination of the acquired data is provided to a medical treatment committee who use the data to individualize therapy for each patient. In this trial, the molecular analysis has to be generated quickly since patients with end-stage progressing disease cannot wait weeks or months for results. In this setting, the entire workflow from tissue receipt to RPMA analysis of dozens of drug targets, including the generation of the molecular report, is achieved in less than 1 week. The trial is investigating and verifying pilot studies showing a statistically significant increase in each patient's progression-free survival interval between assay-informed therapy versus preceding therapy (40). In the NITMEC trial which was recently closed, RPMA data was used to access the activation status of the c-kit/PDGFR pathway in biopsies from metastatic colorectal cancers to the liver in order to select patients likely to benefit from the addition of Imatinib (Glivec or Gleevec) to their standard of care therapy (i.e., an EGFR-targeting antibody).

Until now targeted cancer therapy has been directed in many cases against a single molecular target, but in the future we can imagine targeting an entire set of interconnected (kinase-driven) events along a dysfunctional signaling pathway with a "smart" selection of drugs. Monitoring protein activation will also help to identify treatment-acquired resistance to chemotherapy. As a view to the near-term future, we expect that the use of protein array technology and the clinical implementation and validation of these technology platforms will help to identify the optimal targeted agent and effective dose for each individual patient's disease. Such utilization of array technologies such as the RPMA can have a profound impact on personalized medicine and will allow the monitoring of response and relapse and create ideas on and strategies to circumvent resistance mechanisms.

10. Challenges, Controversies, and Caveats in the RPMA Field

A big challenge lies in transitioning RPMA platforms from research discovery tools to FDA-approved tests that can be used for prognostic or theranostic applications in clinical settings. Some of the technical and regulatory obstacles involved, such as generating data in a CLIA-certified laboratory, inclusion of reliable calibrator reference standards, and obtaining FDA approval, have been discussed in other reviews (6, 9, 17, 41). In contrast to a biomarker used in a clinical setting, a clinical test requires a much higher degree of confidence in measuring and interpreting data generated from an individual's sample. If biomarkers are going to be used to guide therapeutic decisions, then measurements must accurately reflect the true status of signaling pathways in a patient's sample at the time of procurement. Given the robustness of RPMA technology evident by low intra- and inter-assay variability (42), the major roadblock to obtaining accurate biomarker measurements in biopsy specimens appears to be related to how samples are collected (43) and processed (44) prior to running the platform. The relative importance of up-front enrichment for cell populations and procedures for preanalytical handling of tissue to preserve biomarker integrity has generated a large degree of controversy among experts in the RPMA field.

It is well known that patient tumors are heterogeneous with respect to the numbers and types of contaminating nonmalignant cells (e.g., stromal cells, fibroblasts, and leukocytes) present in biopsy samples. These differences are apparent across cancers of multiple histologic type and subtype, although some tumors originating from certain tissues tend to be more heterogeneous than others. For example, in a cohort of primary head and neck cancer biopsies, which were preselected for greater than 65% tumor nuclei by histological staining and microscopic examination, our group found that the percentage of tumor cells actually ranged from 15% to 80% (40% average) when estimated using an ASCAT algorithm (45) for analyzing high-resolution SNP data (unpublished observation). Therefore, even in cases where pathologists choose samples on the basis of high tumor cellularity cutoff values, there can be a large degree of variations and errors when estimating the percentage of tumor cells actually present in a sample.

Given the heterogeneity found in tumors, two schools of thought have emerged with respect to the requirement for up-front cell enrichment prior to performing RPMA. One idea is that clinical tests have to deal with such heterogeneity in routine practice (46), so biomarkers should be chosen ahead of time that are not heavily influenced by these factors. In other words, if biomarkers are discovered using heterogeneous specimens, then they should also work on such samples. For example, if a prognostic proteomic signature is found that more likely originated from infiltrating inflammatory cells, then the results could still be valid because the degree of infiltration could in itself affect the biology of a tumor and generate important predictive value, as in the case of inflammatory breast cancer. Opponents of this approach have argued that some method of up-front enrichment should be performed (44), such as laser capture microdissection (LCM), to separate tumor from other cell types prior to sample preparation. There are merits to both approaches and certainly data using each of these strategies have been successfully used in multiple studies where global analysis has been applied to find biomarkers and predict prognosis (18, 20, 21, 28, 46–48).

Two seminal papers examining the impact of tumor heterogeneity and necessity for microdissection prior to performing RPMA have been published recently (44, 46). Hennessy et al. approached the question by examining RPMA profiles for 82 markers (total and phosphoproteins) using biologic replicates derived by isolating protein from either different pieces of the same tumor samples collected prospectively or different sections from the same biopsies from a previously banked cohort. They found that 74/82 markers showed little variability between different pieces of the same biopsy, and 72% of markers showed strong correlation between biologic replicates derived from different sections of the same sample. As expected there was a high degree of inter-tumor variability (i.e., comparing different patients) for most markers in these cohorts.

The study published by Hennessey et al. (46) concluded that proteomic profiles of frozen samples were consistent in the absence of microdissection. However, several critical points need to be considered: first, the fact that proteomic profiles did not change between biologic replicates might simply mean that the intra-sample variability of the selected proteins was probably low, so that regardless of where the sample was taken the tumor looked the same. This can be explained if tumor samples are homogenous with respect to their heterogeneity. In other words, if different regions of the same tumor sample have similar levels of contaminating stromal or inflammatory cells, then taking different parts would be expected to produce a similar proteomic profile. This is a very important finding because it suggests that effects due to sampling different regions of a tumor may be minimal. On the other hand, it does not demonstrate that the individual profiles observed between patients, which were highly variable, were not heavily influenced by the proportion of contaminating nonmalignant cells. However, RPMA data from several of the endpoints, including levels of p53, cyclin B1, estrogen and progesterone receptors, and phosphorylation of STAT3, could be correlated with disease-free survival in one of the cohorts, arguing that microdissection was not necessary for predicting prognosis (46). Nonetheless, it is possible that the particular biomarkers chosen did not have significant expression in the contaminating noncancerous cells present in breast cancer biopsies (i.e., estrogen or progesterone receptors), and therefore the requirement for microdissection was negligible for these particular markers. This would not necessarily be the case for biomarkers highly expressed in both nonmalignant and cancerous cells.

A more direct approach to address the requirement for upfront cell enrichment was adopted by Silvestri et al. (44), where the proteomic profiles from LCM-enriched cancer epithelium or tumor stroma were compared to patient-matched whole-tissue specimens from 53 primary colorectal biopsies. Slides used for LCM were derived from the same set of contiguous cryostat sections also used for preparation of whole-tissue lysates, and 75 different protein/phosphoprotein endpoints were examined by RPMA for comparison. Two-way unsupervised hierarchical clustering analysis demonstrated that the pathway profiles for the undissected specimens, tumor-enriched cells, and stromal compartments were entirely different within the same patient. Attempts to rescale the tumor signature based upon the estimated percentage of tumor were unsuccessful. Focusing on two markers believed to be important in colon cancer biology, Cox-2 and phosphorylated EGFR (tyrosine 1148), there was little correlation between values derived from undissected material and the LCM-enriched tumor component, regardless of the estimated percentage of tumor cells present. As predicted from the poor correlations, the relative rank order levels for Cox-2 and phosphorylated EGFR dramatically changed in patients when comparing data from the undissected versus microdissected samples. As the LCM process itself does not significantly affect protein expression or activation (49, 50), it is likely that the different signatures obtained were in fact due to the different subpopulations of cells present within the tumor samples.

It could be argued that enrichment for cell types may be a superior strategy for RPMA analysis because of the increased sensitivity in detecting proteomic changes present in both cancerous and noncancerous cell populations. Moreover, there is no loss of data by separating the signatures from each of these populations of cells. Furthermore, if the biomarker pathway detected by RPMA is the therapeutic target, then it is critical to accurately measure activation levels of the drug target within the tumor population. In a clinical diagnostic setting where the percentage of tumor cells may vary drastically between individual patient samples, it is important to be able to analyze correctly every patient's tumor, irrespective of tumor content, which can be accomplished with up-front microdissection.

Another issue in the RPMA biomarker field is the relative impact of tissue handling on protein and phosphoprotein stability. While there is little debate that snap freezing biopsy specimens in liquid nitrogen within minutes of excision is the best way to preserve proteomic biomarker profiles, such a practice is not practical in typical community hospital settings and could even interfere with availability of tissue for pathological diagnosis (43, 51). Often, tissue is sent to a pathologist where it waits in a queue for some time before being stabilized by either freezing or fixation. As tissue remains in a living state for some time following excision, cells can react to the trauma of excision, ischemia, hypoxia, acidosis, accumulation of cellular waste, absence of electrolytes, and changes in temperature (43, 52, 53). It can be expected that cells would react to these changes by mobilization of stress-induced signaling pathways, given that both kinases and phosphatases remain active in tissue following excision and occasionally even during the process of fixation with formalin (51–53).

Several investigators have published RPMA-based studies examining changes in protein biomarker stability during the delay between excision and tissue stabilization (46, 52, 53). Hennessy et al. (46) examined the effects of leaving pieces from the same tumor samples at room temperature for periods of time ranging from 30 min to 24 h before freezing by analyzing 54 total protein and 28 phosphoprotein markers on RPMAs. Surprisingly, they found that only 9/31 (24%) phosphoproteins and 12/66 (18%) total proteins tested showed statistically significant changes over time. Remarkably, two-way unsupervised hierarchical clustering demonstrated that the proteomic signatures from 7/10 breast cancer biopsies remained similar with increasing time to tumor freezing for up to 24 h after resection. Although the impact of delayed processing on proteomic profiles can be potentially minimized by focusing on less labile biomarkers, the Hennessy study nonetheless illustrates that many protein markers do indeed change without some kind of stabilization measures.

The effect of ex vivo delay time on protein biomarkers has also been extensively studied by Espina et al. (52, 53) in two different publications, where the conclusions regarding phosphoprotein stability were far less optimistic. In the larger of these studies, 53 analytes including phosphoproteins, cleaved proteins, and non-phosphorylated proteins were examined by RPMA over a range of post-excision delay times. These experiments were conducted in a large panel of different tissue types that included diseased or non-diseased samples originating from uterus, colon, lung, and ovary (i.e., adenocarcinoma). Time-dependent fluctuations (both decreases and increases) defined by greater than 20% deviation from time zero measurements were observable in 30/53 endpoints, predominately phosphoproteins, that represented cellular signaling pathways involved in hypoxia/ischemia, proliferation/survival, adhesion/cytoskeleton structure, stress/inflammation, or apoptosis from various subcellular compartments. For example, decreases in pAKT (Ser-473) and myristoylated alanine-rich C-kinase substrate (Ser-152/Ser-156) were observed within 90 min post-excision for most tissues, whereas cleaved caspase-3, pIRS-1 (Ser-612), AMP-activated protein kinase Β (Ser-108), GSK3α/β (Ser-21/Ser-9), STAT1 (Tyr-701), and cAMP response element binding protein (Ser-133) increased within 90 min post-procurement. Treatment with phosphatase or kinase inhibitors alone altered protein phosphorylation emphasizing the need to stabilize both kinases and phosphatases to avoid post-excision artifacts.

The issue of biomarker stability transcends the type of platform used. Using IHC, Holzer et al. (54) confirmed significant decreases in phosphorylation of proteins in the PI3K/AKT, ERK/MAPK, and p38 MAPK signaling pathways in as little as 1 h post-excision using human xenograft tumors from mice. Interestingly, they reported that the post-ischemic stability of a given phospho-epitope varied significantly across different tumor types and that within the same tumor type there were also dramatic variations in the stability between different phosphoproteins. Based upon demonstrable fluctuations in protein biomarker stability associated with ex vivo delays, Espina et al. have proposed tissue procurement guidelines for handling clinical specimens (43). According to these guidelines, tissue procurement protocols need to be adopted to take into consideration that tissue is alive and reactive to ex vivo stress; investigators must be aware that post-excision stress can generate false changes in protein biomarkers. Methods to simultaneously block both kinases and phosphatases must be implemented, and tissue should be stabilized as soon as possible after excision preferably within 20 min. Finally, tissue stabilization methods should be compatible with downstream analyses but still preserve morphology and histology vital for clinical diagnosis. Formalin fixation has been the standard preservative for a century, but the procedure has major drawbacks. The extensive cross-linking produced by formalin makes protein extraction difficult, but more importantly the slow penetration time (i.e., millimeters per hour) prevents timely fixation of inner regions of tissue, and the kinetics of phosphatase inactivation are too slow (43, 51, 52). Thus, while it is feasible to extract proteins from formalin-fixed specimens (including overfixed samples) (55, 56) and successfully perform RPMA on proteins from FFPE tissue (38, 47, 57), many of the biomarkers measured may not be truly reflective of their original state owing to the slow and variable kinetics of fixation that are highly dependent on sample thickness.

A new type of chemical fixative known as biomarker and histology preservative (BHP) has been recently developed. It can preserve phosphorylated and non-phosphorylated proteins better than formalin yet still retains cellular/tissue morphology permitting routine tissue histology staining and immunohistochemistry to be performed from paraffin-embedded blocks (51). BHP combines both kinase and phosphatase inhibitors to immediately arrest changes in protein phosphorylation, has permeation enhancers to decrease penetration time into cells, uses a precipitating fixative with reversible cross-linkers, and has an osmotically balanced buffer to maintain tissue/cell morphology during fixation. Total protein yield from BHP-fixed tissue is equivalent to that obtained from snap frozen tissue, phosphoproteins are stable in the fixative over a long time period ranging from hours to at least 7 days, and immunohistochemistry using antibodies to total proteins typically used for clinical diagnosis works as well as they do on formalin-fixed specimens. The efficacy of BHP to preserve proteomic profiles was demonstrated with RPMAs by comparing levels of nine phosphoproteins, six non-phosphorylated proteins, and one cleaved protein in tissue that had been either snap frozen or fixed for 7 days with BHP. Only 1 out of the 16 analytes examined showed a slight alteration after BHP fixation, when compared to snap frozen tissue. Therefore, broad implementation of a chemical fixative with the properties of BHP could circumvent many of the problems associated with faithful preservation of protein biomarkers from biopsies routinely obtained in the clinical setting.

A final caveat in the RPMA field that limits biomarker discovery as well as other uses in drug development is the somewhat limited panel of antibodies that have been validated to date for the platform. Many antibodies that work on western blots have the potential to work with RPMAs as well, because in both platforms proteins are in a denatured state. However, because the molecular weight of analytes detected by RPMA cannot be known, it is imperative to demonstrate the specificity of antibodies, usually measured by the absence of nonspecific reactive bands on western blots developed under similar blocking conditions. A second requirement for antibody validation is proof of principle that an antibody detects quantitative changes in the expected target on an

RPMA, either following serial dilution of the target or after specific modulation by siRNA (58, 59). The number of validated antibodies routinely used for RPMA in publications is currently around a few hundred. A principal reason why the rate of antibody validation has been relatively slow is because there are only a handful of laboratories around the world that routinely perform RPMAs. There is currently no common database that lists all antibodies validated for RPMA. The Antibody Characterization Laboratory at the National Institute of Health (USA) does maintain a public database of antibodies that the agency has validated for a "well-based" reverse phase protein array; however, the methods utilized are substantially different from the RPMA platforms commonly used in research laboratories and publications around the world, thereby limiting its utility. There is currently a great need to validate more antibodies using standard platforms in order to unleash the potential for drug and biomarker discovery that RPMA technology possesses.

References

1. Chua HN, Wong L (2008) Increasing the reliability of protein interactomes. Drug Discov Today 13:652–658
2. Miernyk JA, Thelen JJ (2008) Biochemical approaches for discovering protein-protein interactions. Plant J 53:597–609
3. Smith C (2003) Drug target validation: hitting the target. Nature 422:341
4. Walsh CT, Garneau-Tsodikova S, Gatto GJJ (2001) Protein posttranslational modifications: the chemistry of proteome diversifications. Angew Chem Int Ed 44:7342–7372
5. Paweletz CP, Liotta LA, Petricoin EF (2001) New technologies for biomarker analysis of prostate cancer progression: laser capture microdissection and tissue proteomics. Urology 57:160–163
6. Mueller C, Liotta LA, Espina V (2010) Reverse phase protein microarrays advance to use in clinical trials. Mol Oncol 4:461–481
7. Pierobon M, VanMeter AJ, Moroni N et al (2012) Reverse-phase protein microarrays. Methods Mol Biol 823:215–235
8. Espina V, Wulfkuhle J, Calvert VS et al (2008) Reverse phase protein microarrays for theranostics and patient-tailored therapy. Methods Mol Biol 441:113–128
9. Wilson B, Liotta LA, Petricoin E (2010) Monitoring proteins and protein networks using reverse phase protein arrays. Dis Markers 28:225–232
10. Espina V, Wulfkuhle JD, Calvert VS et al (2007) Reverse phase protein microarrays for monitoring biological responses. Methods Mol Biol 383:321–336
11. Gallagher RI, Silvestri A, Petricoin EF et al (2011) Reverse phase protein microarrays: fluorometric and colorimetric detection. Methods Mol Biol 723:275–301
12. Brase JC, Mannsperger H, Sultmann H et al (2011) Antibody-mediated signal amplification for reverse phase protein array-based protein quantification. Methods Mol Biol 785:55–64
13. Brase JC, Mannsperger H, Frohlich H et al (2010) Increasing the sensitivity of reverse phase protein arrays by antibody-mediated signal amplification. Proteome Sci 8:36
14. Espina V, Woodhouse EC, Wulfkuhle J et al (2004) Protein microarray detection strategies: focus on direct detection technologies. J Immunol Methods 290:121–133
15. VanMeter AJ, Rodriguez AS, Bowman ED et al (2008) Laser capture microdissection and protein microarray analysis of human non-small cell lung cancer: differential epidermal growth factor receptor (EGPR) phosphorylation events associated with mutated EGFR compared with wild type. Mol Cell Proteomics 7:1902–1924
16. Hanahan D, Weinberg RA (2011) Hallmarks of cancer: the next generation. Cell 144: 646–674
17. Sheehan KM, Calvert VS, Kay EW et al (2005) Use of reverse phase protein microarrays and reference standard development for molecular network analysis of metastatic ovarian carcinoma. Mol Cell Proteomics 4:346–355

18. Improta G, Zupa A, Fillmore H et al (2001) Protein pathway activation mapping of brain metastasis from lung and breast cancers reveals organ type specific drug target activation. J Proteome Res 10:3089–3097
19. Agrawal N, Frederick MJ, Pickering CR et al (2011) Exome sequencing of head and neck squamous cell carcinoma reveals inactivating mutations in NOTCH1. Science 333: 1154–1157
20. Frederick MJ, VanMeter AJ, Gadhikar MA et al (2011) Phosphoproteomic analysis of signaling pathways in head and neck squamous cell carcinoma patient samples. Am J Pathol 178(2): 548–571
21. Accordi B, Espina V, Giordan M et al (2010) Functional protein network activation mapping reveals new potential molecular drug targets for poor prognosis pediatric BCP-ALL. PLoS One (Electronic Resource) 5:e13552
22. Lu Y, Muller M, Smith D et al (2010) Kinome siRNA-phosphoproteomic screen identifies networks regulating AKT signaling. Oncogene 30:4567–4577
23. Pierobon M, Calvert V, Belluco C et al (2009) Multiplexed cell signaling analysis of metastatic and nonmetastatic colorectal cancer reveals COX2-EGFR signaling activation as a potential prognostic pathway biomarker. Clin Colorectal Cancer 8:110–117
24. Espina V, Mariani BD, Gallagher RI et al (2010) Malignant precursor cells pre-exist in human breast DCIS and require autophagy for survival. PLoS One (Electronic Resource) 5:e10240
25. Lavezzari G, Lackner MR (2012) Monitoring phosphoproteomic response to targeted kinase inhibitors using reverse-phase protein microarrays. Methods Mol Biol 795:203–215
26. Martiny-Baron G, Haasen D, D'Dorazio D et al (2011) Characterization of kinase inhibitors using reverse phase protein arrays. Methods Mol Biol 785:79–107
27. Gopal YN, Deng W, Woodman SE et al (2001) Basal and treatment-induced activation of AKT mediates resistance to cell death by AZD6244 (ARRY-142886) in Braf-mutant human cutaneous melanoma cells. Cancer Res 70:8736–8747
28. Miller TW, Hennessy BT, Gonzalez-Angulo AM et al (2001) Hyperactivation of phosphatidylinositol-3 kinase promotes escape from hormone dependence in estrogen receptor-positive human breast cancer. J Clin Invest 120:2406–2413
29. Wei X, Guo W, Wu S et al (2009) Inhibiting JNK dephosphorylation and induction of apoptosis by novel anticancer agent NSC-741909 in cancer cells. J Biol Chem 284:16948–16955
30. Espina V, Liotta LA, Petricoin EF (2009) Reverse-phase protein microarrays for theranostics and patient tailored therapy. Methods Mol Biol 520:89–105
31. Stuart D, Sellers WR (2009) Linking somatic genetic alterations in cancer to therapeutics. Curr Opin Cell Biol 21:304–310
32. Hamburg MA, Collins FS (1922) The path to personalized medicine (Erratum appears in N Engl J Med. 2010 Sep 9;363(11):1092). N Engl J Med 363:301–304
33. Macconaill LE, Garraway LA (2010) Clinical implications of the cancer genome. J Clin Oncol 28:5219–5228
34. Pierobon M, Belluco C, Liotta LA et al (2011) Reverse phase protein microarrays for clinical applications. Methods Mol Biol 785:3–12
35. Zha H, Raffeld M, Charboneau L et al (2004) Similarities of prosurvival signals in Bcl-2-positive and Bcl-2-negative follicular lymphomas identified by reverse phase protein microarray. Lab Invest 84:235–244
36. Liotta LA, Kohn EC, Petricoin EF (1914) Clinical proteomics: personalized molecular medicine. JAMA 286:2211–2214
37. Espina V, Dettloff KA, Cowherd S et al (2004) Use of proteomic analysis to monitor responses to biological therapies. Expert Opin Biol Ther 4:83–93 (Review) (95 refs)
38. Petricoin EF, Espina V, Araujo RP et al (2001) Phosphoprotein pathway mapping: Akt/mammalian target of rapamycin activation is negatively associated with childhood rhabdomyosarcoma survival. Cancer Res 67:3431–3440
39. Berg D, Wolff C, Langer R et al (2011) Discovery of new molecular subtypes in oesophageal adenocarcinoma. PLoS One (Electronic Resource) 6:e23985
40. Von H, Stephenson JJ Jr, Rosen P et al (2010) Pilot study using molecular profiling of patients' tumors to find potential targets and select treatments for their refractory cancers. J Clin Oncol 28:4877–4883
41. Liotta LA, Petricoin EF (2012) Regulatory approval pathways for molecular diagnostic technology. Methods Mol Biol 823:409–420
42. Rapkiewicz A, Espina V, Zujewski JA et al (1925) The needle in the haystack: application of breast fine-needle aspirate samples to quantitative protein microarray technology. Cancer 111:173–184
43. Espina V, Mueller C (2012) Reduction of preanalytical variability in specimen procurement for molecular profiling. Methods Mol Biol 823:49–57
44. Silvestri A, Colombatti A, Calvert VS et al (2010) Protein pathway biomarker analysis of

human cancer reveals requirement for upfront cellular-enrichment processing. Lab Invest 90:787–796

45. Van L, Nordgard SH, Lingjaerde OC et al (1928) Allele-specific copy number analysis of tumors. Proc Natl Acad Sci USA 107: 16910–16915
46. Hennessy BT, Lu Y, Gonzalez-Angulo AM et al (2010) A technical assessment of the utility of reverse phase protein arrays for the study of the functional proteome in non-microdissected human breast cancers. Clin Proteomics 6:129–151
47. Berg D, Wolff C, Malinowsky K et al (2012) Profiling signalling pathways in formalin-fixed and paraffin-embedded breast cancer tissues reveals cross-talk between EGFR, HER2, HER3 and uPAR. J Cell Physiol 227:204–212
48. Gonzalez-Angulo AM, Hennessy BT, Meric-Bernstam F et al (2011) Functional proteomics can define prognosis and predict pathologic complete response in patients with breast cancer. Clin Proteomics 8:11
49. Paweletz CP, Charboneau L, Bichsel VE et al (2001) Reverse phase protein microarrays which capture disease progression show activation of pro-survival pathways at the cancer invasion front. Oncogene 20(16):1981–1989
50. Ornstein DK, Gillespie JW, Paweletz CP et al (2000) Proteomic analysis of laser capture microdissected human prostate cancer and in vitro prostate cell lines. Electrophoresis 21:2235–2242
51. Mueller C, Edmiston KH, Carpenter C et al (2011) One-step preservation of phosphoproteins and tissue morphology at room temperature for diagnostic and research specimens. PLoS One (Electronic Resource) 6:e23780
52. Espina V, Edmiston KH, Heiby M et al (2008) A portrait of tissue phosphoprotein stability in the clinical tissue procurement process. Mol Cell Proteomics 7:1998–2018
53. Espina V, Mueller C, Edmiston K et al (2009) Tissue is alive: new technologies are needed to address the problems of protein biomarker pre-analytical variability. Proteomics Clin Appl 3:874–882
54. Holzer TR, Fulford AD, Arkins AM et al (2011) Ischemic time impacts biological integrity of phospho-proteins in PI3K/Akt, Erk/MAPK, and p38 MAPK signaling networks. Anticancer Res 31:2073–2081
55. Wolff C, Schott C, Malinowsky K et al (2011) Producing reverse phase protein microarrays from formalin-fixed tissues. Methods Mol Biol 785:123–140
56. Wolff C, Schott C, Porschewski P et al (2011) Successful protein extraction from over-fixed and long-term stored formalin-fixed tissues (Erratum appears in PLoS One. 2011;6(4). doi:10.1371/annotation/a42e114f-a708-4423-8a3e-a1d8919b9b60). PLoS One (Electronic Resource) 6:e16353
57. Berg D, Malinowsky K, Reischauer B et al (2011) Use of formalin-fixed and paraffin-embedded tissues for diagnosis and therapy in routine clinical settings. Methods Mol Biol 785:109–122
58. Spurrier B, Washburn FL, Asin S et al (2007) Antibody screening database for protein kinetic modeling. Proteomics 7:3259–3263
59. Mannsperger H, Uhlmann S, Korf U et al (2011) Utilization of RNAi to validate antibodies for reverse phase protein arrays. Methods Mol Biol 785:45–54
60. Berg D, Hipp S, Malinowsky K et al (2010) Molecular profiling of signalling pathways in formalin-fixed and paraffin-embedded cancer tissues. Eur J Cancer 46:47–55
61. Espina V, Mueller C, Liotta LA (2011) Phosphoprotein stability in clinical tissue and its relevance for reverse phase protein microarray technology. Methods Mol Biol 785: 23–43
62. Malinowsky K, Wolff C, Gundisch S et al (2010) Targeted therapies in cancer—challenges and chances offered by newly developed techniques for protein analysis in clinical tissues. J Cancer 2:26–35

Part V

Innovative Cell Culture Techniques to Mimic Tissue Microenvironments

Chapter 14

A Cell Culture System That Mimics Chronic Lymphocytic Leukemia Cells Microenvironment for Drug Screening and Characterization

Alessandro Natoni, Michael O'Dwyer, and Corrado Santocanale

Abstract

Chronic Lymphocytic Leukaemia (CLL) is an incurable disease that warrants new therapeutic treatments. CLL cells accumulate in the peripheral blood, in the bone marrow and in secondary lymphoid organs. Unlike circulating CLL cells, CLL cells resident in these last two compartments display high chemoresistance and proliferative capacity. Given the importance of the microenvironment in this disease, strategies that aim to develop new therapeutic agents need to consider this critical factor. Various cell culture conditions have been described that attempt to emulate either the different types of microenvironments in which CLL cells are found or an individual component of a particular microenvironment. Here, a methodology that partially mimics the interaction between CLL cells and the CD3+ CD4+ CD154+ T cells is described. Moreover, within this method, two protocols are presented and compared that may partially recapitulate different physiological states. The methodology can be exploited for target validation and drug development in CLL.

Key words: CLL, Microenvironment, T cells, Chemotherapy, Lymph Nodes, Cell Death

1. Introduction

Chronic lymphocytic leukemia (CLL) represents the most frequent type of leukemia in adults in the Western world (1). This disease is characterized by the accumulation of B lymphocytes in the blood stream, bone marrow, and lymphoid organs such as lymph nodes and spleen. The progressive increase of these clonal B lymphocytes ultimately impairs the function of normal lymphocytes and the bone marrow leading to thrombocytopenia, anemia, and infection.

The microenvironment plays an essential role in CLL (2, 3). Indeed, it is now clear that CLL cells display different phenotypes depending on their localization in the body (4, 5). CLL cells in the

Jürgen Moll and Riccardo Colombo (eds.), *Target Identification and Validation in Drug Discovery: Methods and Protocols*, Methods in Molecular Biology, vol. 986, DOI 10.1007/978-1-62703-311-4_14,

bloodstream are quiescent, while proliferative Ki67 positive CLL cells are found in particular structures of the lymph nodes called proliferation centers or pseudofollicles (6–9). Importantly, CLL cells in this compartment are resistant to apoptosis induced by chemotherapeutic agents that would normally kill the malignant CLL cells in the blood (10). The increased resistance to apoptosis is, in part, due to elevated expression levels of antiapoptotic proteins such as Mcl-1, Bcl-A1, and Bcl-X_L (11–14).

The lymph node microenvironment represents a niche where CLL cells interact with different types of cells including monocyte-derived nurse-like cells (NLC), $CD3^+$ $CD4^+$ $CD154^+$ T cells, mesenchymal stromal cells, dendritic cells, and endothelial cells (15). In addition to cell–cell interactions, CLL cells are also exposed to a variety of soluble factors such as antigens, cytokines, and chemokines (2). It is the combination of such signals that renders CLL cells less susceptible to chemotherapy and promotes clonal evolution and drug resistance. Thus, the role of the microenvironment needs to be carefully considered in order to develop novel and more effective therapies for CLL treatment (16). In particular, the efficacy of new drugs must be evaluated under experimental conditions that recapitulate (or at least partially mimic) the CLL microenvironment.

Despite its complexity, several studies have attempted to partially recapitulate the lymph node microenvironment and in particular the $CD3^+$ $CD4^+$ $CD154^+$ T cell/CLL interaction (16). Activated $CD3^+$ $CD4^+$ $CD154^+$ T cells are in contact with CLL cells in the proliferation centers, and they provide leukemic cells with signaling molecules, including CD154 and interleukin 4 (IL-4), that sustain proliferation and survival (3). In vitro, the interaction between leukemic cells and T cells can be partially mimicked by culturing CLL cells in the presence of CD154 and IL-4 (16–18). CD40 stimulation can be achieved using either soluble CD154 (or anti-CD40 agonistic antibodies) or cell lines engineered to express membrane-bound CD154. The second method of CD154 delivery is usually considered preferable as membrane-bound CD154 displays superior capacity to activate CD40 signaling and also better mimics the in vivo conditions where CD154 is most likely delivered by cell–cell contact (16, 19). Recently the CD154/IL-4 system has been used quite extensively to test the efficacy of new drugs or combinations of drugs for CLL treatment (12, 14, 20–29).

Here we describe a simple method of coculturing CLL cells on a feeder layer of mouse fibroblast cells (NIH3T3) expressing CD154 (NIH3T3 CD154) in the presence of IL-4. This methodology can be applied to screen drugs or combinations of drugs that have shown activity in preliminary screening using either cell lines or primary CLL cells cultured under standard conditions. Two experimental protocols are described that can be used for drug screening (namely, continuous and discontinuous culturing). In the continuous protocol, CLL cells are treated with the drug(s) of choice in the continuous presence of the supporting fibroblasts. In the discontinuous protocol, the CLL cells are first cocultivated on a layer of NIH3T3 cells, then detached and replated in the absence of NIH3T3 and treated with the drug(s) of choice. However, the reader should be aware that alternative methodologies to mimic the CD3⁺ CD4⁺ CD154⁺ T cells/CLL interaction have been developed and reported in the literature (12, 20–29). These methodologies utilize different media, supplements, CD154 expressing cells, source and concentration of IL-4, ratio of CLL–feeder cells, etc. To our knowledge, no comparative studies of the different methodologies have been carried out to date, so one should be cautious when comparing the results obtained with the different CD154/IL-4 systems.

2. Material

Prepare all media and handle cells in aseptic conditions except during the cell death assay.

2.1. Reagents

1. RPMI (ATCC, catalog number 30-2001) (see Note 1).
2. Interleukin-4 (IL-4): IL-4 is available from many suppliers.
3. Fluorescein isothiocyanate (FITC)-conjugated annexin V is available from many different suppliers. In the present study, it was produced in house as previously described (30). Protect from light and store at +4°C.
4. Propidium iodide (PI): prepare a stock solution of 50 μg/ml in water. Protect from light and store at +4°C.
5. PHA-767491 (Tocris, catalog number 3140): prepare a stock solution of 10 mM in dimethyl sulfoxide (DMSO). Protect from light and store at −20°C.
6. Fludarabine (SIGMA, catalog number F2773). Prepare a stock solution of 10 mM in DMSO. Store at −20°C.
7. Bortezomib was kindly provided by Millennium Pharmaceuticals. Prepare a stock solution of 500 μM in Milli-Q sterile water. Store at −20°C.

2.2. Cell Culture Media

1. Complete DMEM: to 500 ml of DMEM add 10% heat/inactivated FBS, 50 μg/ml streptomycin, and 50 U/ml penicillin.
2. Complete RPMI: to 500 ml of RPMI add 10% heat/inactivated FBS, 50 μg/ml streptomycin, and 50 U/ml penicillin.

2.3. Cells

1. Irradiated wild type (WT) and CD154 expressing (CD154) NIH3T3 mouse embryonic fibroblasts (see Note 2): WT and CD154 NIH3T3 cells are cultured in complete DMEM and maintained in a humidified atmosphere of 5% CO_2 at 37°C.

2. Primary CLL cells purified from peripheral blood of volunteer patients (see Note 3): CLL cells are purified by Ficoll gradient centrifugation, by negative selection, or by a combination of both (see Note 4). CLL cells can be used immediately or frozen in 90% heat/inactivated FBS supplemented with 10% DMSO and stored at −80°C or in liquid nitrogen for short- and long-term storage, respectively. CLL cells are cultured in complete RPMI medium and maintained in a humidified atmosphere of 5% CO_2 at 37°C (see Note 5).

3. Methods

3.1. Culturing CLL Cells on WT and CD154 NIH3T3

1. Irradiate WT and CD154 NIH3T3 using an appropriate source of γ radiation and follow all the health and safety procedures of your institution. Usually a dose of 30 Gray is sufficient to block cell proliferation (see Note 6). Cells can be used immediately after irradiation or frozen in 90% heat/inactivated FBS supplemented with 10% DMSO and stored at −80°C or in liquid nitrogen for short- and long-term storage, respectively.
2. Plate 2×10^6 WT or CD154 NIH3T3 irradiated and growth arrested cells on 60 mm tissue culture dishes in 5 ml DMEM and incubate overnight to allow the cells to attach on the surface of the tissue culture dishes.
3. Examine the cells under the microscope to ensure that all the fibroblasts are attached to the tissue culture dishes (Fig. 1a). Wash the layer of fibroblasts twice with 5 ml of pre-warmed PBS. Seed 2×10^7 CLL cells (ratio 10:1; 10 CLL cells per NIH3T3 cell; Fig. 1b) on WT and CD154 NIH3T3 in 5 ml complete RPMI medium supplemented with 10 ng/ml of IL-4 (only in CD154 NIH3T3-coated tissue culture dishes) and incubate for 24 h in a humidified atmosphere of 5% CO_2 at 37°C (see Notes 7 and 8).
4. Examine the CLL cells by microscopy. The cells should lie on top of the fibroblasts (on both the WT and CD154 NIH3T3) (Fig. 1c). Shake the tissue culture dishes gently to verify adherence. Adherent CLL cells should remain attached after gently shaking the tissue culture dish.
5. Detach the CLL cells by gently pipetting up and down using a 1,000 μl pipette tip. Rotate the dishes while pipetting to detach the CLL cells all around the tissue culture dishes. Collect the media containing cells in an appropriate sterile tube. Examine the tissue culture dishes under the microscope to ensure that all the CLL cells are collected and the layer of fibroblasts is still intact (see Note 9) (Fig. 1d).

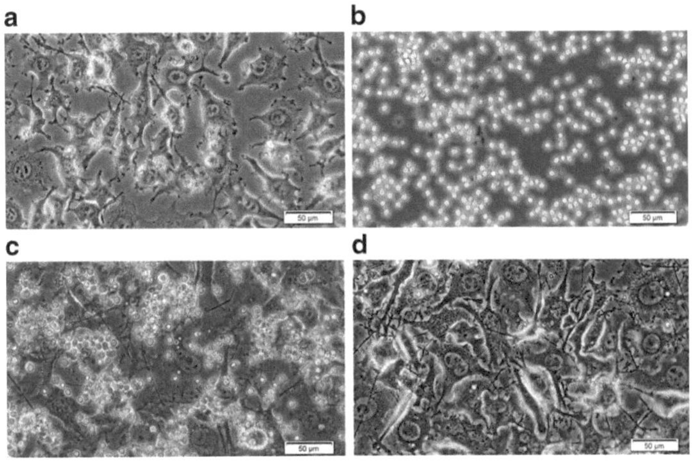

Fig. 1. Culture of CLL cells with CD154 expressing NIH3T3 fibroblasts. Representative images of (**a**) irradiated CD154 NIH3T3 fibroblasts and (**b**) CLL cells cultivated separately or (**c**) after 48 h in coculture. (**d**) Representative image of irradiated CD154 NIH3T3 fibroblasts after removal of CLL cells. Images were acquired at ×10 magnification using an inverted CKX41 Olympus microscope equipped with an Olympus SC30 camera and Olympus U-TV0.5XC-3 camera adaptor (magnification 0.50). Images were analyzed using CellSens Entry 1.5 (Olympus). Bars represent 50 μm.

6. Count the CLL cells, centrifuge at 200 × *g* for 5 min at room temperature (RT), and resuspend in fresh complete RPMI media at 2×10^6 cells/ml.
7. At this point, two types of procedures, namely, continuous (see Subheading 3.1, step 8) and discontinuous (see Subheading 3.1, step 9) coculture may be employed for drug testing.
8. Continuous coculturing:
   (a) For this procedure, WT and CD154 NIH3T3 cells must be seeded the day before in 24 wells/plate at 2×10^5 cells/ml (1 ml per well).
   (b) Wash the layer of fibroblasts in each well twice with 1 ml PBS. Plate the CLL cells from step 6 on the layer of WT and CD154 NIH3T3 fibroblasts at 2×10^6 cells/ml in 1 ml of complete RPMI medium supplemented with 10 ng/ml IL-4 (only in CD154 NIH3T3-coated culture dishes).
9. Discontinuous coculturing:
   (a) Plate CLL cells from step 5 in 24 well plate at 2×10^6 cells/ml in 1 ml of complete RPMI medium.

3.2. Treatment of Cocultured CLL Cells with the Drug(s) of Choice

1. For the purpose of this study, PHA-767491 (a Cdc7/CDK9 inhibitor, final concentration 5 μM), fludarabine (nucleoside analogue, final concentration 10 μM), and bortezomib (proteasome inhibitor, final concentration 30 nM) were chosen as

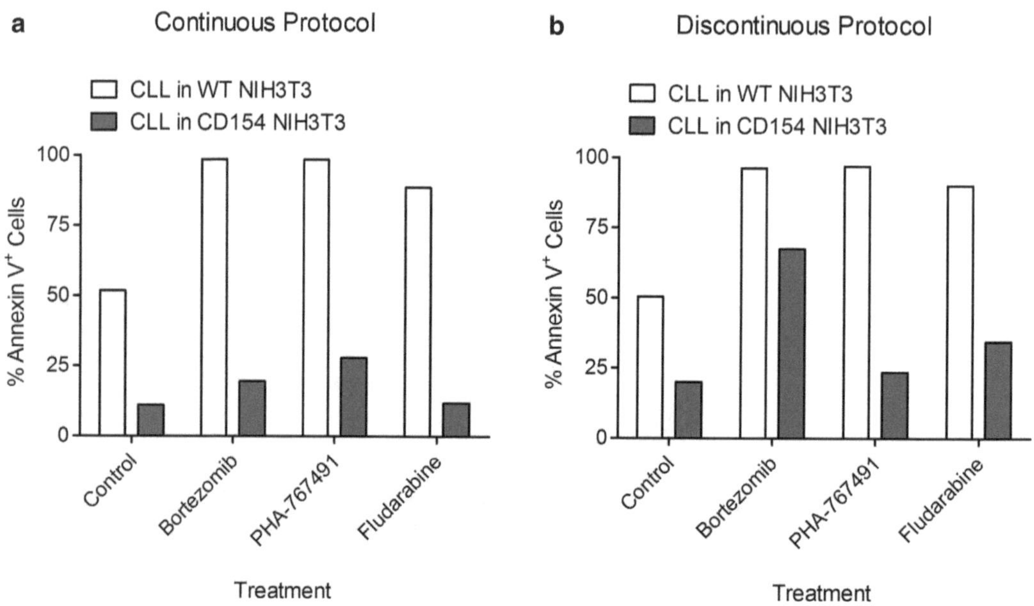

Fig. 2. CD154 and IL-4 stimulation of CLL cells inhibits drug-induced cell death. CLL cells were cultured on WT and CD154 NIH3T3 fibroblasts using either the continuous (**a**) or the discontinuous (**b**) protocol in the presence of bortezomib (30 nM), PHA-767491 (10 μM), fludarabine (10 μM), or vehicle control for 24 h. Cell death was analyzed using a BD FACSCanto I flow cytometer (BD Biosciences) using annexin V/PI staining. Data were analyzed using FlowJo (©Tree Star) and GraphPad Prism software.

they have different targets in CLL cells (see Note 10) (14, 31–33). As a control, include CLL cells treated with vehicle only (DMSO, 1% of the total volume). Incubate for 48 h in a humidified atmosphere of 5% CO_2 at 37°C.

2. Collect the CLL cells. In the continuous coculture procedure, extra care must be taken when detaching the CLL cells from the layer of fibroblasts as the drugs used may render the fibroblasts more prone to detachment (see Notes 9 and 10).

3.3. Cell Death Analysis in Cocultured CLL Cells

1. In this study, annexin V/PI assay was used to detect cell death by flow cytometry (30) (see Note 11).

2. As shown in Fig. 2, seeding CLL cells on CD154 NIH3T3 but not on WT NIH3T3 prevents spontaneous apoptosis of CLL cells in both continuous and discontinuous methods. However, CLL cells cocultured using the continuous protocol (Fig. 2a) become resistant to all treatments, while CLL cells cocultured using the discontinuous protocol (Fig. 2b) remain sensitive to the bortezomib treatment. The continuous protocol of coculturing may better reflect the in vivo condition where the CLL cells are in direct contact and continuously stimulated by the $CD3^+$ $CD4^+$ $CD154^+$ T cells. The discontinuous protocol may represent a situation where CLL cells are disengaged from the $CD3^+$ $CD4^+$ $CD154^+$ T cells and are about to leave the lymph node compartment. The reader should decide which of these

protocols better fulfills the aims of their experiment. The reader should also be aware that interindividual variation (genetic background) between CLL samples obtained from different patients markedly impacts the final outcome of the drug activity in the coculture system. Thus, extensive validation should be performed using multiple CLL samples with different genetic backgrounds.

3.4. Considerations for scaling up

The protocol given in this study may be scaled up. However, it is recommended to maintain the same CLL:NIH3T3 ratio to preserve the efficacy of CLL stimulation by the CD154 NIH3T3 and decrease the chance of contaminating the sample with NIH3T3 cells. Coculturing CLL cells in large culture dishes such as 145–150 mm or in culture flasks may result in loss of CLL cells and/or contamination with NIH3T3 cells due to the complications of pipetting and detaching the CLL cells.

4. Notes

1. It is recommended to purchase RPMI that contains 4.5 g/L glucose (such as the one suggested) as it may better sustain the viability of CLL and NIH3T3 in the coculture conditions.
2. The reader should be aware that other types of cells expressing CD154 have been used to stimulate CLL cells. These include mouse fibroblast L-cells (NTL) (12, 20, 21, 23, 28) and baby hamster kidney cells (27). The CD154 NIH3T3 cells used in this study were kindly provided by Dr. Eldering (Department of Pathology, Academic Medical Center, Amsterdam, The Netherlands).
3. The study must be approved by the local authorities, and blood samples must be obtained from patients after informed consent in accordance with the declaration of Helsinki (34). The present study was approved by the Ethical Review Committee of University College Hospital Galway. Extreme care should be taken when handling human blood sample. The presence in the sample of at least hepatitis B, hepatitis C, and human immunodeficiency virus should be tested. If this is not possible, treat the samples as if they would be infected. The personnel handling the human blood samples should be vaccinated at least for hepatitis B, and they must follow all the health and safety procedures of their institution.
4. Positive selection is not recommended as these procedures may induce signaling pathways in CLL cells due to antibody/antigen interactions. Purity of CLL cells is assessed by flow cytometry using anti-CD19, anti-CD5, and matching isotype control

antibodies. Only samples with ≥95% double CD19 and CD5 positive cells should be used.

5. CLL cells may be cultured in other types of media including Iscove modified Dulbecco's media (22, 24–26, 29) or Dulbecco's modified media (27). The reader should be aware that the type of media may change the results of the experiment.
6. Mitomycin C can be used as an alternative to γ radiation for blocking cell proliferation of the feeder layer of CD154 expressing cells (21, 28). However, to our knowledge, this methodology has not been used for NIH3T3 cells.
7. Different concentrations of IL-4 and CLL:NIH3T3 ratio may be used (12, 20, 21, 27–29). The reader should be aware that the concentration of IL-4 may change the results of the experiment.
8. NIH3T3 and CLL cells may be seeded and treated in 24 well plates avoiding steps 5–8 in Subheading 3.1 when performing continuous coculturing. However, coculturing the CLL cells on the NIH3T3 fibroblasts seeded in the 60 mm tissue culture dishes ensures more homogeneous stimulation of the cells, thereby preventing the variability that could arise from well to well. Moreover, if the discontinuous protocol is used (see Subheading 3.1, step 9), detaching the CLL cells from the NIH3T3 fibroblasts on a single 60 mm tissue culture dish instead of several individual wells of the 24 well plate will be less laborious and reduce possible NIH3T3 cells contamination.
9. WT NIH3T3 cells are more prone to detach than CD154 NIH3T3, so detaching CLL cells from the WT NIH3T3 cells requires extra care. If discontinuities or ruptures in the layer of fibroblasts are noticed, CLL cells may be seeded at this step in 60 mm culture dishes for 2 h to allow possible NIH3T3 cells present in the sample to reattach on the surface of the tissue culture dishes. The tissue culture dishes are then incubated in a humidified atmosphere of 5% CO_2 at 37°C. After incubation, examine the tissue culture dishes for the presence of adherent cells and carefully collect the CLL cells in an appropriate sterile tube.
10. A pilot study is recommended to examine if the drug(s) of choice drastically affects the layer of fibroblasts during the time of treatment. It must be remembered that WT NIH3T3 cells are more prone of detaching from the surface of the culture dishes and the treatment may augment this features.
11. Different methods can be used to detect cell death in CLL cells. Methods based on detection of cell death by flow cytometry are recommended as CLL cells are easily distinguishable by forward and size scatter from possible contamination of

the NIH3T3 fibroblasts. However, other methodologies that allow discrimination between CLL and NIH3T3 cells may be used. In the methodology described below, an annexin V/propidium iodide (PI) assay was used to detect cell death by flow cytometry.

References

1. Lanasa MC (2010) Novel insights into the biology of CLL. Hematology Am Soc Hematol Educ Program 2010:70–76
2. Burger JA (2011) Nurture versus nature: the microenvironment in chronic lymphocytic leukemia. Hematology Am Soc Hematol Educ Program 2011:96–103
3. Ghia P, Chiorazzi N, Stamatopoulos K (2008) Microenvironmental influences in chronic lymphocytic leukaemia: the role of antigen stimulation. J Intern Med 264:549–562
4. Herishanu Y, Perez-Galan P, Liu D et al (2011) The lymph node microenvironment promotes B-cell receptor signaling. NF-kappaB activation, and tumor proliferation in chronic lymphocytic leukemia. Blood 117:563–574
5. Jaksic O, Kardum-Skelin I, Jaksic B (2010) Chronic lymphocytic leukemia: insights from lymph nodes & bone marrow and clinical perspectives. Coll Antropol 34:309–313
6. Chiorazzi N (2007) Cell proliferation and death: forgotten features of chronic lymphocytic leukemia B cells. Best Pract Res Clin Haematol 20:399–413
7. Granziero L, Ghia P, Circosta P et al (2001) Survivin is expressed on CD40 stimulation and interfaces proliferation and apoptosis in B-cell chronic lymphocytic leukemia. Blood 97: 2777–2783
8. Lampert IA, Wotherspoon A, Van Noorden S et al (1999) High expression of CD23 in the proliferation centers of chronic lymphocytic leukemia in lymph nodes and spleen. Hum Pathol 30:648–654
9. Vandewoestyne ML, Pede VC, Lambein KY et al (2011) Laser microdissection for the assessment of the clonal relationship between chronic lymphocytic leukemia/small lymphocytic lymphoma and proliferating B cells within lymph node pseudofollicles. Leukemia 25: 883–888
10. Munk Pedersen I, Reed J (2004) Microenvironmental interactions and survival of CLL B-cells. Leuk Lymphoma 45:2365–2372
11. Smit LA, Hallaert DY, Spijker R et al (2007) Differential Noxa/Mcl-1 balance in peripheral versus lymph node chronic lymphocytic leukemia cells correlates with survival capacity. Blood 109:1660–1668
12. Vogler M, Butterworth M, Majid A et al (2009) Concurrent up-regulation of BCL-XL and BCL2A1 induces approximately 1000-fold resistance to ABT-737 in chronic lymphocytic leukemia. Blood 113:4403–4413
13. Buggins AG, Pepper CJ (2010) The role of Bcl-2 family proteins in chronic lymphocytic leukaemia. Leuk Res 34:837–842
14. Natoni A, Murillo LS, Kliszczak AE et al (2011) Mechanisms of action of a dual Cdc7/Cdk9 kinase inhibitor against quiescent and proliferating CLL cells. Mol Cancer Ther 10: 1624–1634
15. Burger JA, Ghia P, Rosenwald A et al (2009) The microenvironment in mature B-cell malignancies: a target for new treatment strategies. Blood 114:3367–3375
16. Hayden RE, Pratt G, Roberts C et al (2011) Treatment of chronic lymphocytic leukemia requires targeting of the protective lymph node environment with novel therapeutic approaches. Leuk Lymphoma 53(4):537–549
17. Willimott S, Baou M, Huf S et al (2007) Separate cell culture conditions to promote proliferation or quiescent cell survival in chronic lymphocytic leukemia. Leuk Lymphoma 48:1647–1650
18. Willimott S, Baou M, Naresh K et al (2007) CD154 induces a switch in pro-survival Bcl-2 family members in chronic lymphocytic leukaemia. Br J Haematol 138:721–732
19. Neron S, Nadeau PJ, Darveau A et al (2011) Tuning of CD40-CD154 interactions in human B-lymphocyte activation: a broad array of in vitro models for a complex in vivo situation. Arch Immunol Ther Exp (Warsz) 59:25–40
20. Pepper C, Mahdi JG, Buggins AG et al (2011) Two novel aspirin analogues show selective cytotoxicity in primary chronic lymphocytic leukaemia cells that is associated with dual inhibition of Rel A and COX-2. Cell Prolif 44: 380–390
21. Hayden RE, Pratt G, Davies NJ et al (2009) Treatment of primary CLL cells with bezafibrate and medroxyprogesterone acetate induces apoptosis and represses the pro-proliferative signal of CD40-ligand, in part through increased 15dDelta12,14, PGJ2. Leukemia 23:292–304

22. Jak M, van Bochove GG, van Lier RA et al (2011) CD40 stimulation sensitizes CLL cells to rituximab-induced cell death. Leukemia 25:968–978
23. McCaig AM, Cosimo E, Leach MT et al (2011) Dasatinib inhibits B cell receptor signalling in chronic lymphocytic leukaemia but novel combination approaches are required to overcome additional pro-survival microenvironmental signals. Br J Haematol 153(2):199–211
24. Tromp JM, Geest CR, Breij EC et al (2012) Tipping the Noxa/Mcl-1 balance overcomes ABT-737 resistance in chronic lymphocytic leukemia. Clin Cancer Res 18:487–498
25. Jak M, van Bochove GG, Reits EA et al (2011) CD40 stimulation sensitizes CLL cells to lysosomal cell death induction by type II anti-CD20 mAb GA101. Blood 118:5178–5188
26. Tonino SH, van Laar J, van Oers MH et al (2011) ROS-mediated upregulation of Noxa overcomes chemoresistance in chronic lymphocytic leukemia. Oncogene 30:701–713
27. Dietrich S, Kramer OH, Hahn E et al (2012) Leflunomide induces apoptosis in fludarabine-resistant and clinically refractory CLL cells. Clin Cancer Res 18:417–431
28. Hayden RE, Pratt G, Drayson MT et al (2010) Lycorine sensitizes CD40 ligand-protected chronic lymphocytic leukemia cells to bezafibrate- and medroxyprogesterone acetate-induced apoptosis but dasatanib does not overcome reported CD40-mediated drug resistance. Haematologica 95:1889–1896
29. Hallaert DY, Jaspers A, van Noesel CJ et al (2008) c-Abl kinase inhibitors overcome CD40-mediated drug resistance in CLL: implications for therapeutic targeting of chemoresistant niches. Blood 112:5141–5149
30. Logue SE, Elgendy M, Martin SJ (2009) Expression, purification and use of recombinant annexin V for the detection of apoptotic cells. Nat Protoc 4:1383–1395
31. Baou M, Kohlhaas SL, Butterworth M et al (2010) Role of NOXA and its ubiquitination in proteasome inhibitor-induced apoptosis in chronic lymphocytic leukemia cells. Haematologica 95:1510–1518
32. Wright SJ, Robertson LE, O'Brien S et al (1994) The role of fludarabine in hematological malignancies. Blood Rev 8:125–134
33. Montagnoli A, Valsasina B, Croci V et al (2008) A Cdc7 kinase inhibitor restricts initiation of DNA replication and has antitumor activity. Nat Chem Biol 4:357–365
34. World Medical Organization (1996) Declaration of Helsinki. Br Med J 313:1448–1449

Chapter 15

Two-Dimensional vs. Three-Dimensional In Vitro Tumor Migration and Invasion Assays

Miriam Zimmermann, Carol Box, and Suzanne A. Eccles

Abstract

Motility and invasion are key hallmarks that distinguish benign from malignant tumors, enabling cells to cross tissue boundaries, disseminate in blood and lymph and establish metastases at distant sites. Similar properties are also utilized by activated endothelial cells during tumor-induced angiogenesis. It is now appreciated that these processes might provide a rich source of novel molecular targets with the potential for inhibitors to restrain both metastasis and neoangiogenesis. Such therapeutic strategies require assays that can rapidly and quantitatively measure cell movement and the ability to traverse physiological barriers. The need for high-throughput, however, must be balanced by assay designs that accommodate, as far as possible, the complexity of the in vivo tumor microenvironment. This chapter aims to give an overview of some commonly used migration and invasion assays to aid in the selection of a balanced portfolio of techniques for the rapid and accurate evaluation of novel therapeutic agents.

Key words: Migration, Haptotaxis, Chemotaxis, Invasion, Motility, Matrix protein, 3-dimensional cultures, Spheroid

Abbreviations

2-D	Two-dimensional
3-D	Three-dimensional
BBB	Blood–brain barrier
BME	Basement membrane extract
CAM	Chorioallantoic membrane
CAM-DR	Cell adhesion-mediated drug resistance
EC	Endothelial cell
ECM	Extracellular matrix
EGF	Epidermal growth factor
EHS	Engelbreth–Holm–Swarm
EMT	Epithelial-to-mesenchymal transition
FAK	Focal adhesion kinase
FCS	Fetal calf serum

Jürgen Moll and Riccardo Colombo (eds.), *Target Identification and Validation in Drug Discovery: Methods and Protocols*, Methods in Molecular Biology, vol. 986, DOI 10.1007/978-1-62703-311-4_15,

GBM	Glioblastoma
GFP	Green fluorescent protein
HCC	Hepatocellular carcinoma
HGF	Hepatocyte growth factor
HIF-1	Hypoxia-inducible factor-1
HT(S)	High-throughput (screening)
HUVEC	Human umbilical vein endothelial cell
MMP	Matrix metalloproteinase
MMPi	Matrix metalloproteinase inhibitors
PD	Pharmacodynamic
PET	Polyethylene terephthalate
RFP	Red fluorescent protein
s.c.	Subcutaneously
SCC	Squamous cell carcinoma
TIMP	Tissue inhibitor of metalloproteinases
VEGF	Vascular endothelial growth factor

1. Introduction: The Rationale for Studying Tumor Cell Migration and Invasion

The history of anticancer drug development has been characterized by the search for agents that inhibit the (generally accepted) primary "hallmark" of cancer: unrestrained cell proliferation. This has included the use of radiotherapy (mainly for loco-regional control) and systemic cytotoxic chemotherapy to kill rapidly dividing cells—with the inevitable consequence of damage to normal self-renewing tissues such as bone marrow and gut. Recently, attention has turned to the *causes* of malignant transformation. Strategies are being developed to target the aberrant cell signaling pathways that overcome normal growth control processes, with the aspiration of "personalized medicine" where each patient is treated according to the molecular abnormalities driving his or her cancer. Given the enormous heterogeneity of cancers, their ability to escape monotherapy via additional mutational events and the narrow therapeutic window of many agents, it is perhaps not surprising that progress against the major common cancers has been slow. What is more, it is *disseminated* malignant disease that is the cause of over 90 % of cancer deaths, yet identification and validation of therapeutic targets in metastases (and testing of new agents in appropriate experimental models) has so far been limited (1, 2). One exception is the use of bisphosphonates to treat bone metastases (3); nevertheless, these agents are not curative and other sites of metastasis, such as brain, are proving relatively intractable.

We now have a greater appreciation of the multiple hallmarks of cancer (4) and understand that tumors are not simply autonomous factories generating hyper-proliferating cells. Indeed, they depend upon many additional properties, several of which are dependent on the host and which may offer new vulnerabilities for therapeutic targeting (5). This paradigm shift, however, requires new approaches both to validate appropriate molecular targets (in the correct biological context) and to identify novel anticancer agents. There is therefore an increasing demand for higher throughput cell *functional* assays, notably those that accommodate properties of cancer cells beyond mere proliferation.

The aim of this chapter is to focus on two key attributes of malignant cells: the ability to move and to spread across tissue boundaries as essential components of invasion and metastasis. It includes a summary of the most accessible, simple, reductionist (2-D) assays which are appropriate for the higher throughput early stage studies, but also makes the case for the inclusion of more complex (3-D) assays and a description of those that can best be adapted for target validation and drug evaluation.

2. Mechanisms of Cell Motility and Invasion: Opportunities for Intervention and Key Limitations

Cell motility is regulated by extracellular signals in the microenvironment (5). Soluble factors activate cell surface receptors to induce directional migration/chemotaxis. This process mediated, for example, by chemokines, may contribute to tumor cell dissemination in the bloodstream and lodgment at specific secondary sites. Secondly, cells may engage matrix proteins via integrins to enable migration (a process sometimes termed haptotaxis) which may be more allied to local tissue invasion. Cell motility may provide additional, more selective targets than those involved in tumor cell survival and proliferation, since only certain specialized non-epithelial cells in the adult body are highly motile. Inhibitors have different mechanisms of action from (and could be complementary to) conventional cytotoxic drugs. Evidence suggests that there are common elements in the underlying molecular mechanisms of angiogenesis and invasion (6). Similar assays are therefore appropriate to measure both of these critical factors in malignant tumor progression (7), and dual function agents able to inhibit both angiogenesis and tumor invasion have been described (8, 9).

Tumor cells can utilize one or more types of cell motility depending on their environment and may thus escape from inhibitors by virtue of this plasticity. For example, elongated/mesenchymal and rounded/amoeboid motility with different requirements for proteolysis and Rac or Rho/ROCK signaling have been described (10, 11). This suggests that different cell motility assays will be required for comprehensive screening campaigns and that more than one inhibitor (or one that targets an essential common pathway) will be required for effective inhibition of tumor cell spread. These considerations must be taken into account in order to decide which type of assay is appropriate, depending on the intended clinical indication. For example, some tumors cause significant morbidity and mortality by virtue of their local invasive potential (e.g., oral cancers, glioblastoma); in others, the development of multiple benign lesions increases the risk of malignant progression to an invasive metastatic phenotype (e.g., bladder cancer, APC mutant colon cancer). In these examples, an agent that could delay or control motility, matrix, or tissue invasion could be of therapeutic benefit. In contrast, such intervention in highly malignant tumors, in which metastasis is often early and widespread (e.g., melanoma, some lymphomas), would probably not lead to significant survival benefit.

Although much has been learned from in vitro assays, we do not yet fully understand the predominant migratory mechanisms used by cancer cells in vivo. It is important that any molecular mediators (or their inhibitors) identified in one assay are tested in complementary assays and validated in appropriate in vivo models before they can be assumed to play a significant role in invasion and metastasis. There are several examples where a molecule can have either positive or negative regulatory roles in key cellular functions depending on the cellular/microenvironmental context (e.g., tissue inhibitors of matrix metalloproteinases: TIMPs (12)). Thus, care needs to be taken to avoid undesirable activities or, as in the example of some angiogenic inhibitors, compensatory mechanisms that result in adverse events (13).

3. Simple Two-Dimensional Migration Assays

3.1. Haptotaxis Assays

The most popular assay is the "scratch wound" assay, in which a confluent adherent cell monolayer is manually "wounded" using a pipette tip or, in an automated assay, using a silicon stamp or electrical current to avoid cell injury. Wound closure is monitored by periodic imaging or by time-lapse programs. An alternative method is to place cells into a circular ring inside a multiwell plate. On reaching confluence, the ring and unattached cells are removed and adherent cells migrate into the clear area. The assay is terminated by fixation and staining of the cells and quantified using imaging software. The automation of both types of assay has greatly improved, thereby allowing high-throughput screening (HTS). A 384-well wound-healing assay, coupled with automated microscopy and quantification, enabled an RNAi screen for inhibitors of ovarian carcinoma cell motility, illustrating its potential for target identification (14). However, this assay (using an inert substrate) is not very representative of in vivo invasion, and any genes identified will require additional validation in more pathologically relevant systems. Additionally, some types of cells (such as endothelial cells) are relatively translucent and are difficult to quantify using automated imaging systems. Most importantly, in these assays (which normally run over 24 h or more), cell proliferation needs to be taken into consideration. Tracking of single cells enables the differentiation between migration and proliferation; however, subpopulations of cells may have different migratory properties, and cell tracking is time-consuming and may not be adaptable to HTS and the identification/testing of new compounds.

3.2. Chemotaxis Assays

Chemotaxis is directional cell migration in an extracellular gradient of molecules such as chemokines or motogenic factors (15). While the Dunn chamber is considered the "gold standard" for true chemotaxis (since gradients are maintained during migration), it is not practical for routine target validation or drug testing, although it was used effectively to identify the PI3-kinase isoforms responsible for breast carcinoma cell migration (16). More recently, specialized microfluidic imaging chambers have been developed with enhanced imaging potential (17), but the most widely used methods are still based upon the original Boyden chamber design (18). Cells in suspension are added to an upper chamber separated from a lower chamber by a semipermeable membrane. The attractant in the lower chamber depends upon the cell type (e.g., VEGF is appropriate for endothelial cells), although 2–10 % fetal calf serum (FCS) is commonly used, as it provides both adhesive and chemotactic proteins. This works well for non- or weakly adherent cells such as lymphocytes (for which it was designed), lymphoid cells, and leukemic cancer cells, but requires modifications for use with cells that are normally adherent. When testing cell migration towards specific chemotactic factors (such as EGF or HGF), it is generally necessary to coat the membranes with a thin layer of matrix proteins to enable integrin ligation, as signals mediated by both substrate-bound and soluble factors are required for optimal directional migration. Each tumor cell line should be tested with a variety of protein coatings (such as collagens, fibronectin, hyaluronic acid, or laminin) and chemotactic factors, since their responses will depend upon which integrins and cell surface receptors they express (19) (see Chapter 16). Inevitably, their receptor repertoire will impact on the targeted agents to which they will respond.

One of the most practical filter chamber methods utilizes PET (polyethylene terephthalate) membranes that block fluorescence of labeled cells so that cell migration to the lower surface can be quantified in real time (20). Cells can be stained with fluorescent dyes or be genetically engineered to express GFP or RFP, thus allowing differentially labeled cells to be analyzed simultaneously. This is invaluable for comparing the behavior of cells pretreated with test and control siRNAs (for target validation) or different pharmacological inhibitors. It can also indicate whether the presence of one cell type influences the migration rate of another. Some cell lines expressing RFP or GFP are available commercially;

however, such cells are often fast-migrating sarcoma cells (e.g., HT1080) which may use different mechanisms of migration compared with epithelial cells.

4. Invasion Assays: From 2-D Towards 3-D Functional Assays

4.1. Filter-Based Assays

Transwell and Boyden chambers can be modified to test the potential of tumor cells to invade through various 3-D barriers. In the simplest variation, the filters are coated with a thick layer of ECM or proteins of interest which need to be proteolytically degraded prior to cell migration through the pores (21). The ECM is a natural substrate with a complex molecular composition. It is approximated in in vitro cultures by Matrigel™ (a product derived from the murine EHS tumor) or similar natural products (e.g., Cultrex™, ECMatrix). However, Matrigel-type products are closer to basement membranes than extracellular matrix. Individual ECM components or mixtures can also be used. More detailed descriptions of these types of standard assays are provided in the following reviews: (18, 19, 22). Invasion takes longer than chemotaxis, but quantitation can be achieved by the same means. The assay is relatively simple, and using fluorescently labeled cells, real-time dynamic readouts of invasion can be obtained. It also lends itself to target validation (using genetic engineering to overexpress/deplete the molecule of interest) or drug evaluation (compounds can be added to upper and/or lower chambers). However, variations in the thickness of the invasion barrier and nonuniform cell distribution over the surface can present technical issues. Recent studies have also shown the critical importance of the degree of stiffness of the matrix in potentiating (or impeding) invasion (23).

Variations on the filter-based assay have been designed to approximate more physiological contexts. Such assays include tumor cell "invasion" across a confluent cell monolayer (e.g., endothelial cells (EC) as a surrogate for intravasation or extravasation during hematogenous metastasis (24)) and ovarian carcinoma invasion of mesothelial cell monolayers (25). Additionally, 1 mm thick slices of human brain tissue have been used as a tissue barrier on Transwell filters with invasion of GFP-labeled glioma cells measured by confocal microscopy (26).

Other assays are being explored in HTS platforms to identify potential inhibitors of invasion, for example, compounds that affect the shape or morphology of cells or their ability to generate invadopodia. Quintavalle et al. recently described such a method initially using Src-transformed NIH 3T3 fibroblasts grown on 384-well optical plates. Imaging of cellular and nuclear morphology combined with phalloidin-stained F-actin was used to discriminate compounds which reduced (or enhanced) the number of invadopodia, without associated cytotoxicity. Interestingly, paclitaxel *increased* invadopodia and invasion (with potential implications for neoadjuvant therapy), and two novel inhibitors targeting Cdk5 were identified (27). Secondary assays included degradation of a thin film of fluorescently labeled gelatin and validation of some of the molecular targets by siRNA. Such platforms could be very useful for initial screening, although deconvolution with novel compounds of unknown target specificity would be challenging and predictability of in vivo efficacy is uncertain. This leads us to a consideration of other 3-D formats that can be used to measure tumor cell invasion.

4.2. Invasion of Single Cells into 3-D Matrices

In the basic form of this assay, cell suspensions are seeded onto a collagen gel and allowed to invade into the matrix (28). Measurement generally requires confocal microscopy or serial sectioning to determine depth of invasion. More advanced systems add fibroblasts to the gel or generate an asymmetric invasion assay by placing tumor cells (and/or fibroblasts) between two layers of collagen. The latter system was used to show that LOXL2 shRNA inhibited invasion, as measured by perpendicular sectioning of the gels and staining of the cells (29). Prostate carcinoma cells sandwiched between two layers of Matrigel™ developed 3-D structures, underwent EMT, and became invasive. Differences in morphology, gene expression, signaling pathway activation, and effects of drugs compared with 2-D were noted (30). Similarly, oral cancer invasion was demonstrated in an even more sophisticated system in which the tumor cells were separated from a "stroma" (oral mucosa fibroblasts embedded in collagen gel) by a "basement membrane" (Cultrex™ BME) in an OptiCell chamber. This model was used to show the contribution of hypoxia-regulated uPA and PAI-1 (31).

Transparent synthetic ECM hydrogel scaffolds are now available, for example, amino acid-based PuraMatrix synthetic nanofiber scaffolds. Their advantages are that they avoid the use of less defined animal-derived materials and can be altered to fit the purpose (e.g., addition of specific ECM proteins or growth factors or altered stiffness and porosity). These modifications can influence the migratory ability of cells and thus the results obtained in drug screening assays. Cross-linked hyaluronan hydrogels are also relevant to the tumor microenvironment, and this material proved significant in demonstrating the role of hyaluronidase activity in human tumor cell invasion at physiological pH (32).

Recently, a method has been developed to microinject cell–polymer suspensions into 3-D collagen matrices in a 96-well plate format. This was used to perform a proof-of-principle screen of known signaling pathway inhibitors in murine breast cancer cells. The cells were visualized by staining of the actin cytoskeleton and effects of the compounds measured using multiparametric imaging. The system also proved amenable to the use of enzymatically disaggregated tumor xenografts or human sarcoma samples (33). An alternative high-content 3-D tumor cell plate-based migration assay uses an array of microchannels filled with collagen. Cell migration into the channels monitored by imaging reportedly correlated with metastatic potential. Most of the steps could be automated, and the assay is compatible with a variety of multiparametric and/or kinetic readouts (34).

4.3. Invasion of Single Cells into Natural or Surrogate Tissues

Since the implementation of the classic embryonic chick heart invasion assay (35, 36), many alternative tissues have been used in in vitro invasion assays. Such assays reportedly are more predictive of in vivo behavior. For example, primary human uveal melanoma cells differed in their ability to traverse Matrigel™ or EC barriers with the latter more closely correlating with clinical characteristics associated with invasion (37). The skin is a site of invasion/metastasis for several tumor types, primarily melanoma and squamous cell carcinoma, but also other cancers such as breast and lymphoma. Natural and artificial skin equivalents have been devised to study this process (38). Briefly, skin fibroblasts are mixed with type I collagen gels which are then overlaid with the tumor cells of interest before being transferred onto grids covered with a nylon membrane. The grids are transferred to wells containing growth medium or chemoattractant with a tissue–air interface on top of the gel. After 1–2 weeks, the cultures are fixed and processed for immunohistochemical staining to measure the depth of invasion of the tumor cells into the gel. This assay allows comparisons of the behavior of different cell types and can also be used to test the effects of pharmacological agents (or genetic modifications) on tumor and/or stromal cells. Moreover, these organoids can actually be transplanted *s.c.* in mice, allowing true invasion (and also angiogenesis) to be measured using immunohistochemistry to identify tumor and host components. This delivers a useful intermediate between simple in vitro assays and tumor grafts.

Brain microvascular EC monolayers together with an astrocyte-enriched subendothelial collagen gel can be used to simulate the blood–brain barrier (BBB) (39). Such systems are invaluable to screen for compounds able to penetrate the BBB to access brain tumors or metastases. In addition, glioma cell invasion into brain fragments has enabled better understanding of the properties of these highly invasive tumors and identification of potential therapeutic targets (40).

Recently, there has been success in generating cocultures that more faithfully reproduce in vivo metastatic microenvironments. An ex vivo microscale liver perfusion bioreactor was used to assess metastatic seeding, mimicking the salient features of fluid dynamics and functionality of hepatic parenchyma. Invasion and subsequent growth of breast and prostate carcinoma cells were detected by two-photon microscopy of fluorescently labeled cells. Tumors failed to grow in the absence of the 3-D liver microtissue, suggesting paracrine or stromal support functions in "metastasis" establishment (41).

Bone is a significant site of metastasis for several cancer types. Hsaio et al. developed a two-layer microfluidic system comprising DsRed-labeled PC3 prostate carcinoma cells (for confocal fluorescence microscopy tracking), osteoblasts, and endothelial cells. The cells proliferated more slowly than in 2-D culture (a common feature of human prostate cancer), and putative cancer stem-like cells (CD133+) retained viability but remained largely quiescent (42). Others have developed 3-D biomaterials capable of supporting the proliferation of primary osteoblasts and endothelial cells, mainly for tissue engineering and bone repair indications (43), but these are now being exploited in bone metastasis models. Sieh et al. studied the interaction of prostate cancer cells and human osteoblasts in a tissue-engineered bone construct and found that tumor cell–bone matrix interactions led to increased levels of MMPs, steroidogenic enzymes, and PSA, all well-validated markers of prostate cancer progression (44). A specialized bioreactor has also enabled 3-D culture of isolated osteoblasts which generate an osteogenic "tissue" over several months. When metastatic GFP-labeled MDA-MB-231 cells were added, they invaded the thick ECM laid down by the osteoblasts in an "Indian file" pattern similar to that observed in some breast carcinomas (45).

In summary, the majority of these assays are technically challenging and more time-consuming than simple 2-D migration assays, but they are certainly closer to the in vivo situation. Cells moving in 2-D or 3-D are differentially influenced by the substrate and the receptor(s) engaged, and this will influence their sensitivity to inhibitors. Interestingly, differences in the regulation and dynamics of tumor cell motility compared with leukocyte motility in 3-D matrices have been reported, suggesting that the development of compounds that inhibit invasion, but spare immune functions, may be feasible (46).

5. Limitations of 2-D Tumor Cell Cultures and Advantages of 3-D Tumor Cell-Based Assays

A tumor is a highly complex three-dimensional (3-D) organoid, including not only malignant cells but also stromal cells and components of the immune system providing complex interactions and multidimensional signaling (47, 48). In addition, host endothelial cells are critical for the development of a functional blood supply essential for sustained tumor growth, dissemination, and metastasis. Gradients of oxygen and nutrients lead to varying levels of cellular stress and hypoxia which, together with close cell–cell/matrix interactions, are associated with increased resistance to therapeutic agents

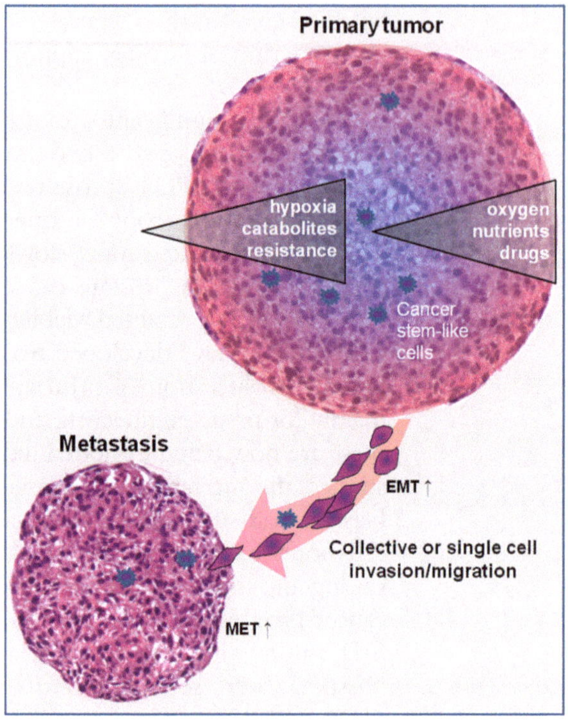

Fig. 1. Microenvironmental factors and the invasive process. The primary tumor is a heterogeneous mix of cell populations, further diversified by gradients of blood-borne nutrients, oxygen, and drugs. Hypoxia contributes to treatment resistance, upregulates pro-angiogenic and pro-invasive molecules, and helps to maintain cancer stem-like cell populations. Tumor cells may undergo epithelial-to-mesenchymal transition (EMT), enter blood vessels, and disseminate to distant sites where they extravasate, invade, and colonize the tissues. Once established, the cells may undergo the reverse program (mesenchymal-to-epithelial transition, MET) and proliferate to form metastases, the major reason for treatment failure.

(49–51) (Fig. 1). Many different pathways are involved, each of which has implications for the tumor cells' response to treatment.

There are challenges in accommodating these important factors in target validation and preclinical drug development, given the relatively high throughput required. In the current preclinical drug development "test cascade," primary in vitro screens generally measure the sensitivity of cells grown as monolayers on plastic in multiwell plates, with advantages of speed and low cost. Cells are uniformly exposed to 20 % oxygen in a nutrient-rich environment, which differs from the stressed, heterogeneous conditions within solid tumors (52). Therefore, current standard models may have very limited predictive power to accurately identify agents that will be efficacious in vivo, especially against invasive, metastatic cancers (53). On the other hand, animal models involve both high costs and ethical concerns and allow only low throughput. This has resulted in efforts to reduce the use of or replace in vivo models with improved and more predictive in vitro drug screening methods (54, 55).

Compounds tested in a more physiological 3-D environment show increased predictability of in vivo responses and help to span the gap between 2-D tissue culture and animal models. There is also evidence that 3-D cultures can identify active compounds that would fail to show their potential in 2-D (56, 57). The complexity of 3-D cultures can be increased by the addition of one or more cell types (fibroblasts, endothelial cells) and/or culture with different media, substrates, or oxygenation conditions. However, a disadvantage is that there is still a lack of simple, standardized, and reliable 3-D protocols that allow their incorporation into the preclinical high-throughput validation and drug evaluation process, although there have been several advances in this area in recent years (30, 58–61).

5.1. Altered Gene and Protein Expression in 3-D Compared with 2-D Tumor Cell Cultures

Target validation depends upon both correlative and functional studies: that is, evidence of the overexpression/activation of a particular molecule in cancers and also proof that by inhibiting its function (by genetic ablation or pharmacological means), the malignant phenotype is suppressed. Similarly, accurate evaluation of inhibitors depends upon the molecular target being expressed at appropriate levels and for the malignant phenotype being dependent on its functional activity. Several important differences in gene expression and signaling pathway activation have been reported in cells cultured in 3-D compared with 2-D, and this has led to significant differences in response to therapeutic agents (see Table 1).

Lai et al. have undertaken comparative transcriptomic and proteomic analyses to determine the simplest 3-D model that still captures the essence of the key in vivo microenvironmental cues. Their analyses of the literature revealed that inflammatory and angiogenic cytokines (often also implicated in cell motility and invasion phenotypes) were most frequently upregulated in 3-D (62) as exemplified by Ghosh (63) and Fischbach (64). Direct explants from human colorectal and lung cancers were shown to retain their native features and gene expression profiles and have proved useful (on a small scale) for pharmacodynamic (PD) profiling of anticancer drugs (65). The next two paragraphs exemplify tumor phenotypes which are more accurately reflected in 3-D cultures.

5.2. Microenvironmental Considerations

The aberrant microenvironment of tumor tissues contributes to the genetic and epigenetic changes that enhance oncogenesis, invasion, and the acquisition of stem-like cell capacities such as self-renewal and drug resistance (66–68). Hypoxia activates genes involved in growth factor signaling, angiogenesis, glycolysis, genetic instability, invasion, and metastasis, thus contributing to tumor growth and progression (69, 70). Hypoxia also mediates

Table 1
Examples of molecular changes and functional consequences of 3-D culture

Cell line	Tumor type[a]	Changes in 3-D compared with 2-D	Functional consequences	Reference
SKBR-3	Breast	ERBB2 mainly formed homodimers and signaled via MAPK–ERK. Increased activated integrin β4 and PAK2	3-D cultures more sensitive to trastuzumab, correlating with clinical results. Enhanced interactions with ECM	(106)
MDA-MB-231	Breast	Invasion dependent on strand-like collective cell migration, involving LIMK1 and 2	LIMK inhibitor effective in 3-D invasion assay, no effect in 2-D haptotaxis assay	(57)
HMT-3522	Breast	Malignant phenotype reversed by anti-integrin blocking mabs	No effect in 2-D	(107)
MMTV-K-Ras transgenic tumors	Murine Breast	Increased production of IL-6 and subsequent STAT3 activation	Matches in vivo expression; linked to drug resistance. STAT3 promotes invasion via EMT	(108)
EMT6	Murine Breast	Cell adhesion-mediated drug resistance (CAM-DR)	Seen in vivo, not in 2-D cultures	(109)
A2058 MM96L A11 D10	Melanoma	Melanospheres: high levels of reciprocal (not co-expressed) BRN2 or MITF	More similar to human melanomas in situ, in 2-D, molecules were co-expressed	(110)
NA8	Melanoma	Increase in multiple cytokines (e.g., CXCL1, 2, 3, IL-8, MIP3a, and angiopoietin-like 4)	Implicated in progression and metastasis	(63)
PC3	Prostate	Invasive (EMT) phenotype developed; STAT/interferon, PI3-kinase, AKT, and integrin signaling pathways became activated	PI3-kinase inhibitors blocked invasive cell growth more effectively in 3-D	(30)

PC3 HCT116 U87MG T47D MDA-MB-231	Prostate Colon Glioma Breast	Enhanced PI3K pathway signaling	Increased inhibition of proliferation and motility by PI3K inhibitors, linked to stronger inhibition of P-AKT	(111)
SMMC-7721	HCC	Enhanced expression of CD147, MMPs, FAK, and paxillin	Proteins implicated in invasion.	(112)
OSC-3 U87MG MDA-MB-231	Oral SCC Glioma Breast	IL-8 upregulated via integrin engagement of RGD adhesion peptides in a hydrogel matrix	IL-8 required for angiogenesis in vivo	(64)
NCI-H226 1° cultures	Mesothelioma Ascites	E-cadherin upregulated	Resistance to SS1P immunotoxin	(113)
HT1080	Sarcoma	Mesenchymal–amoeboid transition in motility with MMPi treatment	Invasion/motility of cells in 3-D and in vivo (intradermal) resistant to MMPi; sensitive in 2-D	(114)

[a]Human unless otherwise stated

resistance to radiotherapy and chemotherapeutic agents such as cisplatin and doxorubicin (71, 72) and therefore has to be taken into account when screening for new therapeutic agents. It has also been shown that hypoxic niches support cancer stem-like cells which can contribute to relapse, metastasis, and resistance to standard therapeutic agents (68, 73, 74).

Cancer stem-like cells have the ability to generate tumor spheroids in suspension cultures using defined (generally serum-free, growth factor-supplemented) media. Indeed, tumor-initiating cells of brain and colorectal tumors (and possibly others) can only be maintained in such specialized nonadherent cultures (75–77). These 3-D structures are generally termed neurospheres (see Chapter 17), colo(no)spheres, or mammospheres when derived from brain, gut, or breast tissue or tumors, respectively, to distinguish them from "spheroids" here defined as tumor cell-derived 3-D cultures grown in standard media. Colospheres derived from surgical specimens accurately reflected the aggressiveness of human colon cancers and correlated with tumorigenic potential in athymic mice. Colospheres were also more invasive in Matrigel™ than standard tumor spheroids and only the former expressed MMPs, which has been linked to invasive potential (78). Since cancer stem-like cells have been implicated in resistance to therapy and as the "seeds" of metastasis, it is critical to ensure that they are represented in cancer target validation and drug evaluation assays. Recently, the power of transgenic animals bearing tissue-targeted oncogenes has been combined with ex vivo culture of organoids (e.g., derived from gut) using Cre recombinase-inducible retrovirus vectors to allow the conditional manipulation of gene expression in primary cultures where stem cells can then be quantified (79).

6. 3-D Models

6.1. Methods of Obtaining 3-D Cultures

Although the focus of this chapter is primarily migration and invasion, it is necessary also to consider how tumor cells may be most effectively grown in 3-D, since these cultures can be used to generate organoids for subsequent functional assays (48, 59). A number of 3-D culture models have been developed: tissue explants, multicellular spheroids/aggregates, cultures using hollow-fiber bioreactors or involving scaffolding, micro-carriers, gels, or matrices (54, 80–85).

In cancer research, the multicellular tumor spheroid model is the most widely used. Spheroids are three-dimensional, often heterogeneous organoids containing gradients of cell proliferation, oxygen, and nutrients. Spheroids may be developed in suspension or in 3-D matrices. The former method has been adapted in various ways to include the use of spinner flasks (86), rotating cell culture systems/rotating wall vessels (82, 87), or liquid hanging drops (88). Low-binding plates (89, 90) or poly-HEMA (poly-2-hydroxyethyl methacrylate) coatings (91) also prevent cell adhesion to a substrate and promote aggregation. Some of these approaches have been successfully applied to high-throughput robotic liquid-handling technologies and automated quantitation of spheroids to determine responses to drugs. For example, Tung et al. used a 384-well hanging drop array to show that 5-FU was significantly less active than in 2-D monolayer cultures, but the hypoxia-activated compound tirapazamine was more potent (60).

For non-suspension cultures, suitable matrices include liquid overlay on agarose (59), Matrigel™, or Cultrex™ (48, 92). More recently, micropatterned arrays have been developed for adherent 3-D spheroid cultures (56) and have been used to show reduced chemosensitivity of colorectal carcinoma cells to irinotecan (58). In some cases, 3-D cultures can be enhanced by the addition of host cells. This increases complexity, but inevitably decreases flexibility and speed of analysis. However, important insights into the role of host cells have emerged: stromal cells modify the gene expression and response of many tumor cell types to chemotherapeutic agents (93) and tumor-associated myofibroblasts can enhance tumor invasiveness (94).

6.2. Motility and Invasion of Cells from 3-D Tumor Structures

Cells may invade individually or collectively; nevertheless, in vivo they emanate from a 3-D solid structure—the primary tumor. The use of migration/invasion assays with enzymatically disaggregated single-cell suspensions is convenient. However, even when the substrates into which these single cells migrate are three-dimensional, they lack the complexity of the solid tumor tissue microenvironment. It is therefore worth considering how this additional dimension may be introduced into an invasion assay in order to increase pathophysiological relevance, although this is likely to be at the expense of throughput.

Rotating culture vessels such as simulated microgravity systems are primarily used to study 3-D tumor growth and differentiation. However, mixed cell populations combined with matrix proteins can be used to generate a complex microenvironment in which cell–cell interactions and "invasion" can be measured (95). A similar system has also been described for the coculture of endothelial cells, myofibroblasts, and tumor cell clusters embedded in Matrigel™. Differential labeling of the cell populations enables their invasion and the effects of inhibitors to be measured (96).

We and others have based invasion assays on preformed tumor spheroids generated either in hanging drops (88) or other nonadherent systems (61). These are then transferred to matrix monolayers or embedded in Matrigel™ and can be analyzed qualitatively or quantitatively, usually by microscopy and image analysis (see also Chapter 16). A 96- or 1,536-well spheroid-based assay was recently used to screen for inhibitors of EMT (considered an important prerequisite of invasion in some systems) using triple-negative breast cancer cells. MDA-MB-231 cells were plated onto a layer of agarose and centrifuged to encourage aggregation. Cells were then covered with Matrigel™, allowed to form spheroids over 4 days, and used to screen a natural products compound library for inhibitors of EMT which yielded four novel hits (97). In this case, EMT was defined by expression of vimentin linked to a luciferase reporter, and cell viability was measured to exclude any cytotoxic compounds.

Bioengineering has contributed a plethora of new porous, synthetic scaffolds in which 3-D tumor behavior can be studied. Interestingly, human oral cancer cells grown in poly-lactide-co-glycolide (PLG) scaffolds showed an enhanced angiogenic profile (with a notable high expression of IL-8) and were highly invasive and drug resistant, much closer to the phenotype of tumors developing in vivo (98). Furthermore, specialized scaffolds have been developed to accommodate the behavior of particular tumor types. For example, gliomas invade through neural tissue, and this was mimicked in vitro by migration of cells from neurospheres or tumor explants along scaffolds formed of poly-ε-caprolactone nanofibers. This activity was associated with increased STAT3 signaling and could be reduced by STAT3 inhibitors which were ineffective in a conventional 2-D migration assay (99).

An interesting autologous invasion assay has been described using human glioblastoma (GBM) spheroids upon coculture in confrontation with dermal spheroids generated from samples taken from the skin incision required for tumor biopsy. Paired spheroids were placed on agar to facilitate contact and then cocultured in a rotating flask (100). Invasion was demonstrated histologically; however, this particular, very demanding setup does not seem to have been followed up for mechanistic studies. An alternative system has utilized 3-D cultures of rat brain cortical cells generated at a liquid–air interface (designated brain Hi-Spots (101)), incorporating GFP-labeled GBM cells. The cultures demonstrated electrophysiological activity and activation of astrocytes, and they were found to be relatively resistant to agents which failed in the clinic (cytosine arabinoside and taxol) despite strong activity in standard 2-D cultures. On the other hand, the standard-of-care agent temozolomide was more active in 3-D than 2-D, although activity was significantly reduced when the tumor cell clusters were well established (101). Thus, these complex cultures mimic several features of clinical GBM. This assay, established in a 48-well format, may prove useful to evaluate novel drugs (and combinations), especially once solid GBM "tumors" have formed in the 3-D cortical cell mass, but higher throughput and automation would be challenging. What is more, the key aspect of GBM (its highly invasive phenotype) is not reflected in this model.

Unfortunately, most of these assays involve several discrete steps, are time-consuming, and are not ideal for large-scale target validation studies (e.g., using shRNA libraries) or drug evaluation. Nevertheless, they more closely approximate solid tumors and can generate interesting insights into tumor biology. The salient features of 2-D and 3-D assays are summarized in Table 2 and representative examples are illustrated in Fig. 2.

6.3. Angiogenesis in 3-D

Additional and versatile systems have been developed including confrontation cultures of tumor spheroids with mouse embryoid bodies which enable a simultaneous assessment of tumor-induced angiogenesis and "tissue" invasion (61, 86). Explanted mouse aortic ring assays have also routinely been used to measure angiogenesis in 3-D, complementing the simpler tube-forming assays utilizing HUVEC or microvascular endothelial cells plated on matrix proteins (7). Bruyere et al. extended this to measure lymphangiogenesis in 3-D using mouse thoracic duct fragments embedded in a collagen gel. The network was analyzed by computerized image analysis and the use of genetically modified mice implicated MMP-2 (but not PAI-1) in the process (102). The system responded to known lymphangiogenic factors which have been implicated in lymphatic metastasis of some tumors and could theoretically be used to screen for inhibitors of lymphangiogenesis.

7. Concluding Remarks

A huge variety of migration and invasion assays are available, and additional modifications and novel technologies introduced in recent years have greatly improved their versatility and potential applications. However, their deployment in HTS target identification, validation, and drug discovery is still limited. How well do these different assays reflect in vivo tumor behavior? Matrigel™ invasion (but not migration through uncoated filters) of a panel of human squamous carcinoma cells was associated with their in vivo aggressiveness (103) and correlated reasonably well with the metastatic potential of many additional tumor cell types (18, 104). Decaestecker critically evaluated available assays of tumor cell migration and concluded that no single assay is sufficient to reliably accommodate all the different cell types and motility mechanisms and to reliably characterize drug efficacy (105). There is a need to improve assay standardization, to develop rapid and accurate quantitation, and to design robust test cascades to enable specific aspects of cell motility/invasion

Table 2
Salient features of common 2-D and 3-D invasion and migration assays

Method/assay	Format	Advantages	Disadvantages	Reference
Scratch wound	2-D	Simple, economical, can be automated for HT	Cell proliferation may contribute, translucency of certain cell types (e.g., ECs) hampers automation, 2-D migration may not correspond to 3-D or in vivo data	(115–118)
Direct-viewing chemotaxis chamber (Dunn, Zigmond, Insall)	2-D	Defined chemotactic gradients, direct cell chemotaxis visualization by microscope (also time-lapse)	Laborious/time-consuming, cannot be used for HT	(119–121)
Radial migration	2-D and 3-D	Well-defined starting area, intact/uninjured cells, colonization (migration + proliferation) in all directions, reproducibility, coating with ECM components possible	Cell proliferation not excluded, potentially time-consuming analysis, low throughput in 6-well plates	(105, 122, 123)
Transwell (Boyden chamber)	2-D and 3-D	2-D migration through filter pores; 3-D invasion through ECM layer and filter pores, amenable to different ECM components / additional cell layers (e.g., ECs) and organotypic tissue slices, migration with or against gravity, potential for HT	Variable migration across the invasion barrier if not carefully controlled for thickness and reproducibility of protein batches; laborious unless using fluorescent cells and automated imaging	(105, 124, 125)
2-D single-cell locomotion	2-D and 3-D	Specific for true migration, software development may enable higher throughput	Computer-assisted single-cell tracking requires specialist equipment, time-consuming, assay in 2-D often does not correspond to 3-D/in vivo data	(105)
Tumor cell migration out of spheroids onto ECM or along aligned nanofibers	2-D from 3-D	Amenable to different ECM components, nanofibers closer to native glioma cell migration, HT 96-well format (12 well for nanofibers)	Necessity of transferable (tight/robust) spheroids, no HT in case of nanofibers	(61, 99, 126)
Microcarrier beads	2-D and 3-D from 3-D	Medium throughput in 24-well format	Laborious, staining, and visual quantification, cannot be used for HT	(127)

CAM invasion	3-D from 2-D	Monitors tissue invasion and metastasis to organs in a physiological microenvironment	Accurate quantification difficult, poor correlation of results with metastasis in patients, radiolabeling of cells, laborious, requires animal tissue, variability, cannot be used for HT	(128, 129)
3-D collagen invasion	3-D from 2-D	Reliable and quantitative assessment of invasion (whole tumor cell population); addition of different ECM proteins possible	Laborious, multiple sections and staining required prior to quantification, low throughput in 8-well plates	(29)
Organotypic, e.g., mimicking skin or mucosa	3-D from 2-D	3-D/in vivo context, imaging using fluorescent cells possible, biological processes can be assessed (invasion, survival, proliferation, differentiation), antibody staining of fixed matrices possible, 24-well format	Slow (8-day contraction of 3-D matrix), invasion 0–21 days, proliferation not excluded, laborious, cannot be used in HT	(31)
3-D invasion of microinjection-derived spheroids in collagen	3-D from 3-D	Reproducible, automated, adaptable to other ECM components or fresh tumor biopsy material. 96-well system containing multiple spheroids/well can be used in HT	Specialist equipment needed, multiple spheroids/well can potentially lead to secreted chemo-attractants	(33)
3-D invasion from spherical structure into Matrigel™	3-D from 3-D	Reproducible, semiautomated, single spheroids/well, ECM components can be varied, HT in 96-well format	Central positioning of spheroids essential for automated quantification, Matrigel™ needs to be added slowly to maintain spheroids' central location	(61, 78, 130, 131)
Invasion from tumor explants into ECM	3-D from 3-D	Use of fresh tumor material recapitulates in vivo features/behavior, various ECM substrates can be used	Laborious, variable due to variation in tumor samples, cannot be used in HT format	(132)
Confrontation cultures between tumor spheroids and solid tissues	3-D from 3-D	Insight into biological processes and involved molecules, mimic of in vivo microenvironment via invasion into embryoid bodies or explanted tissues	Histological analysis and quantification necessary, low throughput	(133, 134)

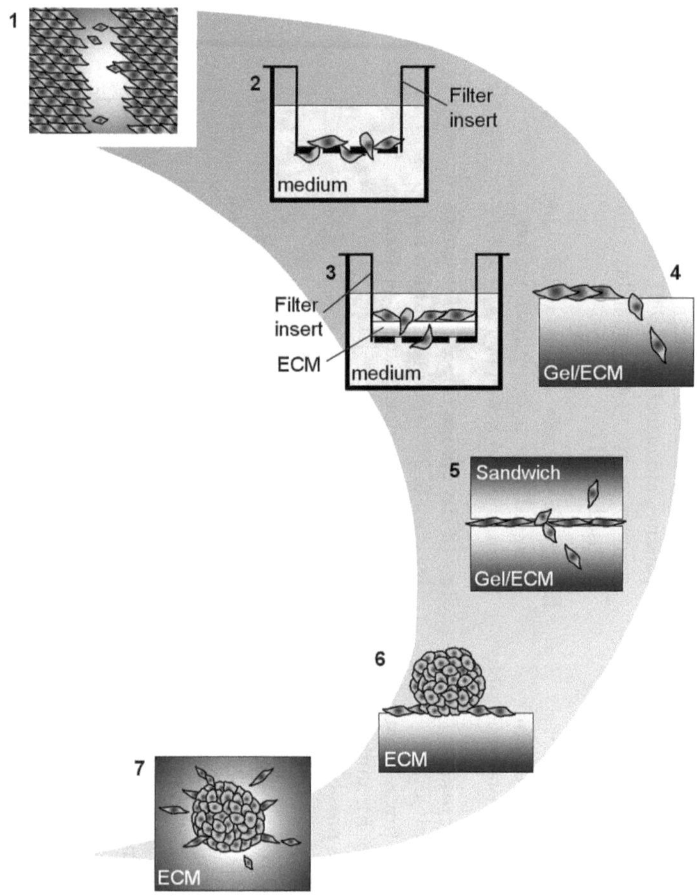

Fig. 2. Summary of key features of migration and invasion assays with differing degrees of two- and three-dimensionality. The scratch wound assay (1) is a 2-D migration assay measuring colonization of a cell-free area, while filter assays (2) can be adapted to incorporate a level of 3-D invasion through the use of a thick ECM layer (3), similar to invasion into a collagen gel (4). The sandwich assay (5) also incorporates a level of three-dimensionality by generating a cell monolayer between two ECM/gel layers. Using a 3-D structure (e.g., a spheroid) as source of cells, migration on (6) and/or invasion into (7) various matrices or tissues can be measured; the latter representing a truly three-dimensional assay (3-D to 3-D).

and associated molecular targets to be routinely assayed. There remain the further challenges of identifying appropriate molecular targets with a key role in inhibiting (rather than simply preventing) metastasis, identifying which subsets of cancer patients will benefit, and developing rapid, sensitive methods of measuring clinical responses. Rational drug development in this important area depends—last but not least—on the intelligent choice of preclinical in vivo model systems (1, 2).

Acknowledgments

The authors are funded by Cancer Research UK grant number C309/A8274 (S.E.) and by the Oracle Cancer Trust (M.Z. and C.B.). We acknowledge NHS funding to the NIHR Biomedical Research Centre. We thank Maria Vinci (funded by the National Centre for the Replacement, Refinement and Reduction of Animals in Research; G1000121 ID no. 94513) and other members of the Tumour Biology and Metastasis group for critical reading of the manuscript and constructive advice.

References

1. Francia G, Cruz-Munoz W, Man S et al (2011) Mouse models of advanced spontaneous metastasis for experimental therapeutics. Nat Rev Cancer 11:135–141
2. Eccles SA, Welch DR (2007) Metastasis: recent discoveries and novel treatment strategies. Lancet 369:1742–1757
3. Coleman R (2011) The use of bisphosphonates in cancer treatment. Ann N Y Acad Sci 1218:3–14
4. Hanahan D, Weinberg RA (2011) Hallmarks of cancer: the next generation. Cell 144:646–674
5. Palmer TD, Ashby WJ, Lewis JD et al (2011) Targeting tumor cell motility to prevent metastasis. Adv Drug Deliv Rev 63:568–581
6. Eccles SA (2004) Parallels in invasion and angiogenesis provide pivotal points for therapeutic intervention. Int J Dev Biol 48: 583–598
7. Eccles SA, Court W, Patterson L et al (2009) In vitro assays for endothelial cell functions related to angiogenesis: proliferation, motility, tubular differentiation, and proteolysis. Methods Mol Biol 467:159–181
8. Brader S, Eccles SA (2004) Phosphoinositide 3-kinase signalling pathways in tumor progression, invasion and angiogenesis. Tumori 90:2–8
9. Eccles SA, Massey A, Raynaud FI et al (2008) NVP-AUY922: a novel heat shock protein 90 inhibitor active against xenograft tumor growth, angiogenesis, and metastasis. Cancer Res 68:2850–2860
10. Friedl P, Gilmour D (2009) Collective cell migration in morphogenesis, regeneration and cancer. Nat Rev Mol Cell Biol 10:445–457
11. Sahai E, Marshall CJ (2003) Differing modes of tumour cell invasion have distinct requirements for Rho/ROCK signalling and extracellular proteolysis. Nat Cell Biol 5:711–719
12. Strongin AY (2010) Proteolytic and non-proteolytic roles of membrane type-1 matrix metalloproteinase in malignancy. Biochim Biophys Acta 1803:133–141
13. Ebos JM, Kerbel RS (2011) Antiangiogenic therapy: impact on invasion, disease progression, and metastasis. Nat Rev Clin Oncol 8:210–221
14. Collins CS, Hong J, Sapinoso L et al (2006) A small interfering RNA screen for modulators of tumor cell motility identifies MAP4K4 as a promigratory kinase. Proc Natl Acad Sci USA 103:3775–3780
15. Roussos ET, Condeelis JS, Patsialou A (2011) Chemotaxis in cancer. Nat Rev Cancer 11: 573–587
16. Sawyer C, Sturge J, Bennett DC et al (2003) Regulation of breast cancer cell chemotaxis by the phosphoinositide 3-kinase p110delta. Cancer Res 63:1667–1675
17. Breckenridge MT, Egelhoff TT, Baskaran H (2010) A microfluidic imaging chamber for the direct observation of chemotactic transmigration. Biomed Microdevices 12:543–553
18. Albini A, Noonan DM (2010) The 'chemoinvasion' assay, 25 years and still going strong: the use of reconstituted basement membranes to study cell invasion and angiogenesis. Curr Opin Cell Biol 22:677–689
19. Eccles SA, Box C, Court W (2005) Cell migration/invasion assays and their application in cancer drug discovery. Biotechnol Annu Rev 11:391–421
20. Maliakal JC (2002) Quantitative high throughput endothelial cell migration and invasion assay system. Methods Enzymol 352: 175–182
21. Marshall J (2011) Transwell((R)) invasion assays. Methods Mol Biol 769:97–110
22. van Roosmalen W, Le Devedec SE, Zovko S et al (2011) Functional screening with a live

cell imaging-based random cell migration assay. Methods Mol Biol 769:435–448
23. Ng MR, Brugge JS (2009) A stiff blow from the stroma: collagen crosslinking drives tumor progression. Cancer Cell 16:455–457
24. Kusama T, Mukai M, Tatsuta M et al (2006) Inhibition of transendothelial migration and invasion of human breast cancer cells by preventing geranylgeranylation of Rho. Int J Oncol 29:217–223
25. Burleson KM, Hansen LK, Skubitz AP (2004) Ovarian carcinoma spheroids disaggregate on type I collagen and invade live human mesothelial cell monolayers. Clin Exp Metastasis 21:685–697
26. Jung S, Kim HW, Lee JH et al (2002) Brain tumor invasion model system using organotypic brain-slice culture as an alternative to in vivo model. J Cancer Res Clin Oncol 128:469–476
27. Quintavalle M, Elia L, Price JH et al (2011) A cell-based high-content screening assay reveals activators and inhibitors of cancer cell invasion. Sci Signal 4:ra49
28. Nystrom ML, Thomas GJ, Stone M et al (2005) Development of a quantitative method to analyse tumour cell invasion in organotypic culture. J Pathol 205:468–475
29. Brekhman V, Neufeld G (2009) A novel asymmetric 3D in-vitro assay for the study of tumor cell invasion. BMC Cancer 9:415
30. Harma V, Virtanen J, Makela R et al (2010) A comprehensive panel of three-dimensional models for studies of prostate cancer growth, invasion and drug responses. PLoS One 5:e10431
31. Duong HS, Le AD, Zhang Q et al (2005) A novel 3-dimensional culture system as an in vitro model for studying oral cancer cell invasion. Int J Exp Pathol 86:365–374
32. David L, Dulong V, Le Cerf D et al (2004) Reticulated hyaluronan hydrogels: a model for examining cancer cell invasion in 3D. Matrix Biol 23:183–193
33. Truong HH, de Sonneville J, Ghotra VP et al (2012) Automated microinjection of cell-polymer suspensions in 3D ECM scaffolds for high-throughput quantitative cancer invasion screens. Biomaterials 33:181–188
34. Echeverria V, Meyvantsson I, Skoien A et al (2010) An automated high-content assay for tumor cell migration through 3-dimensional matrices. J Biomol Screen 15:1144–1151
35. Mareel MM, Van Roy FM, Messiaen LM et al (1987) Qualitative and quantitative analysis of tumour invasion in vivo and in vitro. J Cell Sci Suppl 8:141–163
36. Bracke ME, Boterberg T, Mareel MM (2001) Chick heart invasion assay. Methods Mol Med 58:91–102
37. Woodward JK, Nichols CE, Rennie IG et al (2002) An in vitro assay to assess uveal melanoma invasion across endothelial and basement membrane barriers. Invest Ophthalmol Vis Sci 43:1708–1714
38. Kataoka T, Umeda M, Shigeta T et al (2010) A new in vitro model of cancer invasion using AlloDerm, a human cadaveric dermal equivalent: a preliminary report. Kobe J Med Sci 55:E106–E115
39. Andjelkovic AV, Zochowski MR, Morgan F et al (2001) Qualitative and quantitative analysis of monocyte transendothelial migration by confocal microscopy and three-dimensional image reconstruction. In Vitro Cell Dev Biol Anim 37:111–120
40. Pilkington GJ, Bjerkvig R, De Ridder L et al (1997) In vitro and in vivo models for the study of brain tumour invasion. Anticancer Res 17:4107–4109
41. Yates C, Shepard CR, Papworth G et al (2007) Novel three-dimensional organotypic liver bioreactor to directly visualize early events in metastatic progression. Adv Cancer Res 97: 225–246
42. Hsiao AY, Torisawa YS, Tung YC et al (2009) Microfluidic system for formation of PC-3 prostate cancer co-culture spheroids. Biomaterials 30:3020–3027
43. Unger RE, Halstenberg S, Sartoris A et al (2011) Human endothelial and osteoblast co-cultures on 3D biomaterials. Methods Mol Biol 695:229–241
44. Sieh S, Lubik AA, Clements JA et al (2010) Interactions between human osteoblasts and prostate cancer cells in a novel 3D in vitro model. Organogenesis 6:181–188
45. Mastro AM, Vogler EA (2009) A three-dimensional osteogenic tissue model for the study of metastatic tumor cell interactions with bone. Cancer Res 69:4097–4100
46. Niggemann B, Drell TL IV, Joseph J et al (2004) Tumor cell locomotion: differential dynamics of spontaneous and induced migration in a 3D collagen matrix. Exp Cell Res 298:178–187
47. Joyce JA, Pollard JW (2009) Microenvironmental regulation of metastasis. Nat Rev Cancer 9:239–252
48. Nyga A, Cheema U, Loizidou M (2011) 3D tumour models: novel in vitro approaches to cancer studies. J Cell Commun Signal 5: 239–248

49. Shirinifard A, Gens JS, Zaitlen BL et al (2009) 3D multi-cell simulation of tumor growth and angiogenesis. PLoS One 4:e7190
50. Smalley KS, Lioni M, Herlyn M (2006) Life isn't flat: taking cancer biology to the next dimension. In Vitro Cell Dev Biol Anim 42:242–247
51. Green SK, Francia G, Isidoro C et al (2004) Antiadhesive antibodies targeting E-cadherin sensitize multicellular tumor spheroids to chemotherapy in vitro. Mol Cancer Ther 3:149–159
52. Birgersdotter A, Sandberg R, Ernberg I (2005) Gene expression perturbation in vitro—a growing case for three-dimensional (3D) culture systems. Semin Cancer Biol 15:405–412
53. Cukierman E, Pankov R, Stevens DR et al (2001) Taking cell-matrix adhesions to the third dimension. Science 294:1708–1712
54. Kunz-Schughart LA, Freyer JP, Hofstaedter F et al (2004) The use of 3-D cultures for high-throughput screening: the multicellular spheroid model. J Biomol Screen 9:273–285
55. Takagi A, Watanabe M, Ishii Y et al (2007) Three-dimensional cellular spheroid formation provides human prostate tumor cells with tissue-like features. Anticancer Res 27: 45–53
56. Hirschhaeuser F, Menne H, Dittfeld C et al (2010) Multicellular tumor spheroids: an underestimated tool is catching up again. J Biotechnol 148:3–15
57. Scott RW, Hooper S, Crighton D et al (2010) LIM kinases are required for invasive path generation by tumor and tumor-associated stromal cells. J Cell Biol 191:169–185
58. Hardelauf H, Frimat JP, Stewart JD et al (2011) Microarrays for the scalable production of metabolically relevant tumour spheroids: a tool for modulating chemosensitivity traits. Lab Chip 11:419–428
59. Friedrich J, Seidel C, Ebner R et al (2009) Spheroid-based drug screen: considerations and practical approach. Nat Protoc 4:309–324
60. Tung YC, Hsiao AY, Allen SG et al (2011) High-throughput 3D spheroid culture and drug testing using a 384 hanging drop array. Analyst 136:473–478
61. Vinci M, Gowan S, Boxall F et al (2012) Advances in establishment and analysis of 3D tumour spheroid-based functional assays for target validation and drug evaluation. BMC Biol 10:29. doi:10.1186/1741-7007-10-29
62. Lai Y, Asthana A, Kisaalita WS (2011) Biomarkers for simplifying HTS 3D cell culture platforms for drug discovery: the case for cytokines. Drug Discov Today 16:293–297
63. Ghosh S, Spagnoli GC, Martin I et al (2005) Three-dimensional culture of melanoma cells profoundly affects gene expression profile: a high density oligonucleotide array study. J Cell Physiol 204:522–531
64. Fischbach C, Kong HJ, Hsiong SX et al (2009) Cancer cell angiogenic capability is regulated by 3D culture and integrin engagement. Proc Natl Acad Sci U S A 106:399–404
65. Vaira V, Fedele G, Pyne S et al (2010) Preclinical model of organotypic culture for pharmacodynamic profiling of human tumors. Proc Natl Acad Sci USA 107:8352–8356
66. Keith B, Simon MC (2007) Hypoxia-inducible factors, stem cells, and cancer. Cell 129:465–472
67. Rodriguez-Jimenez FJ, Moreno-Manzano V, Lucas-Dominguez R et al (2008) Hypoxia causes downregulation of mismatch repair system and genomic instability in stem cells. Stem Cells 26:2052–2062
68. Lin Q, Yun Z (2010) Impact of the hypoxic tumor microenvironment on the regulation of cancer stem cell characteristics. Cancer Biol Ther 9:949–956
69. Benita Y, Kikuchi H, Smith AD et al (2009) An integrative genomics approach identifies hypoxia inducible factor-1 (HIF-1)-target genes that form the core response to hypoxia. Nucleic Acids Res 37:4587–4602
70. Lu X, Kang Y (2010) Hypoxia and hypoxia-inducible factors: master regulators of metastasis. Clin Cancer Res 16:5928–5935
71. DeClerck K, Elble RC (2010) The role of hypoxia and acidosis in promoting metastasis and resistance to chemotherapy. Front Biosci 15:213–225
72. Rohwer N, Cramer T (2011) Hypoxia-mediated drug resistance: novel insights on the functional interaction of HIFs and cell death pathways. Drug Resist Updat 14:191–201
73. Li Z, Rich JN (2010) Hypoxia and hypoxia inducible factors in cancer stem cell maintenance. Curr Top Microbiol Immunol 345: 21–30
74. Yu H, Zhang CM, Wu YS (2010) Research progress in cancer stem cells and their drug resistance. Chin J Cancer 29:261–264
75. Ricci-Vitiani L, Lombardi DG, Pilozzi E et al (2007) Identification and expansion of human colon-cancer-initiating cells. Nature 445:111–115
76. Bartholoma P, Impidjati, Reininger-Mack A et al (2005) A more aggressive breast cancer spheroid model coupled to an electronic capillary sensor system for a high-content screening of cytotoxic agents in cancer

therapy: 3-dimensional in vitro tumor spheroids as a screening model. J Biomol Screen 10:705–714
77. Zhang X, Wang W, Yu W et al (2005) Development of an in vitro multicellular tumor spheroid model using microencapsulation and its application in anticancer drug screening and testing. Biotechnol Prog 21:1289–1296
78. Weiswald LB, Richon S, Validire P et al (2009) Newly characterised ex vivo colospheres as a three-dimensional colon cancer cell model of tumour aggressiveness. Br J Cancer 101:473–482
79. Koo BK, Stange DE, Sato T et al (2012) Controlled gene expression in primary Lgr5 organoid cultures. Nat Methods 9:81–83
80. Kim JB (2005) Three-dimensional tissue culture models in cancer biology. Semin Cancer Biol 15:365–377
81. Pampaloni F, Reynaud EG, Stelzer EH (2007) The third dimension bridges the gap between cell culture and live tissue. Nat Rev Mol Cell Biol 8:839–845
82. Mazzoleni G, Di Lorenzo D, Steimberg N (2009) Modelling tissues in 3D: the next future of pharmaco-toxicology and food research? Genes Nutr 4:13–22
83. Haycock JW (2011) 3D cell culture: a review of current approaches and techniques. Methods Mol Biol 695:1–15
84. Prestwich GD, Liu Y, Yu B et al (2007) 3-D culture in synthetic extracellular matrices: new tissue models for drug toxicology and cancer drug discovery. Adv Enzyme Regul 47:196–207
85. Yamada KM, Cukierman E (2007) Modeling tissue morphogenesis and cancer in 3D. Cell 130:601–610
86. Wartenberg M, Donmez F, Ling FC et al (2001) Tumor-induced angiogenesis studied in confrontation cultures of multicellular tumor spheroids and embryoid bodies grown from pluripotent embryonic stem cells. FASEB J 15:995–1005
87. Unsworth BR, Lelkes PI (1998) Growing tissues in microgravity. Nat Med 4:901–907
88. Del Duca D, Werbowetski T, Del Maestro RF (2004) Spheroid preparation from hanging drops: characterization of a model of brain tumor invasion. J Neurooncol 67:295–303
89. Eicher C, Dewerth A, Kirchner B et al (2011) Development of a drug resistance model for hepatoblastoma. Int J Oncol 38:447–454
90. Zhang Q, Nguyen AL, Shi S et al (2011) Three-dimensional spheroid culture of human gingiva-derived mesenchymal stem cells enhances mitigation of chemotherapy-induced oral mucositis. Stem Cells Dev 21(6):937–947
91. Ivascu A, Kubbies M (2006) Rapid generation of single-tumor spheroids for high-throughput cell function and toxicity analysis. J Biomol Screen 11:922–932
92. Lee GY, Kenny PA, Lee EH et al (2007) Three-dimensional culture models of normal and malignant breast epithelial cells. Nat Methods 4:359–365
93. McMillin DW, Delmore J, Weisberg E et al (2010) Tumor cell-specific bioluminescence platform to identify stroma-induced changes to anticancer drug activity. Nat Med 16:483–489
94. Holliday DL, Brouilette KT, Markert A et al (2009) Novel multicellular organotypic models of normal and malignant breast: tools for dissecting the role of the microenvironment in breast cancer progression. Breast Cancer Res 11:R3
95. Rhee HW, Zhau HE, Pathak S et al (2001) Permanent phenotypic and genotypic changes of prostate cancer cells cultured in a three-dimensional rotating-wall vessel. In Vitro Cell Dev Biol Anim 37:127–140
96. Walter-Yohrling J, Pratt BM, Ledbetter S et al (2003) Myofibroblasts enable invasion of endothelial cells into three-dimensional tumor cell clusters: a novel in vitro tumor model. Cancer Chemother Pharmacol 52:263–269
97. Li Q, Chen C, Kapadia A et al (2011) 3D models of epithelial-mesenchymal transition in breast cancer metastasis: high-throughput screening assay development, validation, and pilot screen. J Biomol Screen 16:141–154
98. Fischbach C, Chen R, Matsumoto T et al (2007) Engineering tumors with 3D scaffolds. Nat Methods 4:855–860
99. Agudelo-Garcia PA, De Jesus JK, Williams SP et al (2011) Glioma cell migration on three-dimensional nanofiber scaffolds is regulated by substrate topography and abolished by inhibition of STAT3 signaling. Neoplasia 13: 831–840
100. de Ridder L, Cornelissen M, de Ridder D (2000) Autologous spheroid culture: a screening tool for human brain tumour invasion. Crit Rev Oncol Hematol 36:107–122
101. Biggs T, Foreman J, Sundstrom L et al (2011) Antitumor compound testing in glioblastoma organotypic brain cultures. J Biomol Screen 16:805–817
102. Bruyere F, Melen-Lamalle L, Blacher S et al (2008) Modeling lymphangiogenesis in a three-dimensional culture system. Nat Methods 5:431–437
103. O-Charoenrat P, Rhys-Evans P, Modjtahedi H et al (2000) Overexpression of epidermal growth factor receptor in human head and

neck squamous carcinoma cell lines correlates with matrix metalloproteinase-9 expression and in vitro invasion. Int J Cancer 86: 307–317

104. Kleinman HK, Jacob K (2001) Invasion assays. Curr Protoc Cell Biol Chapter 12:Unit 12 12

105. Decaestecker C, Debeir O, Van Ham P et al (2007) Can anti-migratory drugs be screened in vitro? A review of 2D and 3D assays for the quantitative analysis of cell migration. Med Res Rev 27:149–176

106. Pickl M, Ries CH (2009) Comparison of 3D and 2D tumor models reveals enhanced HER2 activation in 3D associated with an increased response to trastuzumab. Oncogene 28:461–468

107. Weaver VM, Petersen OW, Wang F et al (1997) Reversion of the malignant phenotype of human breast cells in three-dimensional culture and in vivo by integrin blocking antibodies. J Cell Biol 137:231–245

108. Leslie K, Gao SP, Berishaj M et al (2010) Differential interleukin-6/Stat3 signaling as a function of cellular context mediates Ras-induced transformation. Breast Cancer Res 12:R80

109. Kobayashi H, Man S, Graham CH et al (1993) Acquired multicellular-mediated resistance to alkylating agents in cancer. Proc Natl Acad Sci USA 90:3294–3298

110. Thurber AE, Douglas G, Sturm EC et al (2011) Inverse expression states of the BRN2 and MITF transcription factors in melanoma spheres and tumour xenografts regulate the NOTCH pathway. Oncogene 30:3036–3048

111. Howes AL, Chiang GG, Lang ES et al (2007) The phosphatidylinositol 3-kinase inhibitor, PX-866, is a potent inhibitor of cancer cell motility and growth in three-dimensional cultures. Mol Cancer Ther 6:2505–2514

112. Wu YM, Tang J, Zhao P et al (2009) Morphological changes and molecular expressions of hepatocellular carcinoma cells in three-dimensional culture model. Exp Mol Pathol 87:133–140

113. Xiang X, Phung Y, Feng M et al (2011) The development and characterization of a human mesothelioma in vitro 3D model to investigate immunotoxin therapy. PLoS One 6:e14640

114. Wolf K, Mazo I, Leung H et al (2003) Compensation mechanism in tumor cell migration: mesenchymal-amoeboid transition after blocking of pericellular proteolysis. J Cell Biol 160:267–277

115. Pinco KA, He W, Yang JT (2002) alpha4beta1 integrin regulates lamellipodia protrusion via a focal complex/focal adhesion-independent mechanism. Mol Biol Cell 13:3203–3217

116. Keese CR, Wegener J, Walker SR et al (2004) Electrical wound-healing assay for cells in vitro. Proc Natl Acad Sci USA 101:1554–1559

117. Lampugnani MG (1999) Cell migration into a wounded area in vitro. Methods Mol Biol 96:177–182

118. Yarrow JC, Perlman ZE, Westwood NJ et al (2004) A high-throughput cell migration assay using scratch wound healing, a comparison of image-based readout methods. BMC Biotechnol 4:21

119. Zigmond SH (1977) Ability of polymorphonuclear leukocytes to orient in gradients of chemotactic factors. J Cell Biol 75:606–616

120. Zicha D, Dunn GA, Brown AF (1991) A new direct-viewing chemotaxis chamber. J Cell Sci 99(Pt 4):769–775

121. Muinonen-Martin AJ, Veltman DM, Kalna G et al (2010) An improved chamber for direct visualisation of chemotaxis. PLoS One 5:e15309

122. Cai G, Lian J, Shapiro SS et al (2000) Evaluation of endothelial cell migration with a novel in vitro assay system. Methods Cell Sci 22:107–114

123. Pratt BM, Harris AS, Morrow JS et al (1984) Mechanisms of cytoskeletal regulation. Modulation of aortic endothelial cell spectrin by the extracellular matrix. Am J Pathol 117: 349–354

124. Albini A, Iwamoto Y, Kleinman HK et al (1987) A rapid in vitro assay for quantitating the invasive potential of tumor cells. Cancer Res 47:3239–3245

125. Ohnishi T, Matsumura H, Izumoto S et al (1998) A novel model of glioma cell invasion using organotypic brain slice culture. Cancer Res 58:2935–2940

126. Deryugina EI, Bourdon MA (1996) Tenascin mediates human glioma cell migration and modulates cell migration on fibronectin. J Cell Sci 109(Pt 3):643–652

127. Rosen EM, Meromsky L, Setter E et al (1990) Quantitation of cytokine-stimulated migration of endothelium and epithelium by a new assay using microcarrier beads. Exp Cell Res 186:22–31

128. Hart IR, Fidler IF (1978) An in vitro quantitative assay for tumor cell invasion. Cancer Res 38:3218–3224

129. Armstrong PB, Quigley JP, Sidebottom E (1982) Transepithelial invasion and intramesenchymal infiltration of the chick embryo chorioallantois by tumor cell lines. Cancer Res 42:1826–1837

130. An Z, Gluck CB, Choy ML et al (2010) Suberoylanilide hydroxamic acid limits migration and invasion of glioma cells in two and three dimensional culture. Cancer Lett 292: 215–227
131. Stein AM, Demuth T, Mobley D et al (2007) A mathematical model of glioblastoma tumor spheroid invasion in a three-dimensional in vitro experiment. Biophys J 92:356–365
132. Friedl P, Noble PB, Walton PA et al (1995) Migration of coordinated cell clusters in mesenchymal and epithelial cancer explants in vitro. Cancer Res 55:4557–4560
133. Wartenberg M, Finkensieper A, Hescheler J et al (2006) Confrontation cultures of embryonic stem cells with multicellular tumor spheroids to study tumor-induced angiogenesis. Methods Mol Biol 331:313–328
134. Fjellbirkeland L, Bjerkvig R, Laerum OD (1998) Non-small-cell lung carcinoma cells invade human bronchial mucosa in vitro. In Vitro Cell Dev Biol Anim 34:333–340

Chapter 16

Tumor Spheroid-Based Migration Assays for Evaluation of Therapeutic Agents

Maria Vinci, Carol Box, Miriam Zimmermann, and Suzanne A. Eccles

Abstract

Cell migration is a key hallmark of malignant cells that contributes to the progression of cancers from a primary, localized mass to an invasive and/or metastatic phenotype. Traditional methods for the evaluation of tumor cell migration in vitro generally employ two-dimensional (2D), homogeneous cultures that do not take into account tumor heterogeneity, three-dimensional (3D) cell-cell contacts between tumor and/or host cells or interactions with extracellular matrix proteins. Here we describe a 3D tumor spheroid-based migration assay which more accurately reflects the solid tumor microenvironment and can accommodate both extracellular matrix and host cell interactions. It is a rapid and highly reproducible 96-well plate-based technique and we demonstrate its utility for the evaluation of therapeutic agents/drugs with anti-migratory properties.

Key words: Migration, 3D, Tumor spheroids, Extracellular matrix, Drug response, Adhesion, Microenvironment

1. Introduction

Cell migration is a process that occurs in many physiological and pathophysiological events: morphogenesis (e.g., movement of cells in the inner blastocyst that form the embryonic layers), wound healing (e.g., epithelial cells and fibroblasts migrate and proliferate to close the wound), angiogenesis (e.g., endothelial cells that sprout, migrate, and repair the endometrium during the menstrual cycle), inflammation (e.g., leukocyte recruitment in the bloodstream and infiltration into the inflammatory site), and cancer (mobilization of the metastatic cells from the primary site and colonization of the secondary sites) (1–3).

In tumor progression, the acquisition of a motile/migratory phenotype is a process that generally coincides with the switch to a malignant phase. This is achieved either by local invasion/infiltration

Jürgen Moll and Riccardo Colombo (eds.), *Target Identification and Validation in Drug Discovery: Methods and Protocols*, Methods in Molecular Biology, vol. 986, DOI 10.1007/978-1-62703-311-4_16,

into the surrounding tissue, a predominant feature of, for example, glioblastoma (4) and squamous cell carcinoma of the head and neck (SCCHN) (5), or by colonization of distant tissues, for example, breast carcinoma where bone, lung, liver, or brain metastases are common (6). Increased malignancy is also frequently associated with an epithelial–mesenchymal transition (EMT) resulting in enhanced motility (7) and the generation of cancer stem(like) cells (8). In all cases, the fundamental steps of tumor cell invasion/dissemination include (1) changes in adhesion to extracellular matrix (ECM) components and/or changes in cell–cell adhesion, (2) motility which involves primarily the dynamic remodeling of the actin cytoskeleton, and (3) invasion into the surrounding tissue through proteolytic degradation of the ECM (4). Tumor cell migration is driven by responses to hypoxia (9), soluble mediators (e.g., cytokines and growth factors) interactions with stromal cells (10), common features of most solid tumors.

Since cell motility is an essential feature of both invasion and dissemination, in this chapter, we have focused on providing a simple and rapid assay of this phenotype (suitable for target validation and/or evaluation of therapeutic agents) which nevertheless lends itself to more detailed analysis of the molecular and cellular processes involved. The assay consists of transferring preformed tumor spheroids onto ECM-coated 96-well flat-bottomed plates (one spheroid per well) and following the migration/dissemination of tumor cells over time (Fig. 1). Recent advances in the generation of large populations of reproducibly sized tumor spheroids have significantly contributed to the feasibility and reproducibility of such assays (11–15) compared with previously reported methods (16, 17). Using our method, spheroids can be transferred to "migration plates" in a single step by the use of a multichannel pipette (Fig. 1).

Among the different standard 2D chemotaxis and migration assays available (e.g., Transwell/Boyden chambers, wound healing assay), the method described here has the advantage that the tumor spheroid represents a 3D culture model better able to recapitulate the heterogeneous physiological features of a solid tumor (11, 18, 19) (and see chapter 15). The ECM regulates several key cellular processes such as cell differentiation, proliferation, survival, and, last but not least, adhesion and migration (4). Tissue-specific matrices will determine a favorable environment for tumor cells to migrate (4, 16, 20) and metastasize (21). This is well exemplified in Figs. 2 and 3, where two derivatives of CAL 27 (SCCHN cell line), CAL^S and CAL^R, respectively, sensitive or resistant to epidermal growth factor receptor (EGFR) tyrosine kinase inhibitors (TKI), show contrasting migration patterns. CAL^R cells are generally more aggressive in vivo and are more highly motile in vitro than CAL^S on all of the ECM proteins tested, characteristics that were not evident in standard 2D functional assays. We observed that CAL^S migration was most extensive on

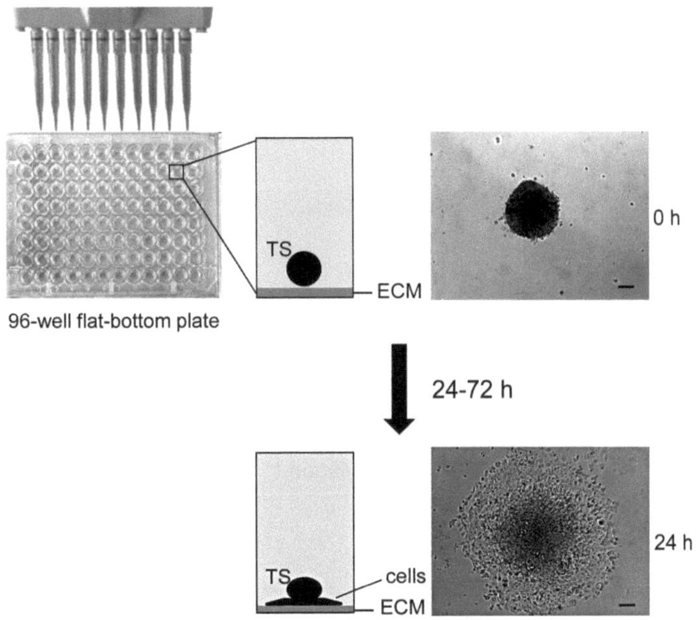

Fig. 1. Schematic overview of the tumor spheroid-based migration assay. Tumor spheroids (TS) are transferred from their culture vessel or plate into a 96-well flat-bottomed migration plate pre-coated with an extracellular matrix (ECM) protein of choice (in this case, gelatin). Digital images of the spheroids are then captured at t = 0 and once every 24 h for a period of up to 72 h, exemplified here by CAL^R spheroids. Image analysis software is used to calculate the spheroid size and extent of migration. Scale bar = 100 μm.

fibronectin while CAL^R cells were more migratory on gelatin or collagen I than on other substrates tested. This illustrates the versatility of the assay to reveal both phenotypic differences and also provide a platform for the exploration of the different (presumably integrin-linked) signaling networks which underlie this behavior and which could potentially be targeted for therapeutic intervention (Fig. 4).

Hypoxia is a common tumor microenvironmental condition that can enhance progression by promoting angiogenesis, invasion, and limiting responses to therapy (9, 22). The LICR-LON-HN4 (HN4) SCCHN cell line (23) is significantly more migratory on a collagen matrix in hypoxia (1% O_2) than in normoxia (20% O_2), at all time points monitored (Fig. 4a). Cisplatin is a standard component of the most effective SCCHN chemotherapy regimens (24) but negatively influenced by hypoxia. This is also reflected in the growth/proliferation of HN4 tumor spheroids when treated in hypoxia compared to normoxia (as determined by the drug concentration inhibiting 50% of growth (GI_{50}): 37.4 μM in hypoxia and 3.7 μM in normoxia). Nevertheless, treatment with cisplatin in hypoxia shows significant inhibition of HN4 cell *migration* at concentrations far below those that inhibit proliferation in 3D, suggesting that cisplatin

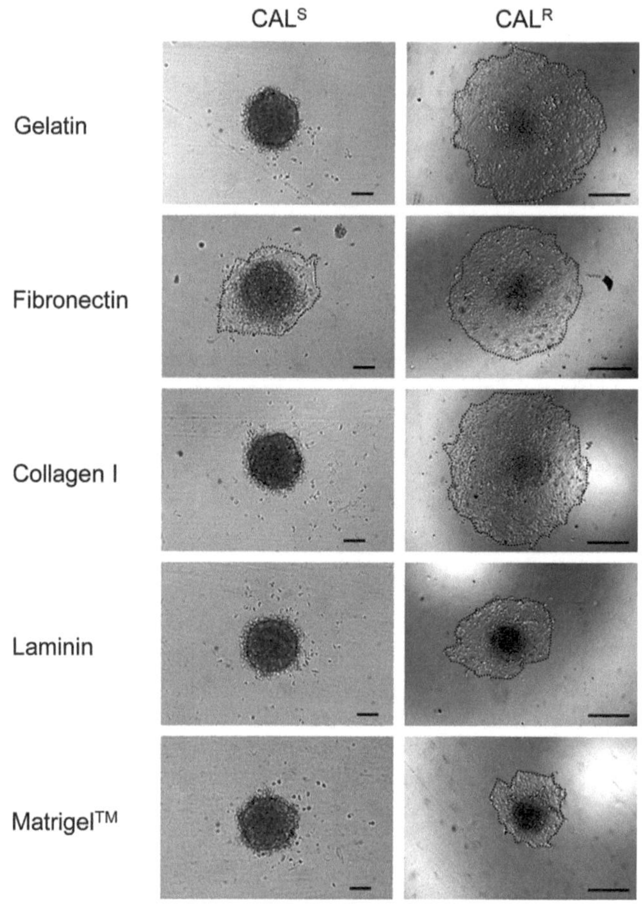

Fig. 2. Representative images of CAL^S and CAL^R tumor spheroids after 48 h of migration in plates coated with different ECM proteins. CAL^S cells are less migratory (images acquired using a ×10 objective, scale bar = 100 μm) than CAL^R (images acquired using a ×4 objective, scale bar = 500 μm) on all substrates. The migrated cell areas have been outlined manually for clarity.

has anti-motility/invasive properties (Fig. 4b). This demonstrates how the spheroid-based migration assay can be a valuable tool for the ongoing evaluation of existing anticancer drugs and also for the characterization of novel agents.

The assay can easily be modified to enable investigations of tumor–host cell interactions through the transfer of spheroids onto fibroblast (25) or endothelial cell (EC) monolayers (26). The latter is particularly relevant when studying glioma cell migration (Fig. 5) since it mimics the ability of these cells to "crawl" along blood vessels to migrate deep into the brain (27). This might be the cause in the notoriously poor response to localized surgery, radiotherapy, and indeed most therapeutic regimens in this disease. In addition, vascular interactions provide an important niche for glioma stem cells, also thought to contribute to

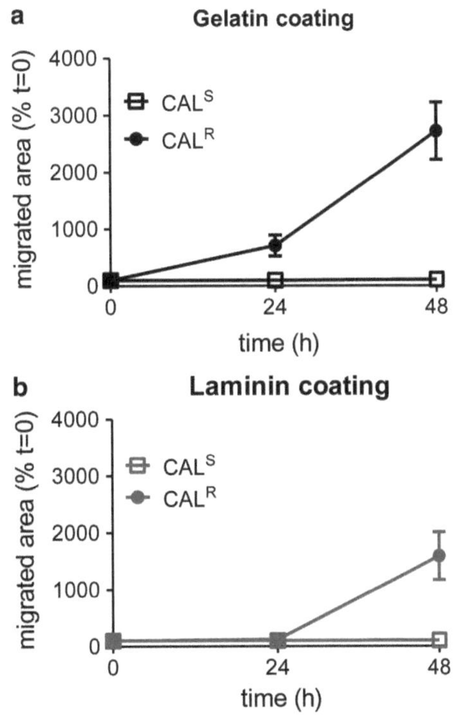

Fig. 3. Quantification of spheroid-based migration highlighting the different migratory properties of CAL^S and CAL^R cells on two different ECM coatings. Representative graphs showing quantification of CAL^S and CAL^R cell migration on (**a**) gelatin and (**b**) laminin, at 0, 24, and 48 h. Data are mean ± SD of replicate spheroids ($n = 6$).

enhanced malignancy and resistance to therapy (28). In this case, for measuring responses to therapeutic agents, we suggest pretreatment of the tumor spheroids prior to transfer onto the cell monolayer to avoid effects on the host cells—unless the specific aim is to test the impact on both populations.

The majority of drugs used to treat cancer are antiproliferative or proapoptotic, but cells that escape the primary tumor and generate metastases are often resistant to therapy. Anti-migratory agents could have an important role in delaying the formation of metastases (10) but have also been shown to sensitize migrating cells to antiapoptotic drugs (29). Deregulated cell migration/invasion is an insidious aspect of cancer progression, and while many screening strategies exist for the identification of cytotoxic compounds (30), few if any high-throughput methods are available to identify and test agents that inhibit this aspect of malignancy (10). Therefore, the significance of the development of rapid, simple, versatile migration assays such as those described here for testing novel agents is clear.

Although some drugs may be able to inhibit *both* cell proliferation and migration (as illustrated here for cisplatin, although

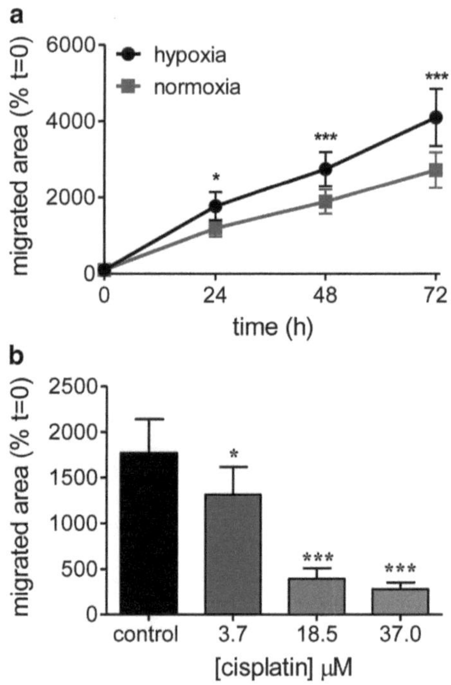

Fig. 4. Evaluation of the impact of tumor microenvironmental factors (e.g., hypoxia) and/or drugs using the tumor spheroid-based migration assay. (**a**) LICR-LON-HN4 spheroids were allowed to migrate on collagen in either normoxia or hypoxia, demonstrating significantly increased migration under reduced oxygen levels (1% 0_2) at all time points (24–72 h). (**b**) The effect of a standard cancer chemotherapeutic drug (cisplatin) on tumor cell migration in hypoxia was evaluated and found to have a significant inhibitory effect, as illustrated by quantification of the 24 h time point. Data are mean ± SD of replicate spheroids ($n = 6$). *$P \leq 0.05$, **$P \leq 0.01$, ***$P \leq 0.001$.

interestingly with different potencies in hypoxic conditions)—there are many signaling pathways unique to motility. It is therefore imperative to investigate the gene expression profiles of invasive/metastatic cells in order to identify such "unconventional" molecular targets. Tumor spheroids generated from cell lines containing stable or inducible knockdown (KD) constructs, or possibly even a screen using shRNA libraries targeting genes of interest (31), could readily be utilized in our migration assay to support target validation.

2. Materials

2.1. Cell Culture

1. Appropriate culture medium for the specific tumor cell line(s) under test (e.g., Dulbecco's Modified Eagle Medium (DMEM) 1× liquid: GlutaMAX™ I, 4,500 mg/L D-glucose, sodium pyruvate; DMEM/F12 (1:1) 1× liquid with GlutaMAX™ I).

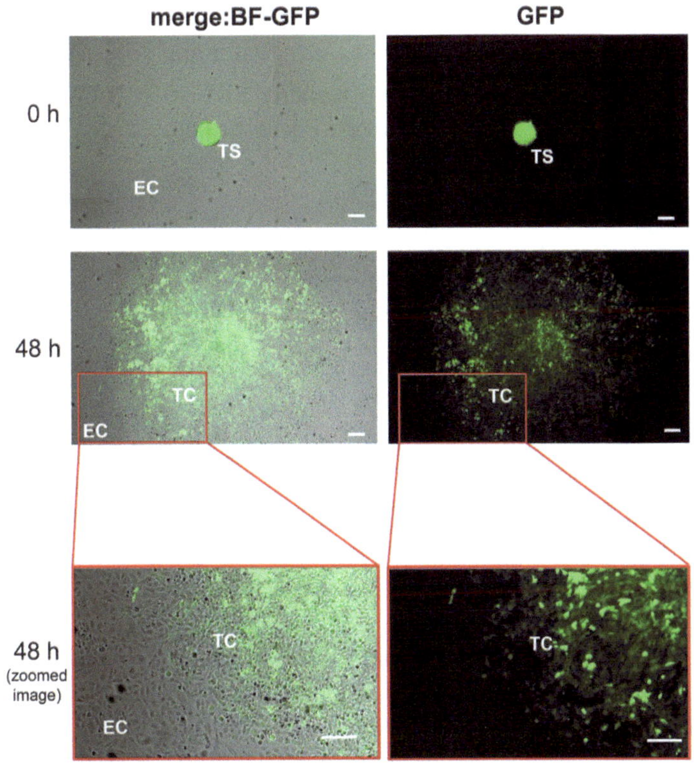

Fig. 5. Visualization of tumor–host cell interactions on an endothelial cell monolayer using the tumor spheroid-based migration assay. Tumor spheroids (TS) of SF188 cells expressing green fluorescent protein (GFP) were placed on an endothelial cell (EC) monolayer. Brightfield (BF) and fluorescence images were acquired over a 48 h period (here using a Celigo™ cytometer). A section of the 48 h image is shown at a higher magnification in the lower two panels where the merged BF and GFP image illustrates how the migrated tumor cells (TC) have adhered to and migrated over the EC layer. Scale bar = 200 μm.

2. Tumor cell lines, preferably characterized by the formation of tight/compact 3D structures (11, 32). Here, we selected three cell lines to exemplify the assay utility:
   (a) CAL^S (EGFR TKI-sensitive) and CAL^R (EGFR TKI-resistant) sublines were derived from CAL 27 (squamous tongue carcinoma, available from American Type Culture Collection®).
   (b) LICR-LON-HN4, a laryngeal SCC described in (23).
   (c) SF188, a pediatric glioblastoma (GBM) cell line described in (33).
3. Human umbilical vein endothelial cells (HUVEC), used to exemplify tumor spheroid-based migration on a cell monolayer.
4. Endothelial cell culture medium (e.g., TCS Cellworks, # ZHM-2953); phosphate-buffered saline (PBS): 140 mM

NaCl, 2.7 mM KCl, 10 mM Na_2HPO_4, 1.8 mM KH_2PO_4, dissolve in water and adjust to pH 7.4 with HCl.

5. Supplement PBS with 0.1% (w/v) EDTA for washing cells prior to detachment.

2.2. ECM Protein Coatings

1. Gelatin.
2. Fibronectin.
3. Laminin.
4. Collagen, type I.
5. Matrigel™ (catalogue number 354234; BD Biosciences).
6. 0.02 N acetic acid.
7. PBS containing 10 mM $CaCl_2$, 10 mM $MgCl_2$.

2.3. Tumor Spheroid-Based Migration Assay

1. Appropriate culture medium for the specific tumor cell line(s) under test.
2. 96-well flat-bottomed tissue culture-treated plates.

2.4. Drug Treatment for Inhibition of Tumor Cell Migration

1. Drug or inhibitor to be tested. Here we use cisplatin (*cis*-diammineplatinum(II) dichloride), stored in aliquots of 100 mM in dimethyl sulfoxide (DMSO) at −20°C for up to 4 weeks.
2. DMSO or appropriate solvent for the agent under test.

2.5. Imaging and Analysis of Tumor Cell Migration

1. Brightfield and/or fluorescence inverted microscope with 10× and 4× objectives.
2. Stage graticule (1 mm) for image calibration.
3. Image analysis software (e.g., Image-Pro Analyzer 6.3, Media Cybernetics Inc.).
4. Scientific graphing and statistics software (e.g., Microsoft Excel or GraphPad Prism5).

3. Methods

General tissue culture procedures are followed as previously described (34). Tumor cells and HUVECs are maintained in culture under standard conditions (37 °C, 5% CO_2, 95% humidity) and passaged when subconfluent.

3.1. Cell Culture

1. Detach cell monolayer and obtain a single cell suspension. Determine cell concentration using a hemocytometer.
2. Adjust the concentration to the appropriate cell density (here we used $0.5–2 \times 10^4$ cells/mL), according to the method of choice, in order to obtain spheroids of at least 300 μm diameter (see Note 1).

3. Initiate tumor spheroids using a method that results in the formation of compact, reproducibly sized spheroids in suspension, for example, ultralow attachment (11), agar-coated (13), poly-Hema-coated 96-well plates (12), or in hanging drops (14, 15) (see Note 2).
4. Spheroids are generally used 4 days after cell seeding when, depending on the cell line doubling time, spheroids reach an average size of 300–500 μm (11) (see Note 1).

3.2. Preparation of Endothelial Cell Monolayer

For the tumor spheroid-based migration assay, prepare the EC monolayer 4 days prior to performing the assay, generally on the same day as tumor spheroid initiation (see Note 3).

1. Coat a 96-well flat-bottomed plate with fibronectin (follow procedure described in Subheading 3.3).
2. Detach HUVECs and obtain a single cell suspension. Determine cell concentration using a hemocytometer.
3. Adjust cell density to 2×10^4 cells/mL and dispense 100 μL/well.
4. Transfer the plate to a cell culture incubator. Four days later, visually confirm that the EC monolayer is confluent and proceed with the migration assay.

3.3. ECM Protein Coating

1. For the preparation of ECM protein stock solutions, storage, and handling, follow the manufacturer's instructions (see Note 4).
2. Dispense 300 μL/well PBS into the outer wells of a 96-well flat-bottomed migration plate.
3. Dilute gelatin stock solution to 0.1% (v/v) in sterile ddH_2O.
4. Dilute fibronectin stock solution to 5 μg/mL in PBS containing Ca^{2+} and Mg^{2+}.
5. Dilute laminin stock solution to 50 μg/mL in sterile ddH_2O.
6. Dilute collagen stock solution to 50 μg/mL in 0.02 N acetic acid.
7. Dilute Matrigel™ to 125 μg/mL (see Note 5) in serum-free medium using precooled pipette tips.
8. Dispense 50 μL/well of one or more of the diluted ECM proteins into the inner 60 wells of the migration plate.
9. We recommend between 6 and 10 replicates per condition.
10. Gently tap the side of the plate to ensure even distribution of the liquid in each well.
11. Incubate at room temperature for 2 h.
12. Aspirate residual unbound ECM proteins from the plate (see Note 6).
13. Wash wells twice with PBS or serum-free medium.

14. Block with 1% (w/v) BSA in PBS, 100 μL/well for 1 h at room temperature.
15. Aspirate residual BSA and proceed with assay or store according to ECM manufacturer's instructions.

3.4. Tumor Spheroid-Based Migration Assay

1. Distribute 200 μL/well of culture medium (see Note 7) supplemented with 2% (v/v) FBS (see Note 8) into the inner 60 wells of the pre-coated migration plate.
2. Using a multichannel pipette (see Note 9), transfer one 4-day-old spheroid in a volume of 100 μL (see Note 10) into each well of the migration plate, resulting in a final volume of 300 μL/well (see Note 11).
3. Dispense the spheroids gently, keeping the tips in the center of the wells.
4. Avoid the formation of bubbles (see Note 12).
5. Gently transfer the plate to the incubator under the culture condition of choice (e.g., normoxia or hypoxia; see Note 13 and Fig. 4).
6. Allow the spheroids to adhere to the coated surface (or EC monolayer) for 30–60 min before imaging for the $t=0$ time point (see Subheading 3.6).

3.5. Evaluation of Test Agents for Inhibition of Tumor Cell Migration

1. For evaluation of potential inhibitors of migration (e.g., cisplatin; Fig. 4), the concentrations used should be ≤ the concentration that inhibits growth by 50% (GI_{50}) (see Note 14). This is important in order to evaluate the effect of the inhibitor more specifically on migration, limiting any inhibitory effects on cell proliferation.
2. Prepare culture medium supplemented with 2% (v/v) FBS and 1.5× the final concentration of the agent under test.
3. Controls contain the equivalent concentration of the appropriate solvent (e.g., DMSO) (see Note 15).
4. Distribute 200 μL/well of medium, containing test agent or vehicle, into 6–10 replicates/condition of the migration plate.
5. Follow the procedure as previously described in Subheading 3.4 (steps 2–6).

3.6. Imaging and Analysis of Tumor Cell Migration

1. Image the plate (see Note 16) at intervals starting from $t=0$ up to 72 h, using an inverted microscope (see Note 17).
2. Take images of a stage graticule (1 mm) using both 10× and 4× objectives.
3. Import graticule images into the image analysis software to perform image calibration (see Note 18). This step is required for the first time only. For subsequent image analysis, simply select the required calibration settings.

4. Import the migration assay images into the image analysis software and select the calibration settings depending on the microscope objective (10× or 4×) used to obtain the images.
5. Measure the area covered by the spheroids at $t=0$ and the area covered by the cells that have migrated from the spheroids at intervals over 24–72 h.
6. Normalize data to the original size of each spheroid recorded at $t=0$ (formula: (migrated area at $t=x$/migrated area at $t=0) \times 100$).

4. Notes

1. Tumor spheroids are an in vitro model that mimic micrometastases or tumor intervascular microregions; thus they can recapitulate some features of in vivo tumor tissue (e.g., gradients of oxygen and proliferation (11, 18, 35)). Considering that in vivo the maximum distance that oxygen can diffuse from capillaries is 100 μm, the spheroid diameter must be above 300 μm in order to reliably include these features.
2. Although all the methods mentioned above can generate reproducibly sized tumor spheroids, the one that we recommend is the method that uses round-bottom ultralow attachment 96-well plates (11). This method is the least time consuming (e.g., plates do not require any coating), reproducible (one centrally located spheroid/well is formed), and suitable for automated imaging. Thus it satisfies important requirements of cell-based HT assays.
3. For a tumor spheroid-based migration assay on EC monolayer, we suggest using a tumor cell line stably transfected with GFP. This enables the migrated tumor cells to be easily distinguished from the EC monolayer and imaging should then be performed using an inverted microscope with fluorescence illumination.
4. Depending on the manufacturer's guidelines, handling, storage, and reconstitution of matrix proteins can be slightly different. Always assess the optimal concentration and coating procedure for each cell line. A few important tips are (a) if gelatin is purchased in powder form, prepare a suspension in H_2O without trying to dissolve; subsequent autoclaving will achieve both sterility and solubility; (b) fibronectin needs to be handled carefully, and vortexing or shaking can cause the fibronectin to precipitate; (c) laminin solution, if warmed too quickly, can form a gel and cannot be reliquefied; and (d) always keep Matrigel™ on ice during preparation of solutions to avoid premature solidification.

5. Matrigel™ total protein concentration is approximately 10 mg/mL, but the precise concentration is specified on the data sheet that accompanies each batch.
6. When removing excess coating, be careful not to scratch the bottom of the well with the pipette tip used. This could impair optimal cell spreading.
7. For the tumor spheroid-based migration assay on EC monolayer, first aspirate the medium from the EC monolayer 96-well plate and then dispense 200 μL/well of fresh, warm EC culture medium. Then follow the procedure described in Subheading 3.4.
8. For highly motile cell lines, the concentration of FBS for the migration assay is kept below that used in standard culture medium (generally 10%). The reasons for this are (1) to slow-down proliferation, thus avoiding effects of proliferation complicating assessment of migration and (2) to reproduce a nutrient deprivation "stress" which tumor cells may be subjected to in vivo.
9. When the number of spheroids to analyze is relatively small (40–60 spheroids), the transfer of one spheroid per well can also be performed using a single channel rather than a multichannel pipette. If using disposable 200 μL pipette tips, it is advisable to cut off the ends using sterile scissors to increase the aperture and reduce the likelihood of disrupting the peripheral rim of cells on the spheroids.
10. Depending on the method of spheroid generation, remove any excess culture medium to leave the spheroid in a volume of 100 μL ready for transfer to the migration plate.
11. The suggested total volume in each well is 300 μL, the maximum total volume/well of standard 96-well tissue culture plates. This helps to reduce the meniscus effect when acquiring images of spheroids, particularly those that are located close to the edge of the well.
12. If bubbles form, use a heated sterile needle to destroy them.
13. For experiments carried out in hypoxia, incubate plates in a humidified hypoxia workstation (H35 Hypoxystation; Don Whitley Scientific, Shipley, UK). This allows oxygen (O_2) levels to be maintained at 1% (hypoxia) instead of 20–21% (normoxia/atmospheric O_2 concentration) which is normally used for cell culture.
14. Cell proliferation GI_{50}s of the test agent should be determined in preliminary studies using assays compatible with 3D cultures (e.g., CellTiter Glo® luminescent assay or WST-1 assay) so that appropriate concentrations can be selected.
15. If the test agent is, for example, diluted 1:1,000 from the stock solution, the same dilution of the vehicle is required to

maintain the vehicle concentration (0.1% (v/v)) across all the drug dilutions tested.

16. For experiments in hypoxia, use Parafilm® to seal the plate during the imaging time (carried out in normoxia) to avoid equilibration of the oxygen levels.
17. Use a 10× objective first and move to the 4× objective only when the migrated cell area is too large to be completely captured using the 10×.
18. Image calibration is required in order to obtain accurate measurements. The objective used (10× or 4×) the units (μm) and the graticule measurements need to be entered into the image analysis software to obtain the correct calibration for the system used.

Acknowledgments

We thank Marta Mendiola for preliminary CAL^R and CAL^S migration data. This work was funded by The National Centre for the Replacement, Refinement and Reduction of Animals in Research (G1000121 ID no. 94513). C.B. and M.Z. are supported by grants from the Oracle Cancer Trust. S.E. is supported by The Institute of Cancer Research and Cancer Research UK grant number C309/A8274. We acknowledge NHS funding to the NIHR Biomedical Research Centre.

References

1. Friedl P, Hegerfeldt Y, Tusch M (2004) Collective cell migration in morphogenesis and cancer. Int J Dev Biol 48:441–449
2. Horwitz R, Webb D (2003) Cell migration. Curr Biol 13:R756–R759
3. Demir R, Yaba A, Huppertz B (2010) Vasculogenesis and angiogenesis in the endometrium during menstrual cycle and implantation. Acta Histochem 112:203–214
4. Lefranc F, Brotchi J, Kiss R (2005) Possible future issues in the treatment of glioblastomas: special emphasis on cell migration and the resistance of migrating glioblastoma cells to apoptosis. J Clin Oncol 23:2411–2422
5. Patel V, Rosenfeldt HM, Lyons R et al (2007) Persistent activation of Rac1 in squamous carcinomas of the head and neck: evidence for an EGFR/Vav2 signaling axis involved in cell invasion. Carcinogenesis 28:1145–1152
6. Weigelt B, Peterse JL, van't Veer LJ (2005) Breast cancer metastasis: markers and models. Nat Rev Cancer 5:591–602
7. Biddle A, Mackenzie IC (2012) Cancer stem cells and EMT in carcinoma. Cancer Metastasis Rev. doi:10.1007/s10555-10012-19345-10550
8. Alison MR, Lim SM, Nicholson LJ (2011) Cancer stem cells: problems for therapy? J Pathol 223:147–161
9. Janssen HL, Haustermans KM, Balm AJ et al (2005) Hypoxia in head and neck cancer: how much, how important? Head Neck 27: 622–638
10. Palmer TD, Ashby WJ, Lewis JD et al (2011) Targeting tumor cell motility to prevent metastasis. Adv Drug Deliv Rev 63:568–581
11. Vinci M, Gowan S, Boxall F et al (2012) Advances in establishment and analysis of 3D tumor spheroid-based functional assays for target validation and drug evaluation. BMC Biol 10:29. doi:10.1186/1741-7007-10-29
12. Ivascu A, Kubbies M (2006) Rapid generation of single-tumor spheroids for high-throughput cell function and toxicity analysis. J Biomol Screen 11:922–932

13. Friedrich J, Seidel C, Ebner R et al (2009) Spheroid-based drug screen: considerations and practical approach. Nat Protoc 4:309–324
14. Foty R (2011) A simple hanging drop cell culture protocol for generation of 3D spheroids. J Vis Exp 51:e2720
15. Kelm JM, Timmins NE, Brown CJ et al (2003) Method for generation of homogeneous multicellular tumor spheroids applicable to a wide variety of cell types. Biotechnol Bioeng 83:173–180
16. Deryugina EI, Bourdon MA (1996) Tenascin mediates human glioma cell migration and modulates cell migration on fibronectin. J Cell Sci 109:643–652
17. Vajkoczy P, Menger MD, Goldbrunner R et al (2000) Targeting angiogenesis inhibits tumor infiltration and expression of the pro-invasive protein SPARC. Int J Cancer 87:261–268
18. Friedrich J, Ebner R, Kunz-Schughart LA (2007) Experimental anti-tumor therapy in 3-D: spheroids—old hat or new challenge? Int J Radiat Biol 83:849–871
19. Hirschhaeuser F, Menne H, Dittfeld C et al (2010) Multicellular tumor spheroids: an underestimated tool is catching up again. J Biotechnol 148:3–15
20. Hood JD, Cheresh DA (2002) Role of integrins in cell invasion and migration. Nat Rev Cancer 2:91–100
21. Buttery RC, Rintoul RC, Sethi T (2004) Small cell lung cancer: the importance of the extracellular matrix. Int J Biochem Cell Biol 36:1154–1160
22. Heo K, Kim YH, Sung HJ et al (2012) Hypoxia-induced up-regulation of apelin is associated with a poor prognosis in oral squamous cell carcinoma patients. Oral Oncol. doi:10.1016/j.oraloncology.2011.1012.1015
23. Easty DM, Easty GC, Carter RL et al (1981) Ten human carcinoma cell lines derived from squamous carcinomas of the head and neck. Br J Cancer 43:772–785
24. Molin Y, Fayette J (2011) Current chemotherapies for recurrent/metastatic head and neck cancer. Anticancer Drugs 22:621–625
25. Brouty-Boye D, Mainguene C, Magnien V et al (1994) Fibroblast-mediated differentiation in human breast carcinoma cells (MCF-7) grown as nodules in vitro. Int J Cancer 56:731–735
26. Offner FA, Wirtz HC, Schiefer J et al (1992) Interaction of human malignant melanoma (ST-ML-12) tumor spheroids with endothelial cell monolayers. Damage to endothelium by oxygen-derived free radicals. Am J Pathol 141:601–610
27. Vajkoczy P, Farhadi M, Gaumann A et al (2002) Microtumor growth initiates angiogenic sprouting with simultaneous expression of VEGF, VEGF receptor-2, and angiopoietin-2. J Clin Invest 109:777–785
28. Kenig S, Alonso MB, Mueller MM et al (2010) Glioblastoma and endothelial cells cross-talk, mediated by SDF-1, enhances tumour invasion and endothelial proliferation by increasing expression of cathepsins B, S, and MMP-9. Cancer Lett 289:53–61
29. Decaestecker C, Debeir O, Van Ham P et al (2007) Can anti-migratory drugs be screened in vitro? A review of 2D and 3D assays for the quantitative analysis of cell migration. Med Res Rev 27:149–176
30. Shoemaker RH (2006) The NCI60 human tumour cell line anticancer drug screen. Nat Rev Cancer 6:813–823
31. Zhang H, Lee MY, Hogg MG et al (2010) Gene delivery in three-dimensional cell cultures by superparamagnetic nanoparticles. ACS Nano 4:4733–4743
32. Ivascu A, Kubbies M (2007) Diversity of cell-mediated adhesions in breast cancer spheroids. Int J Oncol 31:1403–1413
33. Bax DA, Little SE, Gaspar N et al (2009) Molecular and phenotypic characterisation of paediatric glioma cell lines as models for preclinical drug development. PLoS One 4:e5209
34. Box GM, Eccles SA (2011) Simple experimental and spontaneous metastasis assays in mice. Methods Mol Biol 769:311–329
35. Airley R, Loncaster J, Davidson S et al (2001) Glucose transporter glut-1 expression correlates with tumor hypoxia and predicts metastasis-free survival in advanced carcinoma of the cervix. Clin Cancer Res 7:928–934

Chapter 17

The Neurosphere Assay Applied to Neural Stem Cells and Cancer Stem Cells

Rossella Galli

Abstract

The discovery of neural stem cells (NSCs) in the mammalian brain has raised many expectations as these unique cells might recapitulate different neurological diseases, including brain tumors, both from a functional and molecular perspective. Proper in vitro culturing of NSCs has emerged as a critical methodological issue, given that it should preserve the in vivo features of NSCs, with particular emphasis on cell heterogeneity. At the same time, the methodology for NSC culturing should allow the production of large amounts of cells to be exploited not only for prospective clinical applications, but also for drug screening. Direct in vitro selection of NSCs and, very recently, cancer stem cells (CSCs) by means of defined serum-free conditions represents the most reliable methodology to obtain long-term expanding SC lines. Here we describe the methods currently employed to enrich for NSCs/CSCs based on the NeuroSphere Assay (NSA) and their adaptation to specific assays for testing the efficacy of neuroactive compounds.

Key words: Neurosphere assay, Neural stem cell cultures, Cancer stem cell

1. Introduction

The identification of self-renewing and multipotent somatic neural stem cells (NSCs) in the mammalian brain raised many expectations as these unique cells faithfully model different neurological diseases, including brain tumors, both from a functional and molecular standpoint. Likewise, the recent availability of induced pluripotent stem (iPS) cells opens new scenarios for disease-related basic research, as the advent of cell reprogramming enables the generation of NSCs from individual patients, thus leading to the establishment of cellular model systems of neurological diseases. Independent from their origin (somatic neural tissues or iPS cells), NSCs can differentiate in multiple types of neural cells, making cell replacement very effective, in particular for those neurodegenerative diseases that affect specific neural lineages.

Jürgen Moll and Riccardo Colombo (eds.), *Target Identification and Validation in Drug Discovery: Methods and Protocols*, Methods in Molecular Biology, vol. 986, DOI 10.1007/978-1-62703-311-4_17,

Proper in vitro culturing of NSCs has emerged as a critical methodological issue, given that it implies the maintenance of the in vivo characteristics of NSCs, while warranting the generation of large amount of cells ("biomass") to be exploited for drug screening (1). Two main methodologies are currently in use for isolation, enrichment, and propagation of NSCs: (a) the neurosphere assay (2, 3) and (b) the adhesion-mediated enrichment assay (4). Of note, the two in vitro methods have been successfully applied to the isolation of putative cancer stem cells (CSCs) from brain tumors (5, 6) as well as from breast, colon, and prostate cancer (7–9).

The implementation and adaptation of the different technologies developed for the study of NSC proliferation and differentiation might be instrumental for the establishment of screening platforms to identify small molecules with neurogenic and/or antitumor activity.

Some controversies were recently raised concerning the best method to be exploited for NSC isolation and propagation. The NSA was first developed in the early 1990s by the laboratory headed by Reynolds and Weiss (2) and relies on neural cell culturing in a serum-free medium supplemented with two growth factors, namely, epidermal growth factor (EGF) and/or fibroblast growth factor 2 (FGF2) (10–12). By this assay, three-dimensional floating cell clusters, known as neurospheres, are detected in culture after a variable period of time, which depends on cell staging (embryonic, fetal, or adult) and on the species of origin. Neurospheres are highly heterogeneous entities that reproduce the same cellular composition of a typical NSC niche; indeed, they comprise a minority of bona fide stem cells with the remaining bulk of cells containing bipotent progenitors and, although scarcely represented, fully mature neurons, astrocytes and oligodendrocytes. Through many subculturing passages, ideally more than seven, different cell types, which are mostly committed progenitors that persist in the primary culture as a remnant of the tissue of origin, are negatively selected. The neurosphere cell composition reaches finally a steady state equilibrium, in which about 2.5% of the cells within a sphere are true NSCs (3). Neurosphere cultures display an average amplification rate of 50 neurospheres deriving from the dissociation of one single neurosphere. As such, neurosphere expansion in culture might occur quite rapidly, producing relevant numbers of cells to be used for downstream applications. As per definition, NSCs are multipotent and, thus, can differentiate into the three main lineages of the central nervous system (CNS), i.e., neurons, astrocytes, and oligodendrocytes. Several protocols are available to enrich for specific lineages and to simulate in vitro the commitment and differentiation steps taking place in vivo (13, 14).

The second assay currently used for NSC isolation was recently developed by the laboratories directed by Cattaneo and Smith and is based on the culture of putative NSCs as a monolayer, in the presence of EGF and FGF2 on an adhesive substrate such as laminin (4). As opposed to neurospheres, the resultant neural stem (NS) cell cultures constitute homogenous cultures that can be propagated extensively without accompanying spontaneous differentiation. In fact, adherent NS lines obtained by propagation in the presence of mitogens do not show expression of neuronal or glial markers and uniformly express the precursor markers RC2 and nestin. As neurospheres, NS cells differentiate into neurons and astrocytes upon growth factor withdrawal and addition of serum.

Notably, the choice of one of the two methodologies depends on the biological question addressed and on the rationale underlying the study. If the experimental requirement is the availability of highly enriched homogeneous NSC/CSC cell cultures, the appropriate methods should be considered: the adhesion-mediated enrichment assay (4). On the contrary, if the aim of the study entails the preservation of cellular heterogeneity, the NSA may be the optimal choice.

In general, cell cultures derived from neural tissues are considered powerful tools to be exploited for a better understanding of the cellular and molecular mechanisms underlying neural development and function. Moreover, they can also be employed for the screening of potential neuroactive compounds. The development of optimal culture methods for neural and cancer stem cells is a critical foundation for their exploitation in drug screening. To date, very few studies focused on chemical genetic screens for molecules inhibiting neurosphere proliferation (i.e., the closest readout for self-renewal that can be translated in high-throughput settings), revealing that screen-recovered small molecules, in particular neuromodulators, had also potent inhibitory effects on cultures enriched for brain cancer stem cells (15). Others have recently described methods for a simple high-content live-image-based platform, reporting the effect of a panel of several kinase inhibitors on NS cultures (16). Similarly, heterogeneous CSCs isolated from glioblastoma multiforme by the NSA have been exploited for testing the activity of EGFR inhibitors (17).

Here we describe the methods for the isolation of NSCs and CSCs by the NSA and their adaptation to small-scale drug screening applications.

2. Materials

Prepare all solutions using ultrapure cell culture grade water. Filter sterilize and store solutions as indicated below.

2.1. Reagents for Culture Media

1. 10× DMEM-F12 (1:1), Invitrogen.
2. GlutaMax, Invitrogen.

3. 0.2% Heparin sodium salt, grade 1A.
4. 30% D(+)Glucose.
5. 1 M HEPES buffer, Invitrogen.
6. 7.5% $NaHCO_3$ buffer, Invitrogen.
7. 3 mM sodium selenite.
8. 2 mM progesterone.
9. Putrescine dihydrochloride.
10. Apo-transferrin.
11. Insulin.
12. Penicillin/streptomycin.

Growth factors

13. EGF, human recombinant, Peprotech, 100-15.
14. FGF2, human recombinant, Peprotech, 100-18B.

2.2. Reagents for Tissue Digestion

1. EBSS.
2. Papain, Worthington DBA.
3. L-Cysteine hydrochloride monohydrate.
4. EDTA.
5. DNase.

2.3. Other Reagents

1. Matrigel™ (growth factor reduced), Becton Dickinson.
2. Diff-Quick stain set, Dade Behring, Inc.
3. 8 μm Transwell polycarbonate inserts.
4. Trypan blue.
5. 4% Paraformaldehyde.
6. MTT.
7. Vacuum filtration system combining a filter unit with a receiver flask for processing and storing volumes from 150 to 1,000 mL.

2.4. Media and Solutions

For cell media preparation, ultrapure cell culture tissue grade water should be used. Carry out all procedures in a sterile hood.
All media and stock solutions should be prepared only in sterile disposable tubes and/or bottles.

1. 30% glucose. Dissolve 30 g glucose in 100 mL water. Heat the solution at 60°C until complete solubilization. Filter sterilize and store at 4°C.
2. 0.2% heparin. Dissolve 100 mg heparin in 50 mL water. Filter sterilize and store at 4°C.
3. 10× stock solution of DMEM/F12. Dissolve the content of a powder box in 1 L of ultrapure cell culture tissue grade water

under gentle continuous stirring for several hours until the solution is clear. Filter sterilize and store at 4°C.

4. 3 mM sodium selenite. Add 1.93 mL of ultrapure cell culture tissue grade water to a 1 mg vial of sodium selenite. Mix, aliquot into sterile tubes and store at −20°C.
5. 2 mM progesterone. Add 1.59 mL of 95% ethanol to a 1 mg vial of progesterone. Mix, aliquot into sterile tubes and store at −20°C.
6. 10× stock solution of hormone mix (for 400 mL). Mix 40 mL of 10× DMEM/F12, 8 mL 30% glucose, 6 mL 7.5% $NaHCO_3$, 2 mL 1 M HEPES, and 284 mL of water. Dissolve 400 mg of apo-transferrin into the solution. Dissolve 100 mg insulin in 4 mL sterile 0.1 N HCl, mix in 36 mL of water, and add to the hormone mix solution. Dissolve 38.6 mg of putrescine in 20 mL water and add to hormone mix solution. Add 40 μL of 2 mM progesterone and 40 μL of 3 mM sodium selenite. Mix well and filter sterilize. Aliquot in sterile tubes and store at −20°C.
7. Growth factor stock. Reconstitute EGF and FGF2 in order to have a 500 μg/mL and a 100 μg/mL stock solutions, respectively. Aliquot into sterile Eppendorf vials and store at −20°C.
8. Proliferation medium (for 500 mL). Mix 369 mL water, 50 mL 10× DMEM/F12, 10 mL 30% glucose, 7.5 mL 7.5% $NaHCO_3$, 2.5 mL 1M HEPES, 5 mL 200 mM GlutaMax™, 5 mL PenStrep solution, 50 mL 10× hormone mix, 1 mL 0.2% heparin, 20 μL EGF- and 50 μL FGF2-stock (final concentration: 20 ng/mL EGF and 10 ng/mL FGF2).
9. Matrigel™ stock and ready-to-use solution. Thaw a 10-mL vial of Matrigel™ overnight at 4°C. Aliquot into sterile tubes (0.5 mL/aliquot) using refrigerated plastic pipettes and store at −20°C. Dissolve a 0.5 mL aliquot into 150 mL DMEM to obtain the ready-to-use solution.
10. 0.1% DNase stock. Add 10 mL water to 10 mg DNase. Mix well and filter sterilize. Aliquot in sterile tubes (0.5 mL/aliquot) and store at −20°C.

3. Methods

The neurosphere culture paradigm selectively enriches for the stem cell component within neural tissues, including tumors, by taking advantage of the intrinsic property of bona fide stem cells, i.e., their capacity to self-renew long term. When seeded in the presence of the two mitogens EGF and FGF2 with no adhesion substrate, cells from freshly dissociated neural tissues give rise to spherical clusters, which float in suspension and are named "neurospheres."

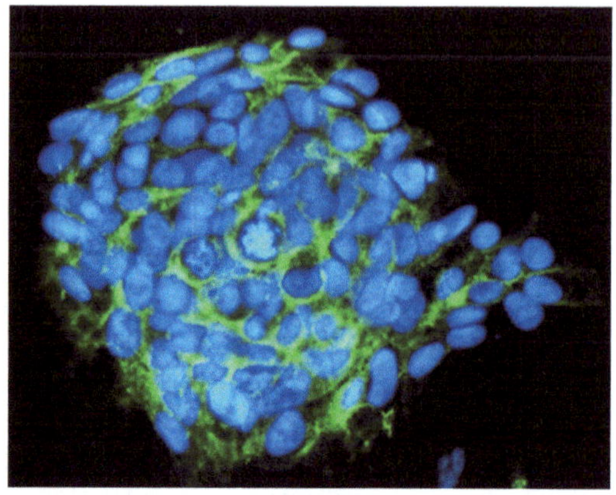

Fig. 1. Only a fraction of cells within a GBM CSC-derived neurosphere express the EGFR. EGFR, *green*; DAPI, *light blue*, nuclei. Magnification 200× (Mazzoleni and Galli, unpublished observations).

Of note, not all the cells within a neurosphere are true stem cells. This concept is highly relevant for CSCs isolated from brain tumors. In fact, multiple subpopulations of tumor-initiating cells have been identified in CSC neurospheres, based on the expression of surface markers, such as EGFR (Fig. 1) and others (17). Neurospheres are routinely subcultured by mechanical dissociation and by replating under the same in vitro conditions. At every subculturing passage, committed cells are selected negatively, whereas NSCs/CSCs do self-renew and generate spheres, allowing the establishment of long-term expanding NSC lines.

3.1. Dissociation of Digested Neural Tissue and Primary Culture

1. Cut the dissected tissue of interest into small pieces (see Note 1). If the neural tissue is obtained from rodents from embryonic day (E) 12 up to postnatal day (P) 7, proceed directly to step 6. If the neural tissue is obtained from rodents after P7, or from tumors, proceed with enzymatic digestion of the tissues. To this end, dissolve 400 U of papain, 5 mg of cysteine and 5 mg of EDTA into 25 mL of EBSS. Vortex until the solution is clear, add 0.5 mL of a 0.1% DNAse stock solution and filter sterilize.
2. Transfer the pieces of tissues into the 15 mL tubes containing the papain mix solution (at least 12 mL) (see Note 2). Transfer the tubes to a rocking device. Incubate at 37°C for 20–60 min, depending on tissue consistence.
3. At the end of the enzymatic incubation, pellet tissues by centrifugation at 100 × *g* for 10 min.
4. Remove almost all the supernatant overlaying the pellets with a pipette.
5. Add 1 mL of fresh EBSS to the tissue.

6. Dissociate by triturating 20–50 times using a sterile 1,000 μL pipette tip. Add 10 mL of fresh EBSS to the cell suspension.
7. Pellet the cells by centrifugation at 100×*g* for 10 min.
8. Discard supernatant and resuspend cells in 200 μL of culture medium. Using a 200 μL Pipetman with the volume set at 180 μL, gently dissociate the pellet for 20–40 times. Dilute a 5 μL aliquot from each sample in 15 μL of Trypan blue and count in a Burker or Neubauer chamber. Seed cells at a density of 8×10^3 viable cells/cm^2 in culture medium containing EGF and FGF2.
9. Incubate at 37°C, 5% CO_2 in a humidified incubator.
10. Cells should proliferate giving rise to primary sphere-shaped clusters, which will float in suspension (see Notes 3 and 4).

3.2. In Vitro Propagation of Neurospheres by Subculturing

1. Pellet the neurosphere suspension by centrifugation at 100×*g* for 10 min (see Note 5).
2. Remove the supernatant leaving approximately a volume of 200 μL. Using a sterilized p200 Pipetman set at a volume of 180–200 μL, gently triturate pellet (see Note 6).
3. Count viable cells by Trypan blue exclusion and seed cells at the appropriate density in culture medium in untreated tissue flasks.

3.3. Assays for Measuring the Effect of Pharmacologically Active Compounds on NSC/CSC Properties

3.3.1. Short-Term Self-Renewal Assay (Clonogenic Assay)

The clonogenic assay is normally exploited to analyze the effect of compounds on the symmetry of division in NSCs/CSCs. In this assay, the number of secondary spheres generated from the dissociation of a single sphere reflects the frequency of NSC/CSCs present in the original primary clone. This analysis also returns an estimate of the relative frequencies between symmetric proliferative (two SCs generated at each cycle) and symmetric differentiative (two differentiated and/or dead cells generated after cell division).

1. Seed the cells derived from the dissociation of clonal single neurospheres into uncoated 96-well plates in proliferation medium with or without the compound under testing. Count the cells in each well 2 h after plating, in order to obtain a "baseline" value to be used for establishing the efficiency of secondary sphere generation.
2. Count the neurospheres generated in each well, 3–10 days after plating, depending on the cell type. Normalize the number of secondary neurospheres to the number of cells originally seeded in each well.

3.3.2. Long-Term Self-Renewal Assay (Growth Curve/Population Assay)

To analyze the effect of compounds on long-term self-renewal in cultures of NSCs/CSCs, the appropriate method is the growth curve/population analysis. Given that NSCs/CSCs under mitogenic stimuli expand in number over serial subculturing, this assay monitors long-term self-renewal that relies on the generation of at least one cell identical to the original clonogenic stem cell at every cell division. Thus, stable stem cell lines can be established that share the same functional properties of the founder cell. In such a system, drug-dependent alterations in NSC/CSC properties will result in the modification of the overall growth rate of the NSC/CSC line under analysis.

1. For growth curve/population analyses, seed the cells at 8×10^3 cells/cm^2 in proliferation medium with or without the compound under testing.
2. Collect the spheres when they reach the proper size. Spin them at $100 \times g$ for 10 min. Remove the supernatant leaving behind approximately 200 μL. Using a sterilized p200 Pipetman set at a volume of 180–200 μL, gently triturate pellet.
3. Count viable cells by Trypan blue exclusion and seed cells at 8×10^3 cells/cm^2 in proliferation medium with or without the compound under testing in new flasks.
4. Repeat steps 1 through 3 for at least ten subculturing passages. At any passage, calculate the relative amplification rate by dividing the amount of counted cells by the number of cells originally seeded. Multiply the amplification rate for the cumulative number of cells obtained at the previous subculturing passages and plot this number in correspondence of every subculturing passage. In this way, a growth curve will be generated, whose slope provides information as to the frequency of putative NSCs/CSCs in each experimental condition (3).

3.3.3. MTT Assay to Determine NSC Proliferation/Survival

The MTT assay allows to measure quantitatively the viability and, as such, the proliferation rate of cells upon drug administration. MTT (3(4,5-dimethylthiazol-2-yl)-2,5-diphenyltetrazolium bromide) is processed by metabolically active cells resulting in the generation of a purple formazan. Release of insoluble formazan by viable cells is promoted by using a solubilization agent as dimethyl sulfoxide (DMSO). The absorbance of the resulting solution is measured at 550 nm wavelength by a spectrophotometer, and it is directly related to cell viability/proliferation.

1. Coat the wells of 96-well plates with Matrigel™ overnight.
2. Generate a titration curve, by plating cells at different densities and by performing MTT assays at the same time point, e.g., 2 h after plating. The absorbance value for each cell density will allow estimating the number of cells survived/proliferated at the end of your experiment.
3. Seed the cells into the required number of 96-well plates at 2×10^4 cells/well in 180 μL of proliferation medium (see Note 7). Allow the cells to adhere to the plates for 2 h. At this point,
   (a) add MTT to "baseline" plates in order to obtain the absorbance value at baseline (see detailed description below);

(b) add the vehicle/compound of interest to the remaining plates to be used for the assay.

4. After the appropriate period of time (normally ranging from 12 to 96 h), add 20 μL of a 5 mg/mL MTT solution in PBS (see Note 8) to each well, which corresponds to a 1:10 dilution. Incubate the cells with MTT for 1 h at 37°C. Next, gently aspirate the medium and add 50 μL of DMSO to each well. DMSO lyses the cells, allowing the insoluble formazan to be released and solubilized (see Note 9). Gently shake the plate on a horizontal shaker for 15 min.
5. Read the plate on a microplate spectrophotometer reader at 550 nm. Plot the data by exploiting the titration curve generated at the beginning of the experiment.

3.3.4. Transwell-Mediated Matrigel™ to Assess NSC Migration/Invasion Potential

Cell migration and invasion are cardinal properties of NSCs and CSCs, respectively. Assessing the effect of drugs on these functional parameters is therefore of paramount preclinical relevance. A Matrigel™ invasion assay, performed in Transwell chambers, represents the most advantageous and efficient method.

1. Coat overnight with Matrigel™ the required number of 6-well Transwell inserts containing a polycarbonate 8.0 μm pore membrane.
2. Include an additional insert for any condition, treatment, or time point, which will serve as internal control, from which non-migrated cells won't be removed, in order to assess the effect of the drug on the overall cell viability (see below for details).
3. Seed cells at a density of 2×10^4 cells/insert onto the layer of Matrigel in proliferation medium with or without the compound under testing for the appropriate time.
4. At the end of the experiment, mechanically remove the cells on the upper side of the inserts (which did not migrate through the membrane) by using a cell scraper or a cotton swab. Do not remove cells from the internal controls.
5. Fix the inserts with 4% paraformaldehyde, and then stain with hematoxylin and eosin (H&E) for 2 min or Diff-Quick staining solutions, according to the manufacturer's instructions.
6. Cut the membrane out of the insert by using a scalpel, transfer the membrane on a glass microscope slide. Glue a coverslip to the microscope slide by using a solvent-free mounting agent or nail polish.
7. The extent of cell migration/invasion can be monitored by standard light microscopy and/or by densitometric analysis (e.g., using ImageJ software). Normalize the extent of migration in each sample to the one of the corresponding internal control.

4. Notes

1. Dissection should be carried out as quickly as possible from tissue collection. If you estimate that more than 2 h are required, keep the tissues at 4°C.
2. The total amount of material to be digested in each tube should not cover more than half of the volume retained in the conic bottom of the tube. If the material is in excess, subdivide it in multiple 15 mL tubes.
3. Rodent NSCs from embryonic stages form neurospheres in about 2–3 days, whereas NSCs from postnatal stages require 5–7 days to give rise to neurosphere.
4. Neurospheres are considered viable and healthy if single cell boundaries can be clearly observed within the cell cluster, if the overall morphology of the neurosphere is regular and round, and if neurosphere appear homogeneously bright. If the center of the neurospheres acquires a dark appearance, the cells within the core are dying, and, as such, the neurospheres need to be subcultured immediately.
5. Subculture neurospheres when their diameter is in the range of 200 and 500 μm. If neurospheres are too small when subcultured, the yield of cells will be very low; if neurospheres are too large, the number of dead cells inside the spheres (that give rise to the "dark core") will be high, mechanical dissociation will be difficult, and viability of the culture will be very low.
6. For achieving maximal disaggregation of neurospheres, slightly tilt the Pipetman and press tip against the bottom of the tube to generate a fair amount of resistance. Gently triturate pellet for 50–70 times for embryonic rodent cells and 60–80 times for adult rodent cells and cancer stem cells.
7. For MTT assays, include a "baseline" control plate for each condition/treatment.
8. MTT solution should be protected from light, when stored at 4°C for up to 2 months.
9. Modified tetrazolium salts such as XTT have recently become available. Viable cells convert these modified compounds to a water-soluble formazan. As a result, the solubilization step can be eliminated from the procedure.

References

1. Winquist RJ, Furey BF, Boucher DM (2010) Cancer stem cells as the relevant biomass for drug discovery. Curr Opin Pharmacol 10:385–390
2. Reynolds BA, Weiss S (1992) Generation of neurons and astrocytes from isolated cells of the adult mammalian central nervous system. Science 255:1707–1710
3. Reynolds BA, Rietze RL (2005) Neural stem cells and neurospheres—re-evaluating the relationship. Nat Methods 2:333–336
4. Conti L, Pollard SM, Gorba T et al (2005) Niche-independent symmetrical self-renewal of a mammalian tissue stem cell. PLoS Biol 3:e283
5. Galli R, Binda E, Orfanelli U et al (2004) Isolation and characterization of tumorigenic, stem-like neural precursors from human glioblastoma. Cancer Res 64:7011–7021
6. Pollard SM, Yoshikawa K, Clarke ID et al (2009) Glioma stem cell lines expanded in adherent culture have tumor-specific phenotypes and are suitable for chemical and genetic screens. Cell Stem Cell 4:568–580
7. Ponti D, Costa A, Zaffaroni N et al (2005) Isolation and in vitro propagation of tumorigenic breast cancer cells with stem/progenitor cell properties. Cancer Res 65:5506–5511
8. O'Brien CA, Pollett A, Gallinger S, Dick JE (2007) A human colon cancer cell capable of initiating tumour growth in immunodeficient mice. Nature 445:106–110
9. Wang S (2009) Anchorage-independent growth of prostate cancer stem cells. Methods Mol Biol 568:151–160
10. Gritti A, Cova L, Parati EA et al (1995) Basic fibroblast growth factor supports the proliferation of epidermal growth factor-generated neuronal precursor cells of the adult mouse CNS. Neurosci Lett 185:151–154
11. Reynolds BA, Weiss S (1996) Clonal and population analyses demonstrate that an EGF-responsive mammalian embryonic CNS precursor is a stem cell. Dev Biol 175:1–13
12. Gritti A, Frolichsthal-Schoeller P, Galli R et al (1999) Epidermal and fibroblast growth factors behave as mitogenic regulators for a single multipotent stem cell-like population from the subventricular region of the adult mouse forebrain. J Neurosci 19:3287–3297
13. Vescovi AL, Reynolds BA, Fraser DD et al (1993) bFGF regulates the proliferative fate of unipotent (neuronal) and bipotent (neuronal/astroglial) EGF-generated CNS progenitor cells. Neuron 11:951–966
14. Parras CM, Galli R, Britz O et al (2004) Mash1 specifies neurons and oligodendrocytes in the postnatal brain. EMBO J 23:4495–4505
15. Diamandis P, Wildenhain J, Clarke ID et al (2007) Chemical genetics reveals a complex functional ground state of neural stem cells. Nat Chem Biol 3:268–273
16. Danovi D, Falk A, Humphreys P et al (2010) Imaging-based chemical screens using normal and glioma-derived neural stem cells. Biochem Soc Trans 38:1067–1071
17. Mazzoleni S, Politi LS, Pala M et al (2010) Epidermal growth factor receptor expression identifies functionally and molecularly distinct tumor-initiating cells in human glioblastoma multiforme and is required for gliomagenesis. Cancer Res 70:7500–7513

Part VI

Genetically Engineered Animal Models to Study Gene Functions

Chapter 18

Genetically Engineered Animal Models for In Vivo Target Identification and Validation in Oncology

Gemma Texidó

Abstract

In vitro approaches using human cancer cell lines aimed to identify and validate oncology targets, have pinpointed a number of key targets and signalling pathways which control cell growth and cell death. However, tumors are more than insular masses of proliferating cancer cells. Instead they are complex tissues composed of multiple distinct cell types that participate in homotypic and heterotypic interactions and depend upon each other for their growth. Therefore, many targets in oncology need to be validated in the context of the whole animal. This review provides an overview on how animal models can be generated and used for target identification and validation in vivo.

Key words: Mouse models, Drosophila, Target identification in vivo, Target validation in vivo, Cancer genes

1. Introduction

The drug development process in oncology includes many steps and requires substantial investments in time and resources (1). The high attrition rate of oncology drugs in clinical development has made many pharmaceutical and biotechnology companies place an early-stage target selection and validation as one of the most crucial steps in drug development. Therefore, pharmaceutical companies have established extensive target identification and validation programs either internally or have outsourced it by expanding collaborations with academia. Although the ultimate validation of a target is when the proof of concept is achieved in human clinical trials, implementing methods able to increase the confidence in a new target at the initial stages of the drug discovery process, and before investing in a full-scale drug discovery program, are expected to reduce attrition rate at the later phases of drug development.

Jürgen Moll and Riccardo Colombo (eds.), *Target Identification and Validation in Drug Discovery: Methods and Protocols*, Methods in Molecular Biology, vol. 986, DOI 10.1007/978-1-62703-311-4_18,

In the last decade, genomic and proteomic approaches analyzing tissue biopsies from different cancers have provided researchers with an array of potential new targets in oncology. While this appears a wealthy resource for the development of new targeted therapies in cancer, the majority of these genomic targets are poorly characterized, and their roles as drivers for the different stages of the disease are often unknown. Many cancer genomes harbor numerous mutations, but the vast majority are likely to be passenger mutations—mutations that do not contribute to the development of cancer but have occurred during the tumorigenesis process—and only a small minority are the critical drivers of the disease. Therefore, for a particular new target to be included in the drug discovery pipeline, it should be validated by ensuring that it is critically involved in the development and progression of human cancers and that its modulation is likely to have a therapeutic effect. In addition, many human cancers develop secondary metastatic tumors. These secondary tumors are subjected to additional mutation and acquire the capability to spread from the original site via the blood or lymphatic system, becoming highly invasive and aggressive (2). Genes involved in these mechanisms can be better identified and validated mainly in an in vivo context. This chapter provides a summary on how animal models can be utilized for this scope. Since the mouse is the most widely used animal model in cancer research, the focus of this chapter is dealing mainly with the opportunities for in vivo target identification and validation of candidate cancer genes using this organism. Strengths of simpler animal models, such as the fruit fly (*Drosophila melanogaster*), for the identification of drug targets in oncology are also illustrated at the end of this chapter.

2. Mouse Models

Mouse models have been an invaluable tool in cancer research. Heterotransplantation of human cancer cell lines or tumor biopsies into immunodeficient mice (xenograft models) and genetically engineered mouse models (GEMMs) are currently being used as the major preclinical in vivo platforms for experimental therapeutics. Each of these models has its intrinsic advantages and shortcomings (3–5); however, both of them have contributed to our understanding of cancer development and treatment.

The advances in recombinant DNA technologies, the improvement of DNA delivery in cells, and the commonly used methodologies for the generation of GEMMs can be exploited for the creation of tailor-made genetically engineered human xenograft mouse models and GEMMs for in vivo target identification and validation. To assess whether a given target has a causal relevance in the tumorigenesis process, we need to evaluate the effects of its modulation in the adequate in vivo model. Since often a selective inhibitor or antibody is not available, a good alternative to modulate the target is by genetic means (i.e., by knocking down or knocking out the target function). This can be achieved by genetically manipulating human cancer cell lines and growing them in the mouse or by genetically manipulating the mouse germ line. In human xenograft mouse models, the target is modulated only in the human tumor-derived xenograft but not in the host, whereas in GEMMs, the modulation of the target can be induced in the whole organism or parts of it. Therefore, the choice of the mouse model is important because it depends on which gene and which tumorigenesis process we want to interrogate. For example, the validation of candidate cancer genes that may play a role in the tumor microenvironment can be best interrogated in GEMMS.

2.1. Genetically Manipulated Human Xenograft Mouse Models and Their Use for Target Validation

The majority of the human cancer cell lines commonly used in cell-based screening platforms are nowadays well characterized with regard to their genetic lesions, and many have the ability to grow in immunocompromised mice as human xenografts (6). Ideally, presumed targets are eliminated or modulated at genomic level in a defined human cancer context, and the resulting effect is subsequently analyzed in vivo. Gene knockout in somatic cells is extremely challenging and only succeeded in a few cell types (7); therefore, the modulation of target genes in cancer cells is normally achieved by knocking down the expression of the gene of interest by means of RNA interference (RNAi) using a specific small interfering RNA (siRNA) or a specific short hairpin RNA (shRNA) (8). Although the involvement of some genes in the process of cancer has been elucidated by intratumoral, systemic, or intraperitoneal injection of siRNAs (9–12), the use of siRNA in vivo encounters the limitations that the knockdown of the target genes is only transient and sometimes incomplete due to oligos instability in serum. Moreover, there are often off target effects and there are no good and reliable delivery systems available yet (13). Hence, a preferred alternative approach is to knock down candidate cancer genes in human xenograft mouse models by means of an shRNA that is stably integrated in the cell genome and can be expressed in a constitutive or inducible manner. This is typically achieved by transfecting or transducing human cancer cell lines with plasmids or viral particles, respectively. The engineered cancer cell lines carrying the shRNA in their genome are subsequently grown in immunocompromised mice, and the effects resulting from the downregulation of a particular target are then evaluated directly in vivo. The knockdown of candidate cancer genes in human xenograft mouse models using this approach has supported the validation of several targets for oncology drug discovery, such as the role of HER3 in HER2-amplified breast cancer (14), the role for Ran GTPase in epithelial ovarian cancer cell survival (15), and the function of phospholipase Cgamma1 (PLCgamma1) in metastasis (16). HER3 knockdown in vitro already suggested a role of HER3 in the proliferation of breast cancer cells that express high levels of HER2, and the role of HER3 in HER2-amplified breast cancer was strengthened by the observation that inducible knockdown of HER3 causes tumor regression in mice (14); similarly, the involvement of Ran GTPase in epithelial ovarian cancer cell survival has been shown in vitro and further corroborated in vivo (15). In the case of PLCgamma1, although in vitro knockdown of this gene in breast cancer cells induced a marked decrease in the number of cell membrane ruffles and impaired cell migration and invasion (16), the conclusive indication that a blockade of PLCgamma1 could have a therapeutic potential to prevent metastasis was obtained by the in vivo data demonstrating that the inducible knockdown of PLCgamma1 in mammary tumors counteracted metastasis dissemination (16).

A detailed protocol for the generation of genetically modified human xenograft mouse models in which the knockdown of a putative oncology target is temporally regulated in vivo using a tetracycline inducible system is found in this book (17).

2.2. Generation of Genetically Engineered Mouse Models

2.2.1. Constitutive GEMMs

There are essentially two basic technical approaches to produce GEMMs. The first is known as classical or conventional transgenesis and consists in the pronuclear microinjection of plasmid DNA or bacterial artificial chromosome (BAC) DNA into the pronucleus of fertilized oocytes, and the successive implantation of the microinjected one-cell-stage embryos in the oviduct of foster mothers. Some of the resulting newborn mice will have the exogenous DNA randomly integrated in their genome. By breeding each transgenic founder to wild-type mice, different transgenic mouse lines are established, each carrying several copies of the transgene in a particular genome position. This method creates what are commonly known as conventional transgenics, and it is normally utilized to insert new genetic information into the mouse genome, usually with the aim to overexpress or knock down genes (Fig. 1a). The first transgenic mouse lines documented were produced in 1981 in the laboratories of Frank Ruddle (Yale) and Frank Constantini and Elisabeth Lacy (Oxford) (18). Classical transgenesis has been extensively used to overexpress oncogenes in the mouse and has produced hundreds of “oncomice” which cover a wide range of cancers affecting most organs of the body (19).

The second approach involves modifying the genome of embryonic stem (ES) cells by homologous recombination (HR) with a DNA construct that carries DNA sequences homologous to the target gene region, leading to targeted genome modification (Fig. 1c, d). Embryonic stem (ES) cells that have recombined the exogenous DNA sequences with the genomic DNA are selected

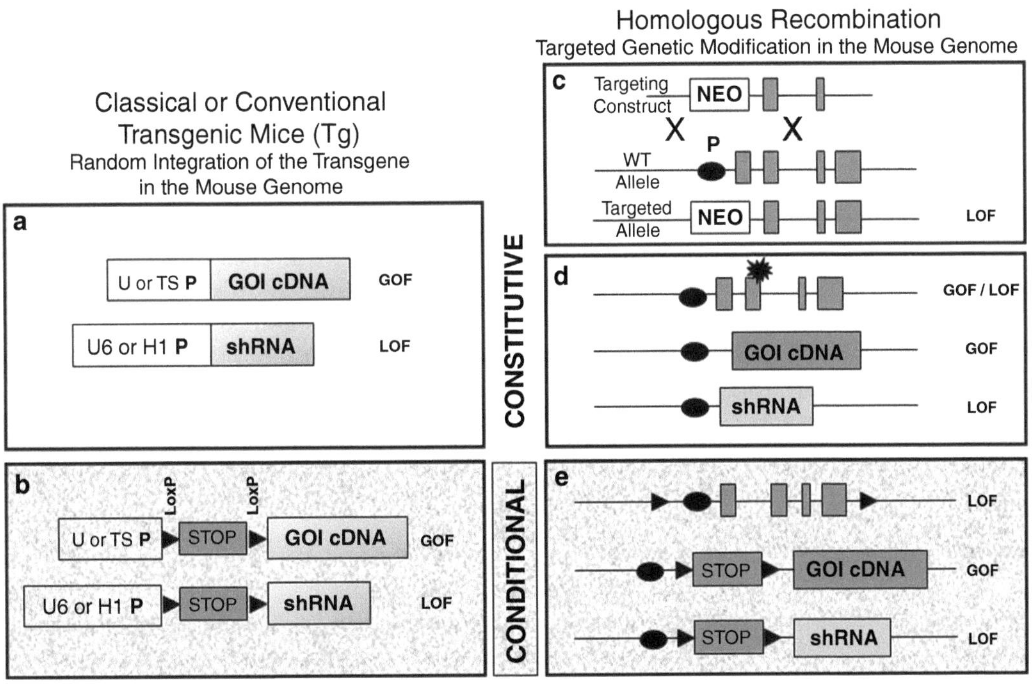

Fig. 1. Genetic modifications of the mouse germ line. This figure shows some examples of genome alterations that can be engineered in mice. (**a**, **b**) Classical or conventional transgenesis technologies result in the random integration of the transgene construct. (**a**) Constitutive gain-of-function (GOF) and RNAi-induced loss-of-function (LOF) mutants result in the overexpression or knockdown of genes, respectively. (**b**) Conditional GOF or LOF mutants can be generated by including a loxP-STOP-loxP cassette upstream of the gene of interest (GOI) cDNA or shRNA sequences in the transgene construct. The GOF or LOF mutation will be induced when an active Cre recombinase is available (see Fig. 2). (**c**–**e**) Homologous recombination in embryonic stem cells results in the targeted integration of the desired genetic modification. (**c**) An example of a homologous recombination strategy is illustrated. The targeting construct is homologous to the desired genomic DNA region to be targeted with the exception that a Neomycin resistance gene (NEO) replaces the promoter region and the first exon in the endogenous wild-type (WT) allele of the gene. In this case, homologous recombination in embryonic stem cells results in a targeted allele that is constitutively nonfunctional, knockout (KO). (**d**) Using similar approaches, subtle mutations (e.g., point mutation) can be introduced in a gene (knock in, KI). Homologous recombination can also be used to target a GOI cDNA or shRNA sequences to any desired locus in the mouse genome (KI), what ensures the expected pattern profile of gene/shRNA expression better than in classical transgenesis. (**e**) By including loxP sites, it is possible to generate conditional silent gene modifications that, upon Cre-induced recombination, will delete the loxP intervening sequences containing key exons of the gene. Similarly, Cre-induced deletion of the STOP signal results in the GOI cDNA or shRNA to be expressed under the control of the locus where they had been targeted. (*P* promoter, *U* ubiquitous, *TS* tissue/cell type specific, ✸ mutation, ▶ LoxP site, ⬬ endogenous promoter).

and then aggregated with mouse morulae or injected into mouse blastocysts, which are subsequently implanted in a foster mouse female. The genetically manipulated ES cells will mix with the morula cells or the blastocyst inner cell mass cells and give rise to a chimeric mouse. The final GEMM will be obtained when the genetically modified ES cells contribute the chimera germ line and the genetic modification is transmitted to the offspring. This method is generally used to manipulate a single gene, in most cases “Knocking” it “Out” (KO) (Fig. 1c), although more subtle genetic modifications can be made, for example, introducing a point mutation (Fig. 1d). Homologous recombination in ES cells also permits to "Knock In" (KI) cDNA sequences of a gene of interest (GOI cDNA) or an shRNA into a preselected locus of the mouse genome, what allows for their expression to be driven by the promoter and enhancers present in that locus (Fig. 1d). The first recorded targeted knockout (KO) mouse was created by Mario R. Capecchi, Martin Evans, and Oliver Smithies in 1989 (20), and this new revolutionary technology was awarded the Nobel Prize for Medicine in 2007. The ablation of suppressor genes by homologous recombination in ES cells has resulted in mice that show a high predisposition to develop tumors (21).

For further information on how GEMMs are generated, I refer the reader to several manuals which describe in detail the methods and technologies most commonly used (22–24).

2.2.2. Conditional GEMMs

Since the first transgenic and the first knockout mouse lines were generated, improved technologies have been developed that allow for the generation of gain-of-function (GOF) and loss-of-function (LOF) mouse mutants in which the mutation is induced in a spatial and temporal fashion by using well-defined tool mice (Figs. 1b, e and 2). These new-generation GEMMs are known as conditional GEMMs and have proven to be of high value for the elucidation of gene function.

In some cases, gene products that function in the adult also play key roles during embryonic development. Mice that are KO for that gene are developmentally lethal, and, as a consequence, adult mice cannot be generated. This limits the studies of gene function to embryonic development but does not allow determining a specific gene function in the normal adult or diseased organism. Conditional/inducible mutation approaches can circumvent this limitation and allow the mouse to develop and mature normally prior to ablation of the gene of interest. These improvements not only solve the problems of embryonic lethality observed in some constitutive GEMMs but also avoid the undesired effects of genetic adaptation by compensatory gene regulation networks which could occur when the mutation is already manifested in the embryo.

The ability to alter gene expression spatially and temporally in the adult mouse is normally achieved by interbreeding different GEMMs with the aim to merge different genetic modifications in the same mouse. In general, one allele consists of a silent mutation that will only be induced when the "inductor" is expressed and activated in a tissue-specific, inducible, or temporal manner (Figs. 1b, e and 2).

A widely used inducible system in the mouse is the bacteriophage P1-derived Cre-LoxP system (25). The site-specific Cre recombinase recognizes specifically 34 bp loxP sites. Generating mice that harbor a DNA sequence flanked by two loxP sites and

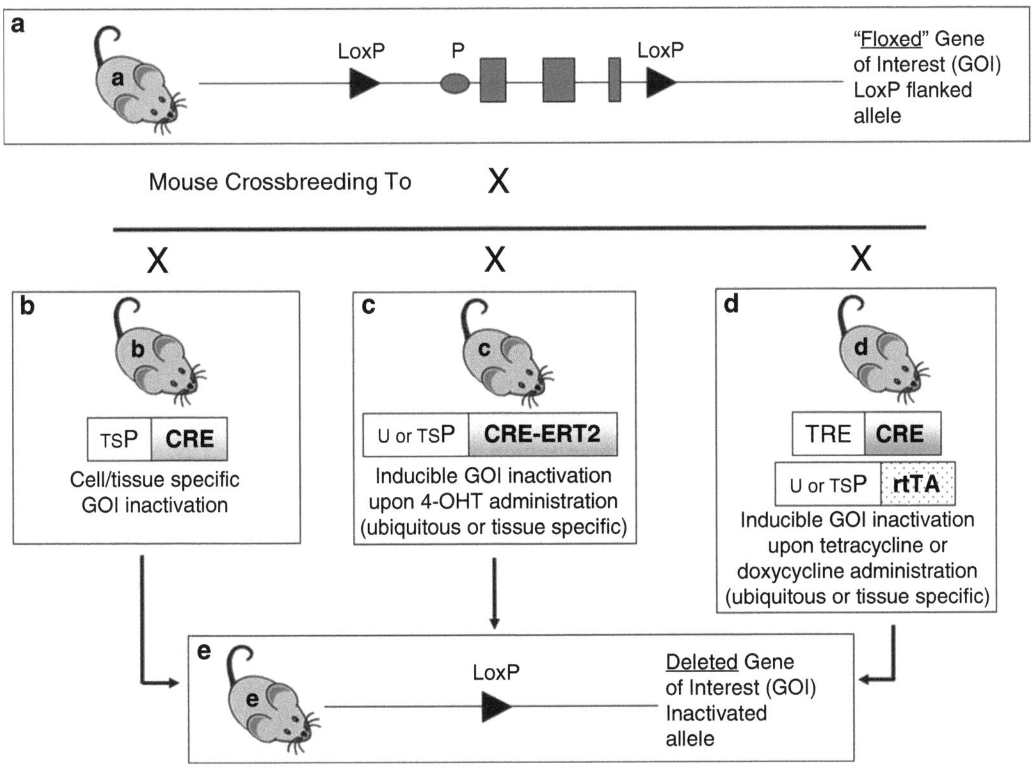

Fig. 2. "Tool" mice for conditional mutagenesis. This figure shows an example on how the inactivation of a gene is achieved conditionally. For this purpose, it is necessary to combine in the same mouse the silent gene alteration (e.g., here by introduction of LoxP sites) (**a**) and the genetic modifications necessary to activate Cre in the desired spatial and temporal fashion (**b**–**d**) to induce a specific mutation (e.g., here a gene deletion) (**e**). (**a**, **b**) Combination of the genetic modifications (**a**) and (**b**) in the same mouse, gene inactivation will be achieved in the tissue/cell type of interest by choosing the appropriate tissue-specific promoter (TSP). (**a**, **c**) Mice with the genetic modifications (**a**) and (**c**) express an inactive form of Cre (i.e., CreERT2) ubiquitously or tissue specifically. Cre will be functional upon 4-hydroxytamoxifen (4-OHT) delivery, what will result in the deletion of the loxP intervening sequences. (**a**, **d**) Crossbreeding of mice (**a**) and (**d**) generates mice in which it is possible to induce Cre expression and targeted gene inactivation upon delivery of tetracycline or doxycycline. (*P* promoter, *U* ubiquitous, *TS* tissue/cell type specific, *TRE* tetracycline response element, *rtTA* tet-on system TetR, ▶ LoxP site, ⬬ endogenous promoter).

crossbreeding them to a Cre transgenic mouse will produce mice that have deleted the intervening "floxed" DNA sequence (Fig. 2a, b, e). As an alternative to crossbreeding the inducible mutant mouse to a Cre mouse line, the Cre recombinase can be sporadically expressed in the organ of interest via engineered adenovirus or lentivirus. This last approach has been, for example, extremely successful for the delivery of Cre to the mouse lungs where it induced the desired mutation (26, 27).

Analogously to the Cre-LoxP system, the 34 bp Flippase recognition target (FRT) sites are recognized by the Flippase recombination enzyme (FLP) derived from *Saccharomyces cerevisiae* (28). The FLP–FRT recombination system is also utilized for the inducible genetic manipulation of the mouse genome (29, 30).

In addition to this spatial control of gene alteration, a temporal control can be added by delivering/activating the inductor at a specific time. This can be accomplished by using a CreERT2 fusion protein in which the Cre enzyme has been fused to a new mutated ligand-binding domain of the human estrogen receptor (ERT2) (31). This ligand-dependent Cre-ERT2 recombinase is inactive until the synthetic estrogen receptor (ER) ligand 4-hydroxytamoxifen (4-OHT) is delivered, for example, by the drinking water, topically or systemically, allowing inducible control of Cre activity and consequently target gene function alteration (Fig. 2a, c, e).

Another commonly used inducible system in GEMMs for the generation of conditional mutations is the tetracycline (Tet)-regulated system (32). This requires two genetic modifications in the same mouse which is usually accomplished by crossbreeding two genetically modified mouse strains. One strain expresses a form of the Tet repressor (rtTA), and the other strain expresses Cre under the control of a tetracycline response element (TRE). Only when rtTA binds tetracycline or its analog, doxycycline, will induce Cre expression and, as a consequence, deletion of the gene of interest (Fig. 2a, d, e).

Given the massive collection of genetically engineered tool mice which express Cre, CreERT2, or rtTA ubiquitously or in a tissue-specific fashion (for a complete list, see www.creline.org, www.nagy.mshri.on.ca/cre) (33), it is nowadays possible to control the expression of a gene in almost any tissue of choice and at any desired time point.

The inducible systems described above, apart from utilizing them for the induction of LOF mutations through gene deletion (Fig. 2), can also be exploited for the inducible expression of cDNAs or shRNAs. This is achieved by crossing mice that have a transcriptional STOP signal, flanked by loxP sites, between the promoter and the GOI cDNA or the shRNA (Fig. 1b, e) to one of the tool mice for conditional mutagenesis (Fig. 2b–d).

An advantage of the Tet system over the Cre and the FRT conditional gene expression systems is that in the Cre and FRT systems, activation or knockout of the gene is irreversible once recombination is accomplished, whereas the Tet system can also be used for a reversible gene activation or knockdown (34), which more closely mimics a drug treatment schedule. This can be achieved by placing the GOI cDNA or shRNA directly under the control of the TRE and switch on and off gene expression in mice by adding or not doxycycline to the mouse drinking water or chow (35).

2.3. Revealing Novel Oncology Targets and Their Validation via Mouse Genetics

GEMMs of cancer have been excellent tools for understanding the mechanisms of oncogenic transformation and advancing our knowledge of tumor biology (36). The technologies used for the manipulation of the mouse genome can be further exploited for the identification and validation of oncology targets relevant in

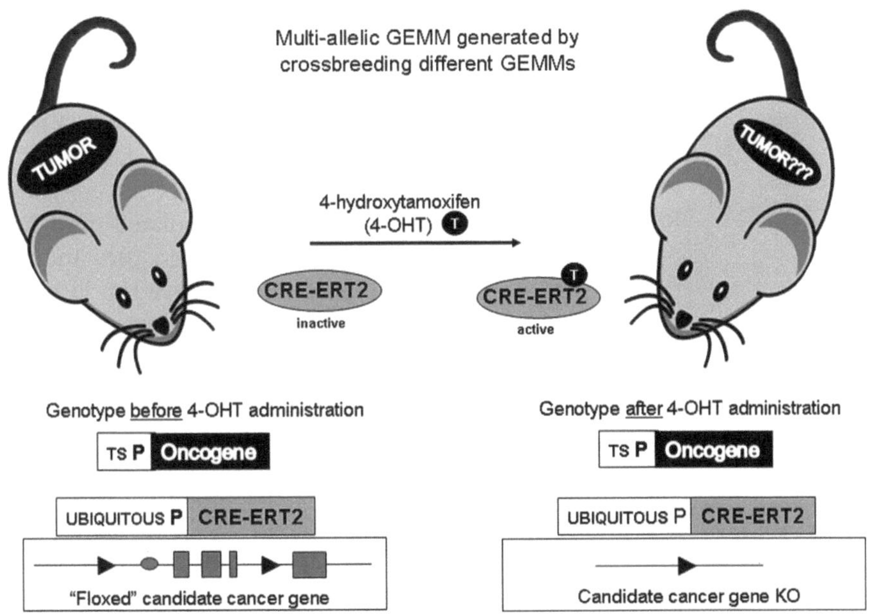

Fig. 3. Target validation in vivo using GEMMs. This figure shows an example of how GEMMs can be exploited for the validation, by genetic means, of candidate cancer genes in vivo. The idea is to combine the genetic modifications necessary to induce modulation of a target in an "oncomouse" background. For simplification the "oncomouse" background is illustrated by only one genetic modification (i.e., oncogene expression—GOF). More complex "oncomice" can be generated by combining GOF of oncogenes and LOF of tumor suppressor genes. To the "oncomouse" genetic background, the genetic modifications that will enable to temporally induce the modulation of the candidate cancer gene are introduced. In this particular example, they consist in a ubiquitous promoter driving the expression of CreERT2 and the candidate cancer gene flanked by loxP sites. Before 4-hydroxytamoxifen (4-OHT) delivery, the GEMM will develop oncogene-driven tumors in a situation in which the candidate cancer gene is expressed normally. However, after 4-hydroxytamoxifen (4-OHT) delivery, the candidate cancer gene will be deleted ubiquitously. This approach allows to interrogate in GEMMs whether a putative cancer gene is a key component in the oncogene-triggered signaling pathway (if yes, its deletion should interfere with the normal tumor growth). Moreover, this approach permits to evaluate what is the effect of the inactivation of this gene in the normal tissues. (***P*** promoter, *TS* tissue/cell type specific, T 4-hydroxytamoxifen).

specific cancer indications. For example, by means of transgenic technologies, it is possible to mutagenize the mouse genome with retroviruses or transposable elements and to assess the causal relationship between a particular gene mutation and cancer (37, 38). Moreover, the use of these technical approaches for the generation of GOF and LOF GEMMs, jointly with selective mouse interbreedings, allows the creation of sophisticated multi-allelic tailor-made GEMMs for target identification and validation via mouse genetics. For example, by combining in the same GEMM a series of genetic lesions necessary for the development of a particular tumor type, and the genetic modifications required for the inducible elimination or downregulation of a gene believed to play a role in that sort of tumor (Fig. 3), it is possible to get insight into the relevance of candidate cancer genes in a defined cancer type.

2.3.1. Target Identification by Retroviral and Transposon-Driven Mutagenesis in GEMMs

Forward genetic screens in the mouse provide a valuable tool for understanding genes and genetic pathways in human cancer. Chemicals and radiation mutagenesis are used to disrupt genes randomly, with the aim to identify gene function and the diseases associated with them, but it can be extremely difficult to pinpoint where in the genome a random mutation has been introduced (39). Retroviruses and transposable element-mediated mutagenesis, on the contrary, have the advantage that when they mutate genes, they provide a "tag" that pinpoints their location in the genomic DNA. The identification of genes that are altered and are driving tumor initiation and maintenance has been instrumental for understanding cancer phenotypic variations and for identifying targets for therapeutic intervention. High-throughput genome-wide screens in vivo to identify tumor suppressors and oncogenes in mammalian models have been advanced by retroviral insertional mutagenesis (39). Slow transforming retroviruses such as mouse mammary tumor virus (MMTV) and Moloney murine leukemia virus (MuLV) do not encode genes that are directly transforming. Instead these viruses are mutagenic when they integrate into a host cell chromosome as a provirus (37). Elements in the proviral genome that regulate the viral transcript also act in cis on cellular gene transcripts. Depending on whether the provirus integrates into or in the vicinity of genes, they can disrupt or enhance normal transcription, inducing LOF or GOF, respectively. Therefore, retroviral-driven mutagenesis allows the identification of novel tumor suppressor genes and oncogenes (37). However, the ability of retroviruses to act as a cancer gene discovery tool has a number of functional limitations. For example, they effectively infect only a limited number of cell types in vivo, infect only dividing cells, have preferred integration sites, and cause effects on distant genes. So far, retroviral insertional mutagenesis has been restricted to elucidate genes involved in hematopoietic and mammary cancers (38).

A more flexible group of insertional mutagens are transposons. Transposable elements are discrete pieces of DNA that can jump around in the genome of a cell. Each transposon system has a specific transposase that mobilizes the transposon from its position within the genome and causes its reintegration elsewhere by a cut-and-paste mechanism (40). Forward genetic screenings using transposon systems have long been used in lower eukaryotes; however, they had to be modified before they could be used as screening tools in mice. A great advantage of the use of transposons for insertional mutagenesis is that they can be modified to carry DNA fragments that are able to alter gene expression (41). An additional plus of such a system is that the mutations caused by transposon insertions are molecularly tagged, greatly facilitating their molecular identification. Another advantage of transposon insertional mutagenesis is the ability to engineer special vectors that can express a reporter gene in a context-dependent manner. Such "gene traps" can be designed to express the reporter gene, if and only if insertion occurs in the right place and orientation. GEMMs bred to carry both the DNA transposon and the corresponding transposase show random mutations due to transposon hops. Transposition can be temporally controlled and targeted to the cell or tissue of interest by controlling the expression of the transposase (41). Analyzing the transposon hop site DNA sequences in mice developing tumors enables the identification of new cancer related genes. The most commonly used transposon systems in mammals are Sleeping Beauty (SB) and the piggyBac transposons, which were originally isolated from the Atlantic salmon and the moth, respectively (42, 43). The first transposons used in mammals were mobilized very seldom and thus could not be properly used for genetic screens. Recently, several groups have improved the efficacy of this genetic tool which is now used as a gene discovery approach in mice (44–46). For example, a more active eukaryotic SB transposition system, T2/Onc2, was generated which is flanked by optimized SB transposase binding sites that increase SB transposition (44). This improved SB-mediated somatic mutagenesis in mice identified a role for Colony Stimulating Factor-1 (CSF1) in the formation of high-grade astrocytomas (47). The identification of CSF1 as a candidate gene for this type of cancer was then further supported by the fact that CSF1 and its receptor, CSF1R, are overexpressed in high-grade human astrocytomas (47). Moreover, a recent study showed that using PLX3397 (a CSF-1R inhibitor which can cross the blood brain barrier) reduced the number of tumor-associated microglia and glioblastoma invasion in an orthotopic mouse model of glioblastoma, further corroborating the relevance of CSF1/CSF1R pathway in high-grade astrocytomas (48). Similarly, a recent insertional mutagenesis screen using this SB transposon system, specifically designed to profile the repertoire of genes that can drive colorectal cancer, has identified several networks of cooperating genes driving intestinal tumorigenesis (49).

2.3.2. Target Validation by Gene Knockdown or Knockout in GEMMs

Selective interbreeding of GEMMs to merge in the same mouse the desired gene mutations offers the attractive possibility of genetically validating targets in vivo and, in parallel, investigating the possible genetic interactions between different proteins and pathways involved in cancer. The idea behind is to firstly create GEMMs in which one or more cancer-relevant genetic lesions are introduced in the same mouse to generate an "oncomouse." Secondly, the introduction of additional genetic modifications will mimic the therapeutic effect of a given target inhibitor by genetically modulating target expression in the whole mouse (Fig. 3). This approach not only elegantly validates the target or pathway with regard to anticancer efficacy but also provides information on any possible mechanism-based safety liability that could be encountered when target function is inhibited. Target modulation in these sophisticated multi-allelic modified mice is commonly achieved genetically through gene silencing by RNAi, which induces gene knockdown, or by gene KO. The technologies used for the generation of GEMMs can also be employed to induce RNAi in mice by introducing validated shRNA in the mouse germ line (50). Alternatively, therapeutic intervention can be mimicked by knocking out the candidate cancer gene in GEMMs (Fig. 3). Sophisticated studies using these strategies have been pursued by several laboratories, which have elegantly identified some of the effector pathways downstream of oncogenic Ras. Using a K-Ras-driven GEMM of non-small cell lung cancer (NSCLC), it was shown that Rac1 function is required for tumorigenesis in this model (51). A similar study demonstrated a critical function of NF-kappaB signaling in lung tumor development and that this requirement was dependent on p53 status (52), identifying the NF-kappaB pathway as a promising therapeutic target in lung adenocarcinoma. This concept was further supported by a recent study in which NF-kappaB pathway inhibitors caused regression in K-Ras and p53 mutant mice (53), although long-term treatment was associated with acquired resistance.

Similarly, GEMMs have led to the identification of synthetic lethal interactions, as for example, the interaction between K-Ras and Cdk4, suggesting that selective pharmacological inhibition of Cdk4 could provide therapeutic benefit for K-Ras-driven NSCLC (54). A study in which another K-Ras-driven GEMM of NSCLC was treated with a Cdk inhibitor which, among other Cdks, inhibits Cdk4 further supported the validation of Cdk4 as a therapeutic target for K-Ras-driven lung cancer (55).

2.3.3. Target Validation by Reversible On/Off Gene

Taking advantage of the reversibility of the tetracycline system (34), it is possible to reveal specific gene requirements for tumor maintenance. The strategy consists in inducing overexpression of cancer-related genes to generate tumors and then switch off gene expression by tetracycline withdrawal. This on/off approach enabled the demonstration that transcriptional extinction of an oncogene such as H-Ras or Myc leads to the almost complete regression of melanoma and lymphoma, respectively (56, 57). Another study using a similar approach revealed that MYC inactivation induces sustained regression of invasive liver cancer. However, the liver cancer cells remain dormant only as long as MYC remains inactivated. MYC reactivation immediately restores their neoplastic features (58). The authors confirmed that the dormant liver cells and the restored tumors retained the identical molecular signature and hence were clonally derived from the original tumor cells. This approach uncovers targets whose inhibition alone may slow down the disease and that the inhibition of additional targets may be necessary for a good therapeutic intervention.

3. Fruit Fly (*D. melanogaster*) as a Model System for Target Identification in Oncology

The mouse has been the primary animal of choice in cancer research. However, other animal models, such as the fruit fly (*D. melanogaster*) and zebrafish (*Danio rerio*), despite their relative simplicity, have made significant contributions to the understanding of the tumorigenesis processes (59, 60). These animal models are easier and cheaper to keep in larger numbers than mice, enabling even medium-sized labs to perform high-throughput screenings. Targets and novel hypothesis identified using these animals models encourage to further proceed with studies in more complex model organisms. This part of the chapter is dedicated to Drosophila as an example of how simpler organisms can also be used for the identification of novel candidate cancer genes.

The Drosophila system combines the advantages of experiments using cell culture with in vivo experiments with higher animals. Short generation time, low breeding costs, and a relatively small genome allow medium or high- throughput screens for the identification of mutations that affect a given biological process. For example, the Drosophila field has been a leader in our understanding of signaling and cell-cycle mechanisms that control epithelial development (59). Studies using Drosophila have identified the hippo signaling pathway, which coordinates cell proliferation and death to achieve normal tissue size, and have provided insight on how oncogenes and tumor suppressor genes cooperate to achieve tumor growth and metastasis (61, 62). Also the WNT, NOTCH, JAK-STAT, and RAS pathways, which are at the center for many cancers in humans, have been genetically dissected in Drosophila (59, 63). The powerful genetic methods in Drosophila and the speed of the fly life cycle continue to uncover new signaling pathways. Considering that the evolutionary conservation of disease-related pathways between *Drosophila* and humans is not limited to individual genes but often also reflects the functional conservation of entire gene networks (64, 65); the findings uncovered in Drosophila can be extrapolated to mammals and be used as a starting hypothesis to perform further studies in other model organisms. For example, studies in Drosophila have identified *raps*, *mira*, *numb*, and *pros* as candidate genes that play a role in cell polarity (66). Loss of polarity in Drosophila might, when affecting asymmetrically dividing stem cells, impair the fate of the daughter cells, rendering them unable to respond to the mechanisms that control proliferation in the wild-type lineage and initiating tumor growth. It will be interesting to interrogate the mammalian counterparts of these genes identified in Drosophila and the pathways where they act, including cancer stem cell signaling. Emerging hypothesis indicating that loss of epithelial cell polarity may have an important role not only at the initiation of tumorigenesis but also at later

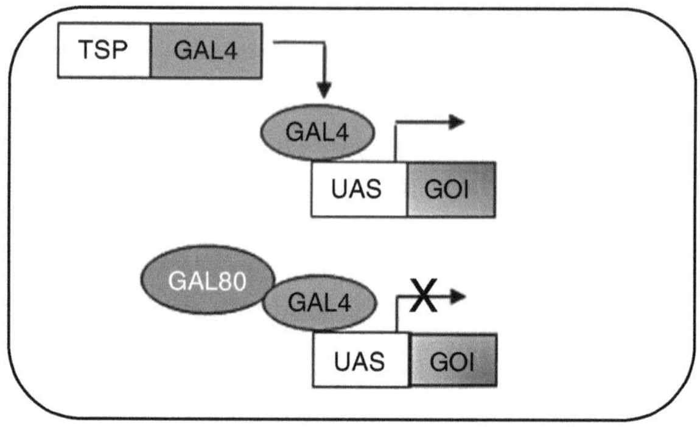

Fig. 4. The UAS/GAL4 system. The UAS/GAL4 system consists of an enhancer/promoter with GAL4 binding sites (UAS sites) driving the expression of the gene of interest (GOI). Specificity of transgene expression in tissues is achieved by the tissue-specific expression of the yeast transcription factor GAL4. An additional level of inducibility is added by the GAL4 inhibitor, GAL80. Binding of GAL8 to GAL4 prevents GAL4-mediated transcriptional activation.

stages of tumor development further encourages these studies. Interestingly, recent observations have indicated that loss of cell polarity and malignant transformation are tightly correlated in human carcinomas (67).

Taking Drosophila as a tool, disease-related pathways can be manipulated. Gene mutations or alterations that specifically interfere with a signal transduction pathway associated with human cancer point to that particular gene product as having an essential function in that pathway. Consequently, it is expected that functional inactivation by low-molecular-weight drugs will attenuate the signaling pathway in much the same way as the mutation in the corresponding gene does.

What is the genetic tool box available in Drosophila that can be used for the identification of candidate cancer genes? Researchers have developed, and continue to develop, methodologies to perform all sorts of genetic manipulations in Drosophila that allows, among others and as described below, GOF and LOF mutants by transgenesis, ability to perform large-scale screens, and the generation of clones, which contain multiple mutations within tissues of phenotypically normal cells (68, 69). And all this is relatively fast and inexpensive.

GOI cDNAs or shRNAs can be expressed in the tissue or cell type of interest and in an inducible manner using the UAS/GAL4 system (70). The system consists of an enhancer/promoter with GAL4 binding sites (UAS sites) driving the expression of the GOI cDNA or shRNA (Fig. 4). Specificity of transgene expression in tissues is achieved by the tissue-specific expression of the yeast transcription factor GAL4. This reverse genetic tool permits to evaluate gene function through GOF or LOF approaches.

Another methodology, and one of the most powerful aspects of Drosophila research, is the ability to perform large-scale screens.

In forward genetic screens, a large number of flies with different mutations are generated and screened for abnormal phenotypes in the biological process of interest. Given the ease with which the Drosophila genome can be saturated for mutations with a specific phenotype, it is possible to identify most, if not all, of the functionally relevant components of a signaling pathway (71). This is further aided by the small genome (72, 73), resulting in a lower degree of functional redundancy. Furthermore, because such screens are performed in living animals, any deleterious or lethal side effects that may be tolerated by cells in culture are not recovered. Therefore, targets that may have liabilities are not identified. Ethylmethanesulfonate (EMS), X-ray, and transposable P or EP elements are the most frequently used mutagens for genetic screens in Drosophila. EMS induces point mutations and small deletions, whereas X-ray generates deletions, chromosomal breaks, inversions, and translocations (74). Transposable elements such as P and EP insert into the genome almost randomly. P elements are transposable elements from *D. melanogaster* P strain (75) and are normally used for the generation of LOF mutations. P elements can be combined with a reporter gene driven by a minimal promoter, in such a way that when the latter comes under the influence of an endogenous enhancer, the reporter will be expressed in the same way as the gene that is being targeted (enhancer trap). EP elements are modified P elements that contain an enhancer/promoter with GAL4 binding sites (UAS sites) which, in combination with the tissue-specific expression of the GAL4 transcription factor, allows the study of over- or miss-expression-related phenotypes of tagged genes in the appropriate tissue. EP elements can be used either for simple GOF screens or can be combined with a modifier screen to search for GOF suppressor mutations in a sensitized genetic background (76). P and EP transposable elements have the advantage that they serve as a molecular tag to isolate flanking sequences, which facilitates the identification of the gene at the integration site. Forward genetic screens in Drosophila have identified and characterized major signaling pathways and cellular processes which are conserved from flies to mammals. Genes identified as suppressors of cancer-related phenotypes caused by overactivation of a signaling pathway (e.g., Ras, WNT) in an EP screen encode potential negative regulators of this pathway and may thus correspond to tumor suppressor genes in humans. Conversely, genes identified in conventional suppressor screens, in which a LOF mutation in heterozygous condition suppresses the phenotype, encode positive regulators of the pathway. Negative regulators may provide diagnostic markers for the classification of a particular disease condition, but the interesting ones as putative drug targets are the positive regulators. Unfortunately, dominant modifier screens cannot identify all the essential components in a signaling pathway. For some proteins which perform an essential function in a given pathway, a 50 % reduction may not be sufficient, even in a sensitized background. Therefore, screening strategies that induce a strong knockdown of positive regulators are needed to identify the majority of the pathway constituents. This can be achieved by LOF screens that use the UAS/GAL4 system for the expression of shRNAs (70, 77). Taking advantage of the available libraries of transgenic Drosophila stocks that conditionally express the interfering shRNAs targeting almost every gene in the Drosophila genome (see: http://stockcenter.vdrc.at/control/main, http://www.shigen.nig.ac.jp/fly/nigfly/index.jsp) (78), LOF screens can be designed for the identification of genes which are relevant for particular steps during the tumorigenesis process.

Another elegant technology in Drosophila consists in the possibility to easily generate mutant cell clones which contain multiple mutations in an otherwise wild-type environment. This adds value to the Drosophila model as a cancer research tool because it permits to explore the interactions among cancer mutations and between cell populations that mimic the tumor microenvironment. Moreover, this clonal analysis permits to elucidate cell competition mechanisms, which could potentially confer malignant cells with a growth advantage over their neighbors. The generation of the mutant cell patches is normally achieved with the MARCM system (Mosaic Analysis with a Repressible Cell Marker) which, by combining the FLP/FRT recombination and the UAS/GAL4 systems, enables to label the homozygous mutant cells, usually by green fluorescent protein (GFP) expression (Fig. 5) (79). The MARCM approach basically consists in the FLT/FRT-mediated mitotic recombination between mutant and nonmutant chromosomes in flies that are in a GAL4; UAS-GFP genetic background, what results in the generation of GFP-marked cell clones that are homozygous mutant for a particular gene, in an otherwise nonmarked and nonmutant environment for that specific gene. To mark the homozygous mutant cells, the MARCM approach takes advantage of GAL80 interfering with the GAL4 induction of gene expression. Binding of GAL80 to the carboxy-terminal 30 amino acids of GAL4 prevents GAL4-mediated transcriptional activation (80) (Fig. 4). Only the cells that are homozygous mutant and have lost the expression of GAL80 will be able to express the GFP gene (Figs. 4 and 5). In these transgenic flies, FLP expression is usually controlled by a heat-shock-inducible promoter that allows for a temporal control of recombination and, as a consequence, generation of mutant clones. The MARCM system allows in vivo tracking of biological processes and identification of new genes involved in these processes by mosaic screening.

The genetic tools in Drosophila described above and selective crosses of the resulting genetically engineered flies allow the generation of flies that show cancer-related phenotypes and, at the same time, permit to modulate the function of a presumed oncology cancer

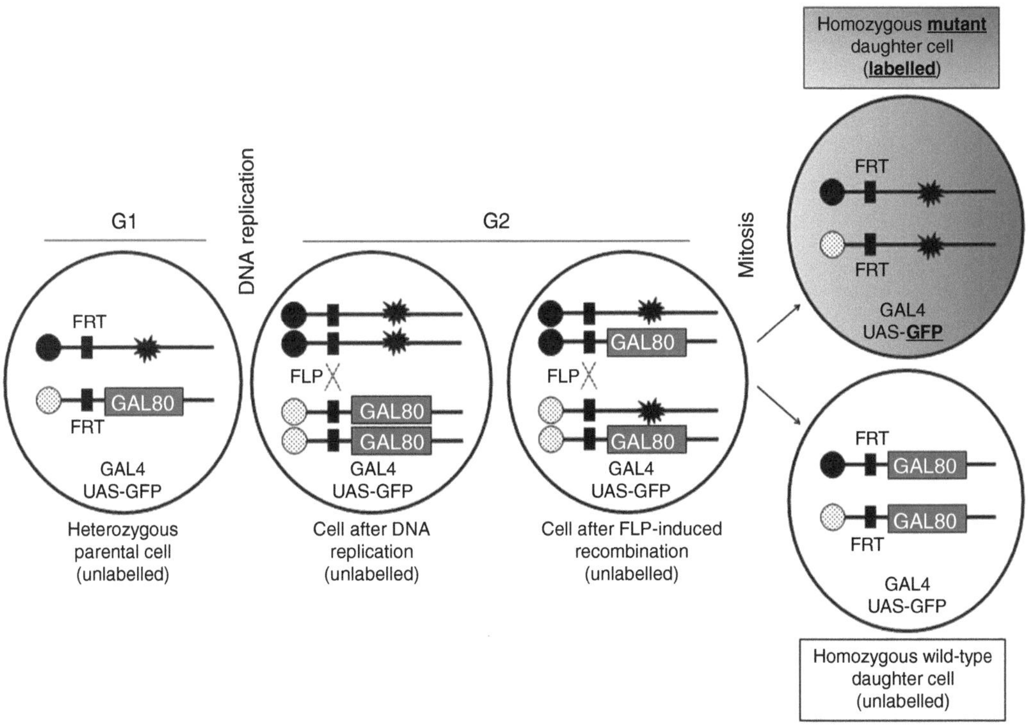

Fig. 5. The MARCM system (Mosaic Analysis with a Repressible Cell Marker). The MARCM system permits to produce clones of green fluorescent protein (GFP)-marked cells that are homozygous mutant in a gene of interest in an otherwise nonmarked, nonmutant environment. The system combines Flippase (FLP)/FRT-mediated mitotic recombination between mutant and nonmutant chromosome, in flies that are in a GAL4; UAS-GFP genetic background. Only the cells that are homozygous mutant will lack GAL80 allowing the expression of GFP (Fig. 4). In these transgenic flies, FLP expression is usually controlled by a heat-shock-inducible promoter, what allows controlling mitotic recombination and, as a consequence, generation of mutant clones (i.e., by warming up the flies). The MARCM system allows in vivo tracking of biological processes and identification of new genes involved in these processes by mosaic screening.

gene in the whole cell population or in cell clones. Therefore, the effects of the putative target modulation can be evaluated using the cancer-related phenotype as a readout. For example, by means of an RNAi genetic screen and clonal analysis, it has been identified that loss of PI3K blocks cell-cycle progression in a Drosophila tumor model induced by loss of polarity and activated Ras signaling (81).

The power of fly genetics, jointly with emerging new assays for the evaluation of cancer-related phenotypes in this organism, is also shedding some light on key genes involved in metastasis (82). Although tumor metastasis has not been observed in wild-type flies, the regulation of cell behaviors such as the migration and invasion of epithelial sheets will probably show mechanistic similarities to those processes involved in the multistep spread of cancer cells. Moreover, assays to discern between benign and malignant growth in Drosophila have recently been developed (83). The method consists in the implantation of the affected tissue, normally referred as allograft or "dauer" culture in a healthy host. Upon implantation, wild-type tissue never overgrows, and benign hyperplasias grow slowly, do not invade other tissues, and retain their capacity to differentiate. Malignant neoplasms, in contrast, display autonomous growth, have the capacity to migrate to and colonize distant organs, and cause lethality to the host. Often, these malignant neoplasms become immortal and can be expanded limitlessly through successive rounds of implantation into healthy hosts. Evaluation of these cancer phenotypes after target modulation uncovers novel players in tumor progression and cell invasion (82).

4. Conclusions and Perspectives

To reduce to a minimum the high attrition rates observed in oncology drug development, it is of tremendous interest to identify real promising targets and to eliminate the failures at an early stage. It is also important to implement an in vivo validation step in the early phases of drug discovery to ascertain that modulation of a target will result in an antitumorigenic effect. Moreover, in vivo target validation will prevent missing targets that are involved in tumorigenesis processes that cannot be assessed in vitro. This chapter has provided a summary on how genetically manipulated animal models can contribute to the discovery and validation of oncology targets in the whole organism. It has been mainly dedicated to the mouse as model system for target identification and validation but has also illustrated how Drosophila is emerging as an inexpensive and easy tool for target identification in cancer research. In conjunction with studies using other model organisms, Drosophila has contributed greatly to our understanding of the mechanisms involved in cancer initiation and progression, revealing previously unknown molecular components and concepts. Although the short-lived fly does not naturally develop cancer manifested by lethal tumor overgrowth and metastasis, as observed in vertebrates, mutations in Drosophila genes that affect cell-cycle control and epithelial cell integrity have been recovered and the pathways where they function dissected genetically (59, 61–63). Reliable data obtained from hypothesis-driven-based screens in Drosophila, suggesting that a given gene may function in a defined tumorigenesis signaling process, encourages further efforts for the investigation on whether this also holds true in mammals. Hence target validation in oncology often includes targets that were initially discovered in flies as being key players in cancer-related processes, such as cell cycle, apoptosis, invasion, and migration.

Although genetic screens using transposons are easily performed in Drosophila and zebrafish (59, 60), this technology is

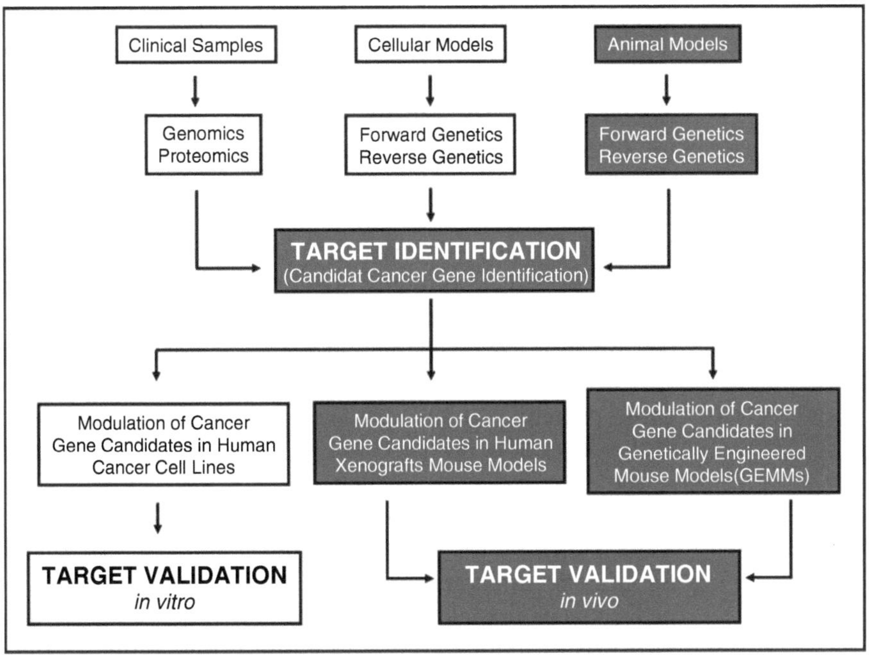

Fig. 6. Target identification and validation—workflow. Putative oncology targets identified by genomic and proteomic approaches applied to clinical samples, by genetic screens using cellular and animal models, constitute the collection of candidate cancer genes that need to be validated before being incorporated into the drug discovery pipeline. These identified candidate cancer genes should be validated by modulating their function in the appropriate cancer context. The shaded path highlights the in vivo strategy for target identification and validation.

now also emerging in mice with the aim to identify genes relevant for the growth of solid tumors (44).

Candidate cancer genes identified in genetic screens using animal models should be added to the collection of putative oncology targets identified by genetic screens using cellular models or by genomic and proteomic approaches applied to clinical samples (Fig. 6). The next step is to validate these emerging targets by modulating their function in the appropriate cancer context and, ideally, using an in vivo strategy (Fig. 6, shaded path).

The mouse is the primary model of choice for the validation of oncology targets in vivo. The technique of homologous recombination in mouse ES cells has revolutionized mouse cancer modeling. It allows to remove genes, to replace genes, or to perform a variety of other genomic alterations that replicate the different types of genetic aberrations or mutations that have been found to be associated with human tumors. We are now able to alter the genome of laboratory mice with great precision and generate “oncomouse” GEMMs in which to test how, when, where, and in which combinations particular gene alterations are involved in the initiation and progression of cancer to invasive, metastatic tumors. These complex “oncomouse” GEMMs can be further crossbred to additional tool

GEMMs with the aim to genetically modulate presumed oncology targets and interrogate them in vivo. Considering the growing collection of GEMMs which harbor conditional mutations of putative oncology targets available, the large collection of genetically modified tool mice (Cre, CreERT2, and Tet mouse lines) and the existing characterized "oncomouse" GEMMs (3, 33), one can potentially explore the relevance of candidate cancer genes in vivo in defined cancer contexts. Moreover, these strategies permit also for the investigation of any possible mechanism-based toxicity and the discovery of genetic interactions between gene products. The last aspect is extremely helpful for the validation of druggable targets in a given cancer signaling pathway triggered by a nondruggable protein. Targets validated genetically in vivo should get a higher level of confidence and hence be prioritized for the feeding of the drug discovery pipeline.

Unfortunately, despite the success of GEMMs in modeling human cancers for biological studies and that the mimicking of therapeutic intervention genetically in mouse models has been proven to be feasible (51–54), the generation and use of germ line GEMMs for high-throughput target validation poses the issues of high expenses and long periods of time for their generation and analyses. Therefore, researchers have started to consider alternatives such as non-germ-line GEMMs (5, 84, 85). For the generation of these models, genetically manipulated ES cells are implanted in mouse embryos which result in chimeras that carry a mixture of predisposed cells that are derived from the ES cells and from wild-type host cells. In these settings, tumors develop in the context of normal tissue, recapitulating human tumorigenesis. Alternatively, non-germ-line GEMMs can be generated by transplanting cells that can be derived either from donor GEMMs or human cancer cells which have first undergone ex vivo genetic engineering.

Although the role of the tumor microenvironment and a competent immune system on tumor progression and growth has been recognized, what has led in the last years to a greater reliance on GEMMs besides classical human xenograft mouse models, the latter have the advantage that the tumor is of human origin and thus reproduces the genetic complexity observed in human tumors. If we want to know whether a human tumor will respond to a specific gene knockdown or a specific therapeutic regimen, it will be more informative to examine the response of a human tumor, not a mouse tumor, to the target modulation (4).

Orthotopic human xenograft mouse models aim at solving the drawbacks of the subcutaneous human xenograft mouse models. The orthotopic implantation of human cancer cells results in the tumor growing at its original organ site, hence reproducing better the original tumor environment. In the past, a challenge presented with orthotopic xenograft models, as compared with subcutaneous xenograft models, was the difficulty of following tumor growth and tumor response to target modulation or therapeutic intervention. However, nowadays, with the use of magnetic resonance imaging and optical imaging technologies, this drawback has been abolished, and tumor behavior, including metastasis, can be followed noninvasively in the whole animal.

Emerging technologies will facilitate the introduction of targeted mutations in human cancer cell lines before they are implanted in the mouse host for the generation of human xenograft mouse models. Recent efforts aimed at creating engineered nucleases hold promise for targeted genome editing in mammalian cells. There are two principal classes of nucleases that can be engineered to recognize specific sequences in the genome: zinc-finger nucleases (ZFNs) and activator-like effector nucleases (TALENS) (86–88). These nucleases are a class of engineered DNA-binding proteins that facilitate targeted editing of the genome by creating double-strand breaks in DNA at user-specific locations. Double-strand breaks are important for site-specific mutagenesis in that they stimulate the cell's natural DNA-repair processes, namely, homologous recombination (HR) and nonhomologous end joining (NHEJ). Using well-established and robust protocols, these cellular processes can be harnessed to generate precisely targeted genomic edits resulting in cell lines, including somatic cell lines, with targeted gene deletions (KO), integrations, or modifications. Both ZNFs and TALENS have shown promising results in the introduction of targeted gene alterations in cells; however, TALENS appear to be much more specific than ZNFs for their target DNA sequence (87–90). Despite in its infancy and although a full confirmation of its value is still needed, the TALENS approach has been awarded the Method of the Year 2011 (91, 92). Several laboratories in the transgenesis field are spending efforts and resources for the generation of TALENS-mediated transgenic animals, and encouraging results have already emerged in rats and zebrafish (93, 94). Moreover, this emerging technology should facilitate the generation of isogenic human cancer cell line pairs, which are genetically identical except for the presence or absence of a key cancer causing gene. These cell lines are very attractive tools for the identification in vitro and in vivo of synthetic lethal interactions, that is, the identification of genes in a pathway that, upon knockdown, is only lethal in a specific cellular context.

After having discussed the advantages and drawbacks of each mouse model, the already available and the newly emerging genetic technologies, and considering that oncology targets are key players in one or several distinct steps of the tumorigenesis process, it is not possible to standardize a protocol for target validation in vivo. For each target, a specific hypothesis-driven-based approach should be designed depending on the available tools. It is important to emphasize that no single mouse model is likely to recapitulate all aspects of the complex genetics and biology of human cancers; hence, we have to recognize the strength and limitations of each model when using them for the validation of specific oncology target candidates. Depending on the hypothesis of what is the specific function of a target, the appropriate model should be chosen.

Another important point to take into account is that target validation using genetic means as described here, that is, through gene KO or gene KD, evaluates the effects of no target expression at all or reduced levels, respectively. We have to be aware that it is not the same to reduce or completely eliminate a protein genetically than to inhibit its activity, for example, by a small molecule, since proteins might have scaffolding functions which are not interrogated by a small-molecule inhibitor. Nevertheless, these surrogate models provide good insights into the in vivo relevance of putative oncology target genes and pathways and encourage further investigations on these target genes by implementing new drug discovery projects. For some targets, for example, kinases, an alternative approach is the expression in vivo of well-characterized dominant negative mutants, which should phenocopy better small-molecule-mediated inhibition (95).

In this new era of genomic advances, including cancer sequencing projects, new candidate cancer genes will emerge which will need to be validated. Implementing an in vivo target validation step early in the drug discovery process will provide more confidence when selecting new drug targets for the feeding of the drug discovery pipeline. Moreover, this action should contribute to reduce the attrition rates for compounds during clinical development.

References

1. Walker I, Newell H (2009) Do molecularly targeted agents in oncology have reduced attrition rates? Nat Rev Drug Discov 8: 15–16
2. Gupta GP, Massague J (2006) Cancer metastasis: building a framework. Cell 127:679–695
3. Singh M, Johnson L (2006) Using genetically engineered mouse models of cancer to aid drug development: an industry perspective. Clin Cancer Res 12:5312–5328
4. Richmond A, Su Y (2008) Mouse xenograft models vs GEM models for human cancer therapeutics. Dis Model Mech 1:78–82
5. Heyer J, Kwong LN, Lowe SW et al (2010) Non-germline genetically engineered mouse models for translational cancer research. Nat Rev Cancer 10:470–480
6. Forbes SA, Bhamra G, Bamford S et al (2008) The catalogue of somatic mutations in cancer (COSMIC). Curr Protoc Hum Genet Chapter 10:Unit 10.11
7. Rago C, Vogelstein B, Bunz F (2007) Genetic knockouts and knockins in human somatic cells. Nat Protoc 2:2734–2746
8. Sandy P, Ventura A, Jacks T (2005) Mammalian RNAi: a practical guide. Biotechniques 39:215–224
9. Futami K, Kumagai E, Makino H et al (2008) Anticancer activity of RecQL1 helicase siRNA in mouse xenograft models. Cancer Sci 99:1227–1236
10. Elez R, Piiper A, Kronenberger B et al (2003) Tumor regression by combination antisense therapy against Plk1 and Bcl-2. Oncogene 22:69–80
11. Verma UN, Surabhi RM, Schmaltieg A et al (2003) Small interfering RNAs directed against beta-catenin inhibit the in vitro and in vivo growth of colon cancer cells. Clin Cancer Res 9:1291–1300
12. Oh BY, Lee RA, Kim KH (2011) siRNA targeting Livin decreases tumor in a xenograft

model for colon cancer. World J Gastroenterol 17:2563–2571
13. Oliveira S, Storm G, Schiffelers RM (2006) Targeted delivery of siRNA. J Biomed Biotechnol 2006:63675
14. Lee-Hoeflich ST, Crocker L, Yao E et al (2008) A central role for HER3 in HER2-amplified breast cancer: implications for targeted therapy. Cancer Res 68:5878–5887
15. Barres V, Ouellet V, Lafontaine J et al (2010) An essential role for Ran GTPase in epithelial ovarian cancer cell survival. Mol Cancer 9:272
16. Sala G, Dituri F, Raimondi C et al (2008) Phospholipase Cgamma1 is required for metastasis development and progression. Cancer Res 68:10187–10196
17. Mazzoletti M, Texidó G. (2013) In vivo target validation by inducible RNAi in human xenograft mouse models, methods in molecular biology. Target identification and validation in drug discovery. Eds. Jürgen Moll and Riccardo Colombo.
18. Costantini F, Lacy E (1981) Introduction of a rabbit beta-globin gene into the mouse germ line. Nature 294:92–94
19. Hanahan D, Wagner EF, Palmiter RD (2007) The origins of oncomice: a history of the first transgenic mice genetically engineered to develop cancer. Genes Dev 21:2258–2270
20. Capecchi MR (1989) Altering the genome by homologous recombination. Science 244: 1288–1292
21. Taneja P, Zhu S, Maglic D et al (2011) Transgenic and knockout mice models to reveal the functions of tumor suppressor genes. Clin Med Insights Oncol 5:235–257
22. Nagy A, Gertsenstein M, Vintersten K et al (2003) Manipulating the mouse embryo—a laboratory manual. Cold Spring Harbor Laboratory Press, New York
23. Hofker MH, van Deursen J (2002) Transgenic mouse: methods and protocols, methods in molecular biology, transgenic mouse: methods and protocols, vol 209. Humana Press, New Jersey
24. Pease S, Saunders TL (2011) Advanced protocols for animal transgenesis: an ISTT manual. Springer Protocols, Springer-Verlag, Berlin, Heidelberg
25. Sauer B, Henderson N (1988) Site-specific DNA recombination in mammalian cells by the Cre recombinase of bacteriophage P1. Proc Natl Acad Sci USA 85:5166–5170
26. Jackson EL, Willis N, Mercer K et al (2001) Analysis of lung tumor initiation and progression using conditional expression of oncogenic K-ras. Genes Dev 15:3243–3248
27. DuPage M, Dooley AL, Jacks T (2009) Conditional mouse lung cancer models using adenoviral or lentiviral delivery of Cre recombinase. Nat Protoc 4:1064–1072
28. Zhu XD, Sadowski PD (1995) Cleavage-dependent ligation by the FLP recombinase. Characterization of a mutant FLP protein with an alteration in a catalytic amino acid. J Biol Chem 270:23044–23054
29. Kwan KM (2002) Conditional alleles in mice: practical considerations for tissue-specific knockouts. Genesis 32:49–62
30. Lee CL, Moding EJ, Huang X et al (2012) Generation of primary tumors with Flp recombinase in FRT-flanked p53 mice. Dis Model Mech. doi:10.1242/dmm.009084
31. Leone DP, Genoud S, Atanasoski S et al (2003) Tamoxifen-inducible glia-specific Cre mice for somatic mutagenesis in oligodendrocytes and Schwann cells. Mol Cell Neurosci 22:430–440
32. Hillen W, Berens C (1994) Mechanisms underlying expression of Tn10 encoded tetracycline resistance. Annu Rev Microbiol 48:345–369
33. Montoliu L (2011) Global resources: including gene trapped ES cell clones—is your gene already knocked out? In: Pease S, Saunders TL (eds) Advanced protocols for animal transgenesis. Springer Protocols, Springer-Verlag, Berlin, Heidelberg, pp 25–42
34. Schonig K, Bujard H, Gossen M (2010) The power of reversibility regulating gene activities via tetracycline-controlled transcription. Methods Enzymol 477:429–453
35. Cawthorne C, Swindell R, Stratford IJ et al (2007) Comparison of doxycycline delivery methods for Tet-inducible gene expression in a subcutaneous xenograft model. J Biomol Tech 18:120–123
36. Marks C (2009) Mouse models of human cancers consortium (MMHCC) from the NCI. Dis Model Mech 2:111
37. Uren AG, Kool J, Berns A et al (2005) Retroviral insertional mutagenesis: past, present and future. Oncogene 24:7656–7672
38. Collier LS, Carlson CM, Ravimohan S et al (2005) Cancer gene discovery in solid tumours using transposon-based somatic mutagenesis in the mouse. Nature 436:272–276
39. Kool J, Berns A (2009) High-throughput insertional mutagenesis screens in mice to identify oncogenic networks. Nat Rev Cancer 9: 389–399
40. Curcio MJ, Derbyshire KM (2003) The outs and ins of transposition: from mu to kangaroo. Nat Rev Mol Cell Biol 4:865–877
41. Weiser KC, Justice MJ (2005) Cancer biology: sleeping beauty awakens. Nature 436: 184–186
42. Ivics Z, Hackett PB, Plasterk RH et al (1997) Molecular reconstruction of Sleeping Beauty, a

Tc1-like transposon from fish, and its transposition in human cells. Cell 91:501–510
43. Ding S, Wu X, Li G et al (2005) Efficient transposition of the piggyBac (PB) transposon in mammalian cells and mice. Cell 122:473–483
44. Copeland NG, Jenkins NA (2010) Harnessing transposons for cancer gene discovery. Nat Rev Cancer 10:696–706
45. Dupuy AJ, Rogers LM, Kim J et al (2009) A modified sleeping beauty transposon system that can be used to model a wide variety of human cancers in mice. Cancer Res 69:8150–8156
46. Yusa K, Zhou L, Li MA et al (2011) A hyperactive piggyBac transposase for mammalian applications. Proc Natl Acad Sci U S A 108:1531–1536
47. Bender AM, Collier LS, Rodriguez FJ et al (2010) Sleeping beauty-mediated somatic mutagenesis implicates CSF1 in the formation of high-grade astrocytomas. Cancer Res 70: 3557–3565
48. Coniglio SJ, Eugenin E, Dobrenis K et al (2012) Microglial stimulation of glioblastoma invasion involves EGFR and CSF-1R signaling. Mol Med. doi:10.2119/molmed.2011.00217
49. March HN, Rust AG, Wright NA et al (2011) Insertional mutagenesis identifies multiple networks of cooperating genes driving intestinal tumorigenesis. Nat Genet 43:1202–1209
50. Kleinhammer A, Wurst W, Kuhn R (2011) Constitutive and conditional RNAi transgenesis in mice. Methods 53:430–436
51. Kissil JL, Walmsley MJ, Hanlon L et al (2007) Requirement for Rac1 in a K-ras induced lung cancer in the mouse. Cancer Res 67:8089–8094
52. Meylan E, Dooley AL, Feldser DM et al (2009) Requirement for NF-kappaB signalling in a mouse model of lung adenocarcinoma. Nature 462:104–107
53. Xue W, Meylan E, Oliver TG et al (2011) Response and resistance to NF-kappaB inhibitors in mouse models of lung adenocarcinoma. Cancer Discov 1:236–247
54. Puyol M, Martin A, Dubus P et al (2010) A synthetic lethal interaction between K-Ras oncogenes and Cdk4 unveils a therapeutic strategy for non-small cell lung carcinoma. Cancer Cell 18:63–73
55. Degrassi A, Russo M, Nanni C et al (2010) Efficacy of PHA-848125, a cyclin-dependent kinase inhibitor, on the K-Ras(G12D)LA2 lung adenocarcinoma transgenic mouse model: evaluation by multimodality imaging. Mol Cancer Ther 9:673–681
56. Chin L, Tam A, Pomerantz J et al (1999) Essential role for oncogenic Ras in tumour maintenance. Nature 400:468–472
57. Felsher DW, Bishop JM (1999) Reversible tumorigenesis by MYC in hematopoietic lineages. Mol Cell 4:199–207
58. Shachaf CM, Kopelman AM, Arvanitis C et al (2004) MYC inactivation uncovers pluripotent differentiation and tumour dormancy in hepatocellular cancer. Nature 431:1112–1117
59. Vidal M, Cagan RL (2006) Drosophila models for cancer research. Curr Opin Genet Dev 16:10–16
60. Liu S, Leach SD (2011) Zebrafish models for cancer. Annu Rev Pathol 6:71–93
61. Kango-Singh M, Singh A (2009) Regulation of organ size: insights from the Drosophila Hippo signaling pathway. Dev Dyn 238:1627–1637
62. Staley BK, Irvine KD (2012) Hippo signaling in Drosophila: recent advances and insights. Dev Dyn 241:3–15
63. Moon RT, Bowerman B, Boutros M et al (2002) The promise and perils of Wnt signaling through beta-catenin. Science 296:1644–1646
64. Reiter LT, Potocki L, Chien S et al (2001) A systematic analysis of human disease-associated gene sequences in Drosophila melanogaster. Genome Res 11:1114–1125
65. Chien S, Reiter LT, Bier E et al (2002) Homophila: human disease gene cognates in Drosophila. Nucleic Acids Res 30:149–151
66. Caussinus E, Gonzalez C (2005) Induction of tumor growth by altered stem-cell asymmetric division in Drosophila melanogaster. Nat Genet 37:1125–1129
67. Royer C, Lu X (2011) Epithelial cell polarity: a major gatekeeper against cancer? Cell Death Differ 18:1470–1477
68. Blair SS (2003) Developmental biology: boundary lines. Nature 424:379–381
69. Brumby AM, Richardson HE (2005) Using Drosophila melanogaster to map human cancer pathways. Nat Rev Cancer 5:626–639
70. Brand AH, Perrimon N (1993) Targeted gene expression as a means of altering cell fates and generating dominant phenotypes. Development 118:401–415
71. Wang H, Chattopadhyay A, Li Z et al (2010) Rapid identification of heterozygous mutations in Drosophila melanogaster using genomic capture sequencing. Genome Res 20:981–988
72. Adams MD, Celniker SE, Holt RA et al (2000) The genome sequence of Drosophila melanogaster. Science 287:2185–2195
73. Boutros M, Perrimon N (2000) Drosophila genome takes flight. Nat Cell Biol 2:E53–E54
74. Carroll PM, Dougherty B, Ross-Macdonald P et al (2003) Model systems in drug discovery:

chemical genetics meets genomics. Pharmacol Ther 99:183–220
75. O'Hare K, Rubin GM (1983) Structures of P transposable elements and their sites of insertion and excision in the Drosophila melanogaster genome. Cell 34:25–35
76. Rorth P (1996) A modular misexpression screen in Drosophila detecting tissue-specific phenotypes. Proc Natl Acad Sci USA 93: 12418–12422
77. Duffy JB (2002) GAL4 system in Drosophila: a fly geneticist's Swiss army knife. Genesis 34:1–15
78. Dietzl G, Chen D, Schnorrer F et al (2007) A genome-wide transgenic RNAi library for conditional gene inactivation in Drosophila. Nature 448:151–156
79. Lee T, Luo L (1999) Mosaic analysis with a repressible cell marker for studies of gene function in neuronal morphogenesis. Neuron 22: 451–461
80. Lue NF, Chasman DI, Buchman AR et al (1987) Interaction of GAL4 and GAL80 gene regulatory proteins in vitro. Mol Cell Biol 7:3446–3451
81. Willecke M, Toggweiler J, Basler K (2011) Loss of PI3K blocks cell-cycle progression in a Drosophila tumor model. Oncogene 30: 4067–4074
82. Miles WO, Dyson NJ, Walker JA (2011) Modeling tumor invasion and metastasis in Drosophila. Dis Model Mech 4:753–761
83. Gonzalez C (2007) Spindle orientation, asymmetric division and tumour suppression in Drosophila stem cells. Nat Rev Genet 8: 462–472
84. Zhou Y, Rideout WM 3rd, Zi T et al (2010) Chimeric mouse tumor models reveal differences in pathway activation between ERBB family- and KRAS-dependent lung adenocarcinomas. Nat Biotechnol 28:71–78
85. Huijbers IJ, Krimpenfort P, Berns A et al (2011) Rapid validation of cancer genes in chimeras derived from established genetically engineered mouse models. Bioessays 33: 701–710
86. Isalan M (2011) Zinc-finger nucleases: how to play two good hands. Nat Methods 9:32–34
87. Miller JC, Tan S, Qiao G et al (2011) A TALE nuclease architecture for efficient genome editing. Nat Biotechnol 29:143–148
88. McMahon MA, Rahdar M, Porteus M (2012) Gene editing: not just for translation anymore. Nat Methods 9:28–31
89. Zhang F, Cong L, Lodato S et al (2011) Efficient construction of sequence-specific TAL effectors for modulating mammalian transcription. Nat Biotechnol 29:149–153
90. Hockemeyer D, Wang H, Kiani S et al (2011) Genetic engineering of human pluripotent cells using TALE nucleases. Nat Biotechnol 29: 731–734
91. Editorial (2012) Method of the year 2011. Nat Methods 9:1
92. Barker M (2012) Gene-editing nucleases. Nat Methods 9:23–26
93. Tesson L, Usal C, Menoret S et al (2011) Knockout rats generated by embryo microinjection of TALENs. Nat Biotechnol 29:695–696
94. Sander JD, Cade L, Khayter C et al (2011) Targeted gene disruption in somatic zebrafish cells using engineered TALENs. Nat Biotechnol 29:697–698
95. Weiss WA, Taylor SS, Shokat KM (2007) Recognizing and exploiting differences between RNAi and small-molecule inhibitors. Nat Chem Biol 3:739–744

Chapter 19

Target Validation in Mice by Constitutive and Conditional RNAi

Aljoscha Kleinhammer, Wolfgang Wurst, and Ralf Kühn

Abstract

Gene silencing by RNA interference (RNAi) has become a standard method for the characterization of gene function in mammalian cells. Short hairpin (sh) RNAs expressed from stably integrated vectors mediate gene knockdown both in cultured cells and in mice, presenting a fast alternative to gene knockout approaches. We describe three strategies to control gene silencing in mice that can be applied to any transcript of interest. This shRNA based approach enables either i) constitutive body-wide knockdown, ii) cell type-specific knockdown controlled by Cre recombinase, or iii) inducible body-wide knockdown controlled by doxycycline. For reliable expression the shRNA vector of interest is inserted into a Rosa26 docking site of ES cells by a site-specific recombinase. These ES cells can then be used to generate shRNA transgenic mice. This technology enables the production of adult knockdown mice within 11 months for an expedite in vivo validation of drug targets.

Key words: RNAi, Conditional, ShRNA, Cre recombinase, TetR/O, Transgenic mice, Rosa26, RMCE

1. Introduction

Silencing of gene expression by RNA interference (RNAi) is used as standard tool for functional genomics in cultured mammalian cells. RNAi is a sequence-specific gene silencing process that occurs at the messenger RNA (mRNA) level. In mammalian cells short dsRNAs (<30 bp) trigger the specific knockdown of mRNAs in mammalian cells without interferon activation (1). Such synthetic siRNAs can be introduced into cultured cells and induce the transient knockdown of target mRNAs. Alternatively, expression vectors driven by RNA polymerase III enable the permanent production of small dsRNAs in mammalian cells (2, 3). Such vector-derived transcripts are designed to contain an antisense region, complementary to a selected mRNA sequence, and the

Jürgen Moll and Riccardo Colombo (eds.), *Target Identification and Validation in Drug Discovery: Methods and Protocols*, Methods in Molecular Biology, vol. 986, DOI 10.1007/978-1-62703-311-4_19,

corresponding mRNA sense region. These transcripts fold into a stem-loop structure and form short hairpin RNAs (shRNAs) that are processed by Dicer in a similar way as siRNAs. Expression vectors for shRNAs can be stably integrated into the genome and allow permanent gene silencing in cell lines and mice.

Like in cultured cells gene silencing can be transiently induced in mice by delivering siRNA to somatic tissues. Direct siRNA delivery is frequently used in experimental settings requiring only a short term knockdown of gene function, e.g., to study the replication of hepatitis C virus in the liver (4). In clinical studies, local delivery of siRNAs has already proved of value as disease gene inactivating drugs (5–7). In vitro screening procedures using genome-wide libraries of RNAi vectors enable to identify important regulatory proteins that may be suited as drug targets (8–12).

Alternatively, permanent gene silencing is achieved in mice or rats by integrating shRNA transgenes into the genome. The efficiency of gene silencing in transgenic mice varies and can be as high as 90% and sometimes even higher. In several instances, corresponding knockdown and knockout phenotypes were compared and found to be identical or very similar (4, 13, 14). As compared to somatic siRNA delivery, for which the uptake rate is critical, transgenic shRNA can be expressed in all cells of the organism using an appropriate promoter. shRNA transgene delivery to mice can be achieved by pronucleus injection, viral embryo infection, or targeted insertion (knock-in) into the genome of ES cells (14–17).

In this article we summarize our approach for a streamlined production of shRNA transgenic mice using targeted transgenesis in ES cells (13, 16, 18, 19). This approach involves the insertion of shRNA expression vectors into the ubiquitously active *Rosa26* locus of ES cells by recombinase-mediated cassette exchange (RMCE), followed by the generation of knockdown mice. The targeted insertion of shRNAs into *Rosa26* ensures reproducible and ubiquitous shRNA expression. A single shRNA vector copy expressed within the *Rosa26* locus is sufficient to elicit constitutive, body-wide gene silencing (16, 18). However, tissue-specific knockdowns require conditional vectors equipped with a molecular switch. This switch is provided by the Cre/loxP recombination system enabling both a spatial and temporal control of shRNA expression (16). Alternatively the doxycycline-controlled tet repressor/tet operator (tetR/O) system can be used to reversibly induce gene silencing in a temporally and spatially restricted manner (20). We describe here three RNAi approaches in transgenic mice to control gene knockdown: (1) constitutive and body-wide gene silencing, (2) temporally and spatially controlled gene silencing based on Cre/loxP recombination, and (3) reversible gene silencing based on the doxycycline-controlled tetR/O system.

In all of these cases the shRNA expression construct is stably integrated into the genome of ES cells and subsequently transferred to the mouse germline. In these constructs the shRNA of interest is expressed from an RNA polymerase III promoter, which mediates ubiquitous expression and thus enables for body-wide knockdown. To obtain the double-stranded RNA configuration required for short hairpin RNA formation from single-stranded RNA, we link a 19–22 bp "sense" siRNA sequence to its complementary "antisense" sequence via an 8 bp loop region and combine it with a poly T termination signal (Fig. 1a). Upon transcription the complementary RNA strands hybridize to form dsRNA that is required for Dicer recognition and gene silencing. The easiest way to obtain a cloneable DNA fragment yielding a sense-loop-antisense configuration is by annealing of two complementary oligonucleotides shA and shB of about 60 bp. Oligonucleotides shA and shB are flanked by restriction enzyme recognition sequences that form restriction enzyme compatible overhangs after hybridization (Fig. 1b).

For stable integration of a shRNA expression vector into the ES cell genome we use RMCE. Figure 1c illustrates an example using the *C31*Int-/*att*B-/*att*P-mediated cassette exchange at a pre-engineered *Rosa26* docking site. Since *Rosa26* is an ubiquitously active locus in all cell types, this docking site insures the stable and ubiquitous expression of any integrated shRNA transgene. Since shRNA donor cassettes include a promoterless neomycin resistance gene (neo) which is activated only after correct cassette exchange at the *Rosa26* acceptor locus (Fig. 1c), RMCE positive cell lines can be conveniently selected using the aminoglycoside geneticin (G418). After stable integration, vectors expressing shRNAs constitutively will directly induce gene silencing.

For the conversion of constitutive shRNA vectors into Cre-inducible vectors, a loxP-STOP cassette which disrupts its function is inserted into a unique *Hin*dIII restriction site engineered into the shRNA loop region (Fig. 2c, d). Given the diversity of the existing Cre and Cre-ER transgenic driver lines (21), Cre-inducible shRNA transgenes enable detailed gene analysis in the context of an entire organism.

Unlike the Cre/loxP system, the tetR/O switch is not dependent on enzyme-based DNA rearrangements but is controlled by the small synthetic compound doxycycline (Dox). In the absence of Dox the tetracycline repressor (tetR), expressed from a constitutive promoter, binds to the tet operator (tetO) within a modified H1 promoter and blocks shRNA transcription (Fig. 3b). Exposure to Dox dissociates the tetR from tetO which enables shRNA expression from the H1 promoter (22). Thus, gene silencing is induced by Dox addition (20, 22, 23). This system is successfully in use to validate the potential of candidate drug targets in vivo (24–28). A commercial service for the production of Dox inducible knockdown mice and rats is available from Taconic (see: www.taconic.com).

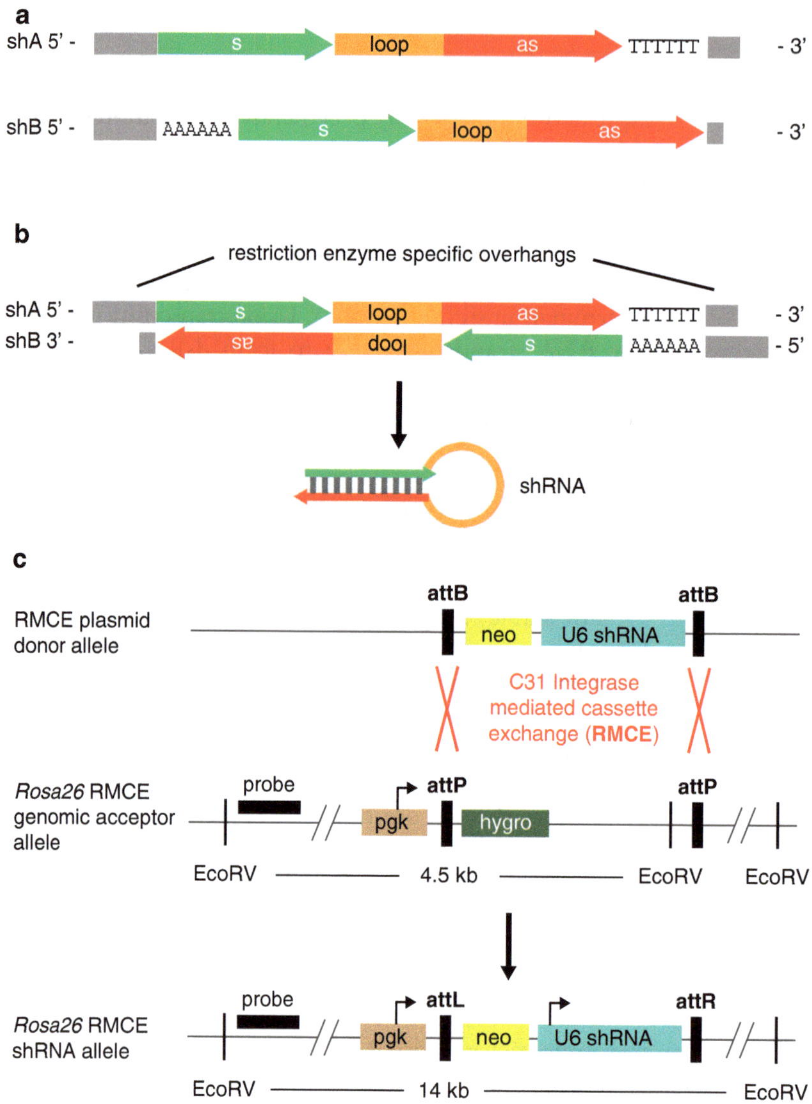

Fig. 1. Design principle of shRNAs and genomic integration of shRNA vectors by RMCE in ES cells. (**a**) For each selected shRNA target sequence a pair of complementary oligonucleotides shA and shB is designed. The "sense" sequence (19–29 bp, "s" *arrows*) represents the CDS (coding sequence) sense target sequence, and the "antisense" sequence (19–29 bp, "as" *arrows*) is its complementary sequence. Both sequences are linked by an 8 bp loop region followed by a poly T termination signal. (**b**) Annealing of oligonucleotides shA and shB results in a cloneable double-stranded DNA fragment with restriction enzyme compatible overhangs. The loop region enables bending of the correspondent single-stranded RNA transcript. Hybridization of "sense" and "antisense" sequences results in a double-stranded short hairpin (sh) RNA. (**c**) shRNA constructs are integrated into the genomic *Rosa26* RMCE acceptor allele of IDG26.10-3 ES cells by RMCE, as exemplified for the constitutive knockdown construct. The RMCE donor allele contains the shRNA expression cassette (U6 promoter and shRNA sequence) and a promoterless neomycin resistance gene (neo), both flanked by *att*B sites. C31 integrase mediates the cassette exchange between donor and acceptor allele by recombination of *att*B and *att*P sites. As a result the hygromycin resistance gene becomes exchanged against the neomycin resistance gene and the shRNA expression cassette. Cells that underwent correct recombination express both the shRNA and the neomycin resistance gene; the latter can therefore be used as selection marker. Positive clones can be identified via PCR (*see* Subheading 3.1.3, step 4 for details) and by Southern blot. Using EcoRV digestion and pRosa26-5′-probe, unrecombined *Rosa26* RMCE acceptor allele produces a band of 4.5 kb, whereas correctly recombined clones exhibit a band of 14 kb. Note that the fragment size of correct *Rosa26* RMCE shRNA clones depends on the insert, for Cre-inducible constructs the size is 15 kb and for Doxycycline-inducible constructs the size is 17.7 kb.

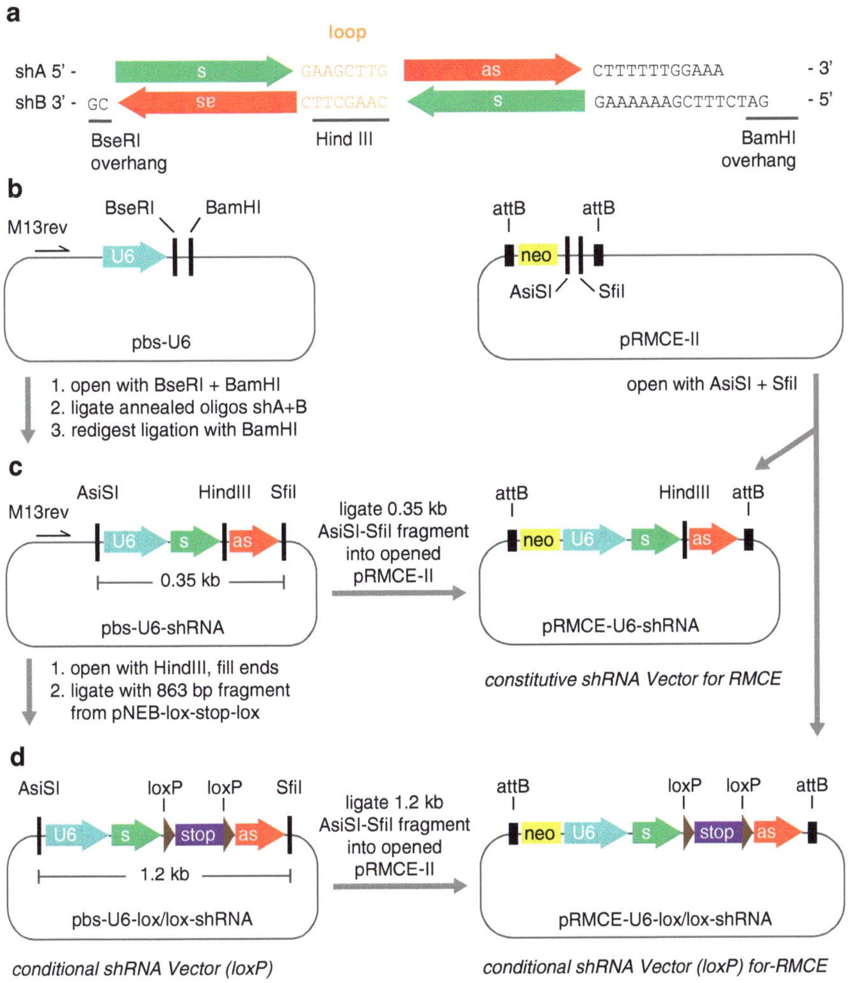

Fig. 2. Generation of constitutive and Cre-inducible shRNA vectors for RMCE. (**a**) A pair of complementary oligonucleotides shA/B is designed and annealed resulting in a cloneable double-stranded DNA fragment with *Bse*RI and *Bam*HI compatible ends. The specific target gene siRNA sense and antisense sequences are to be inserted as indicated by the "s" and "as" *arrows*. As the vector contains the G nucleotide that is required by U6 promoter at position +1, the first base of the sense sequence of shA represents the second transcribed nucleotide (+2) of the shRNA. The loop region contains a unique *Hin*dIII site and poly T serves as termination signal for Pol-III. (**b**) For cloning of a constitutive shRNA, vector-annealed oligonucleotides shA/B are ligated into *Bse*RI/*Bam*311HI opened vector pbs-U6 containing an U6 promoter. Empty vector is removed by redigestion with *Bam*HI. Vector pRMCE-II, containing the elements necessary for RMCE (neomycin resistance and the *att*B sites), is opened in parallel by digestion with AsiSI and SfiI. (**c**) A 0.35 kb AsiSI/SfiI fragment from vector pbS-U6-shRNA is ligated into opened vector pRMCE-II resulting in a constitutive shRNA expression vector compatible for RMCE. (**d**) For cloning of a Cre-inducible shRNA vector a removable floxed stop cassette is inserted into the loop region of the shRNA sequence. Therefore, vector pbs-U6-shRNA (**c**) is opened with *Hin*dIII, ends are filled up, and a 863 bp MlyI fragment from pNEB-lox-stop-lox is inserted by ligation. From the resulting vector pbs-U6-lox/lox-shRNA a 1.2 kb AsiSI/SfiI fragment is isolated and ligated into AsiSI/SfiI opened vector pRMCE-II (**b**). The resulting vector pRMCE-U6-lox/lox-shRNA can be induced by removal of the floxed stop cassette by Cre recombinase and can be used directly for RMCE.

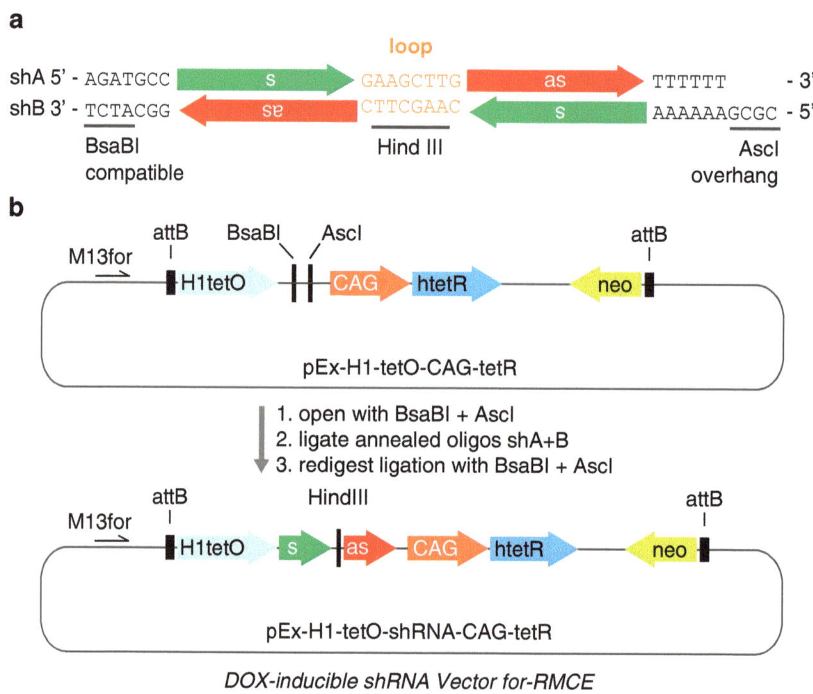

Fig. 3. Generation of doxycycline-inducible shRNA vectors for RMCE. (**a**) A pair of complementary oligonucleotides shA/B is designed and annealed resulting in a cloneable double-stranded DNA fragment with *Bsa*BI and *Asc*I compatible ends. The specific target gene shRNA sense and antisense sequences are to be inserted for the "s" and "as" *arrows* as indicated. The loop region contains a unique *Hin*dIII site and poly T serves as termination signal for Pol-III. (**b**) For cloning of a doxycycline-inducible shRNA vector, annealed oligonucleotides shA/B are ligated into *Bsa*BI/*Asc*I opened vector pEx-H1-tetO-CAG-tetR containing tetR/O regulatory elements. Empty vector is removed by redigestion with BsaBI and AscI. The generated plasmid pEx-H1-tetO-shRNA-CAG-tetR can be used directly for recombinase-mediated cassette exchange (RMCE) as it contains a promoterless neomycin resistance gene (neo) and C31 integrase recognition sites (*att*B).

Alternatively, following the same principle, constitutive or conditional knockdown ES cell lines and mice can be obtained by following the protocols described below. The shRNA vector approach described in this chapter enables to elicit body-wide or cell type-specific gene silencing in most organs of adult mice. In the brain of shRNA vector transgenic mice the levels of a target gene mRNA and protein can be reduced by 90% or more, comparable to the utility of RNAi in cultured cell lines. However, to achieve a satisfying level of gene silencing in vivo it is instrumental to identify well working RNAi target sequences through initial in vitro screening. This can be done within 2–3 months by comparing the gene silencing efficiencies of predicted candidate shRNA sequences in ES cells. As compared to the generation of knockout mice by gene targeting, the shRNA vector approach is technically less demanding, since vector construction and vector insertion into ES cells are simple and highly reproducible. The insertion of shRNA vectors into the *Rosa26* locus as universal docking site enables the production of transgenic mice with reproducible expression properties. Upon the initial characterization of the basic parameters of this approach, any newly generated strain can be expected to represent a functioning knockdown mouse strain. In addition, the generation of knockdown mice is relatively fast—a first group of adult mutants can be generated within a period of 11 months.

Alternatively, shRNA expression can be controlled by doxycycline as inducing compound. This system enables the reversible and body-wide expression of shRNAs in mice and rats (20, 28, 29) and offers the opportunity to reverse the induced gene knockdown at a given time. This property of the shRNA system offers unique applications to study gene function in mice that cannot be achieved with knockout technologies. Therefore, an emerging application of transgenic RNAi is in the field of drug target validation in order to predict the therapeutic potential of small molecule inhibitors (30–32). Several recent publications, describing mice produced by the Dox inducible shRNA system from Taconic, highlight the utility of this approach (24–28). The Dox inducible shRNA vector (Subheading 3.3) developed by us independently follows the same principle.

As a rule of thumb, the gene silencing performances of at least five predicted shRNAs per target are recommended to be evaluated in ES cells before creating stable knockdown ES cell lines. Moreover, at least two transgenic mouse lines expressing two different sh-R NAs against the same target are required to rule out off-target effects. Mouse lines are deemed appropriate for further analysis if the knockdown phenotypes of the two shRNA mouse lines are comparable (14).

2. Materials

2.1. Plasmids

1. pbs-U6	Contains U6 promoter
2. pCAG-*C31*Int	Vector for constitutive expression of C31 integrase
3. pEx-H1tetO-CAG-tetR	Contains doxycycline-inducible H1 promoter and elements for constitutive expression of tetR
4. pNeb-lox-stop-lox	Contains a stop cassette flanked by loxP sites
5. pRMCE-II	Contains a promoterless neomycin resistance flanked by *att*B sites allowing for recombinase-mediated cassette exchange
6. pRosa26-5′ probe	Contains Southern blot probe for identification of right clones

DNA and sequence information for all vectors is available from the authors.

2.2. Cell Lines and Bacteria

1. ES cell line IDG26.10-3.
2. ES cell line IDG3.2.
3. Feeder cells: G418 resistant murine embryonic fibroblasts.
4. Bacteria: *Escherichia coli* DH5a.

ES cell lines, feeders, and bacteria are available from the authors.

2.3. Primers

1. Neo: (5′-GTT GTG CCC AGT CAT AGC CGA ATA G-3′).
2. Pgk (5′-CAC GCT TCA AAA GCG CAC GTC TG-3′).
3. Hyg-1 (5′-GAA GAA TCT CGT GCT TTC AGC TTC GAT G-3′).
4. Hyg-2 (5′-AAT GAC CGC TGT TAT GCG GCC ATT G-3′).
5. Rosa-5′ (5′-CGT GTT CGT GCA AGT TGA GT-3′).
6. Rosa-3′ (5′-ACT CCC GCC CAT CTTCTA G-3′).

2.4. Reagents

1. *Doxycycline*: Doxycycline hyclate (Sigma). Dissolve in water (1 mg/ml) and store in single-use aliquots at –20°C.
2. *F1 ES cell medium*: 500 ml Dulbecco's Modified Eagle's Medium (DMEM, Invitrogen), supplemented with 75 ml ES-qualified FCS (PAN Biotech), 9 ml 200 mM L-glutamine (Invitrogen), 12 ml 1 M HEPES (Invitrogen), 6 ml 100× MEM nonessential amino acids (Invitrogen), 1.2 ml 50 mM β-mercaptoethanol (Invitrogen), 90 microliter LIF (10 to the seven (10 million) U/ml, Chemicon).
3. *G418*: Geneticin solution with an active concentration of 50 mg/ml (Invitrogen). Store in single-use aliquots at –20°C.
4. *PBS*: 1× PBS for cell culture, sterile.
5. *Gelatine*: 0.1% gelatine in water, autoclaved.
6. *TE buffer*: 10 mM Tris–HCl, 1 mM EDTA.

3. Methods

3.1. Generation of Constitutive Knockdown Mice

3.1.1. Cloning of Constitutive shRNA Vectors

1. Select five shRNA target sequences for your gene of interest using siRNA prediction tools as outlined below. For the constitutive and the Cre-inducible knockdown vectors the U6 gene promoter is used. Make sure to use only predicted siRNA target sequences with guanine as the first nucleotide of the siRNA sequence because it is required for efficient transcription from the U6 promoter. Free siRNA prediction programs are available at the Web pages of Ambion, Invitrogen, or Qiagen and the University of Tokyo (http://genomics.jp/sidirect/index.php?type=fc), University of Vienna (http://rna.tbi.univie.ac.at/cgi-bin/RNAxs), and the Whitehead Institute

(http://jura.wi.mit.edu/bioc/siRNAext/home.php). Design a pair of complementary oligonucleotides shA and shB for each siRNA target sequence and follow the design shown in Fig. 2a (see Note 1). As the vector already contains the first G (+1) of the corresponding shRNA transcript, omit the first G of the selected siRNA sequences and start with its second base (+2) when designing the shA oligonucleotides (see Note 2).

2. Anneal complementary oligonucleotides shA/B (1 μg/μl) by incubating 1 μl shA and 1 μl shB in 98 μl TE at 100°C for 5 min and subsequent slow cooling down to room temperature (see Note 3). The resulting ends are compatible with *Bse*RI and *Bam*HI restriction enzymes.
3. Digest the pbs-U6 plasmid DNA with *Bse*RI and *Bam*HI. Ligate 2 μl hybridization reaction with 100 ng opened, gel-purified vector pbs-U6 in a 10 μl standard ligation reaction with 1 μl T4 DNA-Ligase and 1 μl T4-ligase buffer at 16°C overnight (Fig. 19.2b). Redigest the ligation with *Bam*HI to eliminate self-ligated empty vector.
4. Transform the ligation reaction into *E. coli* cells (e.g., DH5α), and digest miniprep DNA with PstI and *Hin*dIII. Correctly ligated clones show two bands of 310 bp and 2.9 kb. Plasmids without insert show only the 2.9 kb vector band. M13 forward and M13 reverse primers can be used for sequence confirmation. Prepare Maxipreps with one confirmed clone for each of the five predicted target sequences and use the five generated plasmids of pbs-U6-shRNA for pretesting their knockdown efficiency in ES cells.

3.1.2. Pretesting Knockdown Efficiency by Transient Transfection into ES Cells

1. Sterilize plasmid DNA from step 4 of the previous section by mixing with 1/10 volume of 3 M sodium acetate (pH 5.3) and 2.5 volumes of 100% ethanol. Incubate for 2 h to overnight at –20°C and pellet the DNA by centrifugation. After washing with 70% ethanol, dissolve the DNA pellet in sterile water to a final concentration of 1 μg/μl while working under a sterile flow. Plasmids are now ready for transfection (see Note 4).
2. As predicted siRNA sequences vary greatly in their gene silencing efficiency, all five sequences should be pretested by transient transfection (see Note 5). Cultivate wild-type IDG3.2 ES cells on gelatine for at least 2 days. On transfection day (day 0) trypsinize 2 or 3 10 cm dishes with IDG3.2 ES cells; one dish usually yields $1–2 \times 10^7$ cells. For each transfection pellet 5×10^6 cells and wash with PBS. Discard the supernatant and resuspend the pellet in 800 μl PBS containing 80 μg plasmid DNA (step 5). Transfer the suspension into an electroporation cuvette and electroporate at 330 V for 3 ms, time constant (see Note 6). Incubate for 5 min at room temperature, then plate the cells on a gelatine-coated 10 cm dish in 10 ml medium.

At day 1 supply with fresh medium, at day 2 harvest cells by trypsinization. After washing with PBS, the pelleted cells can be used directly for analysis or stored at −20°C.

3. Determine knockdown efficiency for each target sequence by analyzing the expression level of your target gene by Western blot or real time PCR. Select the two most efficient sequences generating a knockdown efficiency of at least 70% and use the correspondent plasmids for the following cloning steps.

3.1.3. Stable Integration of shRNA Vectors into ES Cells by RMCE and Generation of Knockdown Mouse Lines

1. The U6-shRNA expression cassette is ligated into the pRMCE-II vector that provides elements (neomycin resistance, *att*B sites) required for RMCE into *Rosa26*. Digest the pbs-U6-shRNA plasmid with AsiSI and SfiI and the pRMCE-II vector with AsiSI and SfiI and dephosphorylate the ends (Fig. 2c). Ligate the 350 bp AsiSI–SfiI U6-shRNA fragment derived from pbs-U6-shRNA into the AsisSI/SfiI pRMCE-II vector (4.3 kb).
2. Transform into *E. coli* cells, prepare minipreps and identify correct clones by digesting with SalI and *Hin*dIII. Correctly ligated clones show two bands of 1.6 and 3.0 kb, plasmids without insert show only the 4.3 kb vector band. M13 forward, M13 reverse, and bpA-for primers can be used for sequence confirmation. Prepare Maxipreps and use two independent plasmids of pRMCE-U6-shRNA for stable transfection by RMCE in ES cells.
3. For stable integration of the two shRNA vectors selected in the previous step coelectroporate 25 μg of the shRNA plasmid and 25 μg of C31 integrase expression plasmid into IDG26.10-3 ES cells cultured on feeder cells. All plasmids have to be sterilized before electroporation as described in step 1 of Subheading 3.1.2. IDG26.10-3 ES cells provide a modified Rosa26 locus enabling RMCE (see Subheading 1 and Fig. 1c). Transfect both selected shRNA plasmids in parallel. For each transfection use 5×10^6 pelleted cells washed with PBS. Resuspend the cells in 800 μl PBS containing the shRNA vector and the C31 integrase vector. Transfer into an electroporation cuvette and electroporate at 300 V for 2 ms, time constant. After an incubation of 5 min at room temperature plate the cells in G418-free medium on two 10 cm dishes containing neomycin-resistant feeders. After 24 h (day 1 after transfection) supply with G418-free medium. From day 2 to day 8 supply daily with fresh medium containing G418 (140 μg/ml) for selection of neomycin-resistant cells. On day 9 after transfection pick at least 12 colonies per transfected shRNA-RMCE construct (see Note 7). Expand clones on feeders for freezing and on gelatine for DNA isolation. After isolation of genomic DNA screen for positive clones by PCR or Southern blot.

4. For identification of positive clones by PCR use the primers Pgk and Neo (Tm: 65°C, Subheading 2.3) which bind in the Pgk-promoter and the neomycin resistance gene, respectively. Correct clones show a band of 280 bp. If using feeders harboring a Pgk-promoter-driven neomycin resistance gene, an additional band of 160 bp will appear. Use a second PCR with the Hyg-1 and Hyg-2 primers (Tm, 65°C; product size, 550 bp; Subheading 2.3) to identify and exclude Hyg-positive clones with unrecombined *Rosa26* loci (Fig. 1c), representing mixed clones, partial recombination, or random integration. Primers Rosa-5′ and Rosa-3′ (Tm: 57°C; product size: 536 bp Subheading 2.3) detect the *Rosa26* wild-type allele. Southern blot verification of correct clones is mandatory for exclusion of partially recombined alleles or chromosomal rearrangements. For identification of positive clones by Southern blot digest genomic DNA with EcoRV and hybridize with the 450 bp EcoRI fragment from plasmid pRosa26-5′-probe (Subheading 2.1). Clones positive for the RMCE allele exhibit a band of 14 kb due to the integration of the U6-shRNA and neo gene cassettes. *Rosa26* wild-type locus gives rise to a band of 11.5 kb, whereas the modified *Rosa26* locus of the IDG26.10-3 ES cells produces a band of 4.5 kb (Fig. 1c). According to our experience, ~40% of the neomycin-resistant clones undergo complete RMCE.
5. Efficiency of stably integrated shRNA vectors should be confirmed in vitro. Therefore, expand two positives clones per shRNA construct on feeders, freeze aliquots, and cultivate the cells feeder-free on gelatine-coated 10 cm dishes for at least 2 days. Harvest cells and analyze the four clones for knockdown efficiency as described in steps 2 and 3 of Subheading 3.1.2. Determine the best performing clone for each siRNA target sequence.
6. Use confirmed clones for generating mouse lines by blastocyst injection or tetraploid embryo aggregation as described in ref. 33. Generate two knockdown mouse lines for the same target gene but with different siRNA target sequences to minimize the probability of off-target effects as phenotype-causing factors.

3.2. Generation of Cre-Inducible Conditional Knockdown Mice

3.2.1. Cloning of Cre/loxP Conditional shRNA Vectors

1. Open the vector pbs-U6-shRNA (see Subheading 3.1.1, step 4, and Fig. 2c) with *Hin*dIII, fill in ends, and dephosphorylate the 5′-overhanging ends using standard procedures (vector fragment size is 3.3 kb).
2. Prepare the 863 bp lox-STOP-lox cassette from the plasmid pNEB-lox-stop-lox plasmid by digesting with MlyI.
3. To generate plasmid pbs-U6-lox/lox-shRNA, ligate the fragment containing the lox-STOP-lox cassette from step 2 into the pbs-U6-shRNA plasmid from step 1, and eliminate religated vector background by digesting with *Hin*dIII.

4. Transform the ligation product into *E. coli* cells and screen minipreps for inserts by digesting with EcoRI and SacII. Correctly ligated clones show an 850 bp band for the insert and a 3.4 kb vector band. Inversely oriented inserts produce 150 bp and 4 kb bands. Clones without insert show only the 3.4 kb vector band. Use M13 forward and M13 reverse primers for sequence confirmation. Prepare Maxipreps for one confirmed clone and test the functionality of the lox-STOP-lox cassette as described below.

3.2.2. In Vitro lox-STOP-lox Cassette Excision and Testing shRNA Knockdown Efficiency in ES Cells

1. Delete the lox-STOP-lox cassette in vitro by incubating 1 μg conditional shRNA plasmid DNA with 4 units of Cre recombinase (New England Biolabs) in a total volume of 50 μl 1× Cre buffer at 37°C for 30 min. Inactivate the recombinase for 10 min at 70°C. Eliminate nonrecombined plasmids by EcoRI digestion (see Note 8).
2. Transform the reaction into *E. coli* cells and confirm recombination by digesting miniprep DNA with XbaI. Cre-recombined plasmids that lost the stop cassette produce two bands of 460 bp and 2.9 kb. Nonrecombined plasmids produce bands of 1.3 and 2.9 kb. After sequence confirmation using the M13 forward and reverse primers, prepare a Maxiprep from one confirmed clone.
3. Pretest these activated shRNA plasmids by transiently transfecting the Maxiprep DNA into ES cells as described in Subheading 3.1.2.
4. Use the correspondent constructs in the conditional state, which still contain the floxed stop cassette for ligation into RMCE vectors and subsequent stable transfection into ES cells by RMCE.

3.2.3. Stable Integration of Cre/loxP Conditional shRNA Vectors into ES Cells by RMCE and Generation of Knockdown Mouse Lines

For construction of Cre-inducible shRNA vectors compatible for RMCE follow the protocol for the construction of constitutive shRNA-RMCE vectors (Subheading 3.1.3). Here we point out only those steps that differ to the protocol in:

- *Subheading 3.1.3*, step 1: Ligate the U6-shRNA-lox-stop-lox fragment (*1.2 kb*) into the opened pRMCE-II vector (4.3 kb).
- *Subheading 3.1.3*, step 2: Identify correct clones by plasmid digestion with SalI and *EcoRI*. Correctly ligated clones show two bands of *1.7* and 3.0 kb, plasmids without insert show only the 4.3 kb vector band.
- *Subheading 3.1.3*, step 4: In Southern blot analysis, the wild-type *Rosa26* locus can be identified by a band of 11.5 kb. The parental IDG26.10-3 ES cells show an additional band of 4.5 kb derived from the modified *Rosa26* allele before RMCE (Fig. 19.1c). Positive RMCE clones show an additional band of

15 kb due to the integration of the U6-shRNA-lox-stop-lox and neo gene cassettes.

- *Subheading 3.1.3*, step 5: The knockdown efficiency of stably integrated vectors should be tested in ES cells with the additional necessity of activating the integrated vectors by Cre before testing them (see Note 9).
- *Subheading 3.1.3*, step 6: Use only confirmed ES cell clones for generating mouse lines by blastocyst injection or tetraploid embryo aggregation. Resulting mice need to be mated to *Cre deleter strains* (21) for knockdown activation.

3.3. Generation of Doxycycline-Inducible Knockdown Mice

In this section, we describe how to generate a doxycycline-inducible vector for reversible expression of shRNA in a single cloning step. The shRNA sequence is introduced into the pEx-H1tetO-CAG-tetR vector, which contains a tetO-modified H1 promoter and a codon-optimized tetR gene under the control of the constitutive CAG promoter. Two flanking *att*B sites and a promoterless neomycin resistance gene enable direct usage of the cloning product for RMCE in IDG26.10-3 ES cells (see Subheading 1).

3.3.1. Cloning of Doxycycline-Inducible shRNA Vectors

1. Select five shRNA sequences for your target gene by using Web resources (see Subheading 3.1.1, step 1). As the H1 promoter is used, siRNA sequences may start with any nucleotide. Design a pair of shA and shB complementary oligonucleotides for each siRNA sequence.
2. Anneal the complementary nucleotides shA and shB (1 μg/μl) by incubating 1 μl shA and 1 μl shB in 98 μl TE at 100°C for 5 min and subsequent slow cooling down to room temperature (see Note 3). The resulting ends are compatible with *Asc*I and *Bsa*BI (Fig. 3).
3. Open the plasmid pEx-H1tetO-CAG-tetR by digestion with *Asc*I and *Bsa*BI and ligate annealed shA/B into the vector. Remove self-ligated empty vector by redigestion with *Asc*I and *Bsa*BI. Heat inactivate enzymes by incubation for 15 min at 75°C.
4. Transform the ligation product into competent *E. coli* cells (in our lab DH5α) and check resulting minipreps by digestion with *Pst*I and *Hin*dIII. Correctly ligated clones show bands of 0.2, 1.0, 2.8, and 3.3 kb. Empty plasmids show bands of 0.2, 1.0, and 6.1 kb. Confirm correct sequence for one positive clone using M13 forward primer. Prepare Maxipreps with one confirmed clone per target sequence and use the five generated plasmids of pEx-H1tetO-shRNA-CAG-tetR in the following steps.
5. Pretest shRNA knockdown efficiency by transient transfection in ES cells as described in Subheading 3.1.2, with the difference

that the knockdown has to be induced by cultivating the cells for 2 days before harvesting with medium containing 1 μg/ml doxycycline. Select the two most efficient sequences generating a knockdown efficiency of at least 70% and use the correspondent plasmids for stable transfection by RMCE.

3.3.2. Stable Integration of Doxycycline-Inducible shRNA Vectors into ES Cells by RMCE and Generation of Knockdown Mouse Lines

For stable integration by RMCE of the two doxycycline-inducible shRNA vectors selected in the previous section, follow the protocol as described in Subheading 3.1.3, steps 3–6. Here, we point out only those steps that differ to the protocol in:

- *Subheading 3.1.3*, step 4: In Southern blot analysis, the wild-type *Rosa26* locus can be identified by a band of 11.5 kb (Fig. 1c). The parental IDG26.10-3 ES cells show an additional band of 4.5 kb derived from the modified *Rosa26* allele before RMCE (Fig. 1d). Clones positive for the pEx-H1tetO-shRNA-CAG-tetR RMCE allele exhibit a band of 17.7 kb.
- *Subheading 3.1.3*, step 5: The knockdown has to be induced by cultivating the cells for 2 days with medium containing 1 μg/ml doxycycline before harvesting (see Notes 10 and 11).
- *Subheading 3.1.3*, step 6: In contrast to the constitutive system, full induction of shRNA production via the tetR/O system in the mouse requires 2 mg/ml doxycycline in the drinking water. The extend of the resulting knockdown can be regulated by adjusting the doxycycline concentration. Add 10% sucrose for covering the aversive taste of doxycycline. Refresh water every second day to prevent microbial contamination and protect water bottles from light.

4. Notes

1. The purchase of expensive phosphorylated oligonucleotides is unnecessary as the vector ends remain phosphorylated after the following ligation reaction.
2. Note that the complementary C nucleotide of the corresponding antisense sequences has to be omitted as well.
3. Slow cooling down to room temperature guarantees proper annealing and is therefore an important factor. It can be achieved by simply switching off the thermoblock after 5 min incubation at 95°C and removing samples after the block has cooled down to approximately room temperature.
4. For sterilization all, or only a part of the Maxiprep, can be used. Determine the concentration of the DNA after the resuspension in sterile water as a loss of DNA is to be expected. Always store the DNA at −20°C.

5. Using positive and negative control plasmids is strongly recommended. A good positive control plasmid will express constitutively a shRNA directed against your target protein or any other protein expressed in ES cells. For negative control use a plasmid established for the knockdown of, e.g., luciferase or any other protein that is NOT expressed in ES cells.
6. We use the BioRad Genepulser Xcell; however, it might be necessary to adapt the parameters to individual lab conditions. As a general rule, lower voltage and shorter time settings result in better cell survival but yield a lower percentage of transfected cells. Poor transfection efficiency can be an explanatory factor if no gene silencing is observed in transient transfection assays. We recommend therefore the usage of controls (see Note 5), coelectroporation of reporter plasmids, or the comparison of all five candidate shRNA sequences by stable integration.
7. Within the first days under selecting conditions you should observe cell death or an appearance change in a part of the colonies. This serves as an indicator that G418 treatment is deleting the cells which are not expressing the neomycin resistance gene.
8. In our experience, the in vitro removal of the floxed stop cassette by this procedure works quite efficiently; we usually obtain ~90% positive clones.
9. The floxed stop cassette of the integrated construct can be deleted in ES cells by transient transfection with a Cre-expressing plasmid (available, e.g., at http://www.addgene.org/) and subsequent low-density plating as selection for recombined clones is impossible. After electroporation, we plate 500–2,000 cells on feeders in a 10 cm dish and analyze 24 clones for recombination by PCR or Southern blot. At least two recombined clones should be analyzed for knockdown efficiency.
10. Use as negative control nonrecombined IDG26.10-3 ES cells for assessing knockdown efficiency of your induced shRNA ES cell lines. Use as a second negative control IDG26.10-3 ES cells containing your shRNA construct that were cultured without the inducer doxycycline. Compare the expression level of your target gene in the uninduced to the one in the unrecombined IDG26.10-3 ES cells. This will give you information about the leakiness of the construct. However, in our studies we could never detect leakiness of the pEx-H1tetO-shRNA-CAG-tetR RMCE allele.
11. The extent of the resulting knockdown can be regulated by adjusting doxycycline concentration. As starting dilutions try, e.g., 1 ng up to 10 μg doxycycline per ml medium.

References

1. Elbashir SM, Harborth J, Lendeckel W et al (2001) Duplexes of 21-nucleotide RNAs mediate RNA interference in cultured mammalian cells. Nature 411:494–498
2. Brummelkamp TR, Bernards R, Agami R (2002) A system for stable expression of short interfering RNAs in mammalian cells. Science 296:550–553
3. Paddison PJ, Caudy AA, Bernstein E et al (2002) Short hairpin RNAs (shRNAs) induce sequence-specific silencing in mammalian cells. Genes Dev 16:948–958
4. Kuhn R, Streif S, Wurst W (2007) RNA interference in mice. Handb Exp Pharmacol 178:149–176
5. Dykxhoorn DM, Chowdhury D, Lieberman J (2008) RNA interference and cancer: endogenous pathways and therapeutic approaches. Adv Exp Med Biol 615:299–329
6. Dykxhoorn DM, Lieberman J (2006) Knocking down disease with siRNAs. Cell 126:231–235
7. Vaishnaw AK, Gollob J, Gamba-Vitalo C et al (2010) A status report on RNAi therapeutics. Silence 1:14
8. Falschlehner C, Steinbrink S, Erdmann G et al (2010) High-throughput RNAi screening to dissect cellular pathways: a how-to guide. Biotechnol J 5:368–376
9. Seyhan AA, Rya TE (2010) RNAi screening for the discovery of novel modulators of human disease. Curr Pharm Biotechnol 11:735–756
10. Seyhan AA, Varadarajan U, Choe S et al (2011) A genome-wide RNAi screen identifies novel targets of neratinib sensitivity leading to neratinib and paclitaxel combination drug treatments. Mol BioSyst 7:1974–1989
11. Tiedemann RE, Zhu YX, Schmidt J et al (2010) Kinome-wide RNAi studies in human multiple myeloma identify vulnerable kinase targets, including a lymphoid-restricted kinase, GRK6. Blood 115:1594–1604
12. Zhu YX, Tiedemann R, Shi CX et al (2011) RNAi screen of the druggable genome identifies modulators of proteasome inhibitor sensitivity in myeloma including CDK5. Blood 117:3847–3857
13. Delic S, Streif S, Deussing JM et al (2008) Genetic mouse models for behavioral analysis through transgenic RNAi technology. Genes Brain Behav 7:821–830
14. Kunath T, Gish G, Lickert H et al (2003) Transgenic RNA interference in ES cell-derived embryos recapitulates a genetic null phenotype. Nat Biotechnol 21:559–561
15. Hasuwa H, Kaseda K, Einarsdottir T et al (2002) Small interfering RNA and gene silencing in transgenic mice and rats. FEBS Lett 532: 227–230
16. Hitz C, Wurst W, Kuhn R (2007) Conditional brain-specific knockdown of MAPK using Cre/loxP regulated RNA interference. Nucleic Acids Res 35:e90
17. Rubinson DA, Dillon CP, Kwiatkowski AV et al (2003) A lentivirus-based system to functionally silence genes in primary mammalian cells, stem cells and transgenic mice by RNA interference. Nat Genet 33:401–406
18. Seibler J, Küter-Luks B, Kern H et al (2005) Single copy shRNA configuration for ubiquitous gene knockdown in mice. Nucleic Acids Res 33:e67
19. Steuber-Buchberger P, Wurst W, Kuhn R (2008) Simultaneous Cre-mediated conditional knockdown of two genes in mice. Genesis 46:144–151
20. Seibler J, Kleinridders A, Küter-Luks B et al (2007) Reversible gene knockdown in mice using a tight, inducible shRNA expression system. Nucleic Acids Res 35:e54
21. Nagy A, Mar L, Watts G (2009) Creation and use of a cre recombinase transgenic database. Methods Mol Biol 530:365–378
22. van de Wetering M, Oving I, Muncanet V et al (2003) Specific inhibition of gene expression using a stably integrated, inducible small-interfering-RNA vector. EMBO Rep 4:609–615
23. Wiznerowicz M, Szulc J, Trono D (2006) Tuning silence: conditional systems for RNA interference. Nat Methods 3:682–688
24. Acehan D, Vaz F, Houtkooper RH et al (2011) Cardiac and skeletal muscle defects in a mouse model of human Barth syndrome. J Biol Chem 286:899–908
25. He G, Luo W, Li P et al (2010) Gamma-secretase activating protein is a therapeutic target for Alzheimer's disease. Nature 467:95–98
26. Jordan SD, Krüger M, Willmes DM et al (2011) Obesity-induced overexpression of miRNA-143 inhibits insulin-stimulated AKT activation and impairs glucose metabolism. Nat Cell Biol 13:434–446
27. Koch L, Wunderlich FT, Seibler J et al (2008) Central insulin action regulates peripheral glucose and fat metabolism in mice. J Clin Invest 118:2132–2147
28. Kotnik K, Popova E, Todiras M et al (2009) Inducible transgenic rat model for diabetes mellitus based on shRNA-mediated gene knockdown. PLoS One 4:e5124
29. Herold MJ, van den Brandt J, Seibler J et al (2008) Inducible and reversible gene silencing

by stable integration of an shRNA-encoding lentivirus in transgenic rats. Proc Natl Acad Sci USA 105:18507–18512

30. Christoph T, Bahrenberg G, De Vry J et al (2008) Investigation of TRPV1 loss-of-function phenotypes in transgenic shRNA expressing and knockout mice. Mol Cell Neurosci 37:579–589
31. Sacca R, Engle SJ, Qin W et al (2010) Genetically engineered mouse models in drug discovery research. Methods Mol Biol 602: 37–54
32. Yang R, Castriota G, Chen Z et al (2011) RNAi-mediated germline knockdown of FABP4 increases body weight but does not improve the deranged nutrient metabolism of diet-induced obese mice. Int J Obes 35:217–225
33. Reid SW, Tessarollo L (2009) Isolation, microinjection and transfer of mouse blastocysts. Methods Mol Biol 530:269–285

Chapter 20

In Vivo Target Validation by Inducible RNAi in Human Xenograft Mouse Models

Marco Mazzoletti and Gemma Texidó

Abstract

Proper target selection and validation are crucial to the discovery of new anti-cancer agents. Since tumors depend on a suitable microenvironment for their growth, once a putative target has been identified, its validation should be performed whenever possible in vivo. This chapter deals with the generation of human xenograft mouse models genetically modified to induce the modulation of cancer-related genes as an approach to validate oncology targets.

Key words: RNAi, Tet-inducible, Target validation in vivo, Xenograft, Cancer genes

1. Introduction

The choice of the target is a key step to develop powerful anticancer drugs aimed to prevent tumor growth and cancer progression. One of the most widely used approaches for target validation is to down-regulate the gene of interest via gene silencing by RNA interference (RNAi) technology to prevent its translation and to evaluate the resulting phenotype (1). Transient transfection of custom-designed ready-to-use siRNAs or plasmids encoding siRNA usually lead to a short-lasting downregulation of the target in cells. Limits of the transient approach are the short-term modulation and the inter-experiment variability. To overcome this problem, a stable integration of the RNAi construct into the genome of cells can be pursued by using a RNAi cassette in which a shRNA is expressed under a constitutive or inducible promoter. While constitutive inhibition does not represent a physiological condition mimicking closely drug treatment regimens and may exert effects besides those strictly associated to the inhibition of the target. The use of inducible expression vectors has a series of advantages, including the prevention

Jürgen Moll and Riccardo Colombo (eds.), *Target Identification and Validation in Drug Discovery: Methods and Protocols*, Methods in Molecular Biology, vol. 986, DOI 10.1007/978-1-62703-311-4_20,

of undesired effects on cell growth eventually associated with constitutive expression of a specific shRNA. Inducible expression opens the opportunity for a time-dependent regulation and the possibility to mimic in vivo tumor progression and the effects of targeted therapy (2).

In particular cases the downregulation of the target through the constitutive expression of a shRNA could be a useful alternative to assess specific aspects of tumor nature, as those related with the growth and spreading of the tumoral cells from their original location towards other distant sites (3). Theoretically a cell clone lacking a particular protein involved in migration or tissue adhesion could be prevented to settle in other body sites and form metastases when injected in the blood flow of a mouse model.

It should be noted, however, that a siRNA leads to the disappearance of a protein from the cell, while a small molecule leaves an inactive protein inside the cell, which might still exert other functions. For example, PI3K p110gamma exerts a scaffold function not related with catalytic activity, which is also mediated by a kinase-dead mutant (4).

In vitro target validation using RNAi in human cancer cell lines has already provided some indications on promising good targets in oncology (5). However, tumors are complex and require a suitable microenvironment for their growth (6). Therefore, once a putative cancer target is identified, its validation in vivo should be considered before investing in a full drug discovery program.

In genetically engineered mouse models (GEMMs) (7), shRNA-based gene knockdown has evolved as an alternative approach to the classical gene knockout, yet the process of generating, breeding, and characterizing these knockdown GEMMs typically takes several years and is expensive. These obstacles, together with the large amount of new target candidates that are emerging from large-scale cancer genomics, prohibit the use of such mouse models for systematic target validation in vivo. Therefore, alternative strategies have to be considered at this initial stage of the drug discovery process.

Conventional human xenograft mouse models commonly employed to evaluate the therapeutic activity of drug candidates can also be exploited for an early target validation in vivo. Human xenograft mouse models are typically created by subcutaneous or orthotopic transplantation of human cancer cell lines in immunocompromised mice. Knockdown of potential oncology targets in these models can provide insight on whether the modulation of a given target is likely to have a therapeutic effect in human disease. Promising results have been achieved by delivering siRNA directly into the tumors or systemically in human xenograft mouse models (8–11); however, this approach often encounters the issue that si-R NAs do not reach all the organs of interest and are not stable in serum (12). Therefore delivery systems need to be improved before this technology can be widely applied.

An alternative approach to induce the knockdown of presumed cancer targets is to express shRNA against the target of interest in the human tumor cell line before their growth in mice. The assessment of genes knockdown effects can then be performed in vivo in the resulting mouse model. This has been used for inducible knockdown of cancer-related genes supporting the validation of targets such as Ran, HER3, and PLCγ1 (13–16).

Considering the large collections of available human cancer cell lines and that basically any gene can be knocked down by means of specific shRNAs, one could essentially downregulate any relevant target in any pertinent cancer context. An additional advantage of using human xenograft mouse models is that many of the most commonly used human cancer cell lines have been engineered to express luciferase genes (17, 18), which allows to trace their growth, and therefore the tumors that they generate can be followed noninvasively in vivo using optical imaging technologies, reducing the number of mice used in each study. Moreover this imaging approach allows to assess easily metastases and other processes associated with tumorigenesis (19, 20).

Xenograft models have often been regarded as inferior to the GEMMs because the tumor xenografts grow in immunocompromised mice whereas tumors that develop in GEMMs do so in an immunocompetent organism. On the other hand, GEMM-derived tumors often cannot fully reproduce the genetic complexity of human tumors (21). By now, it is well accepted that each model has its own advantages and drawbacks, and it is important to know a priori what we want to ask to a specific model and how to select the most appropriate cell line.

In this chapter we describe the procedures for the generation of xenograft models with a temporally induced knockdown of a specific target candidate gene in a particular human cancer context. This approach allows mimicking therapeutic intervention in vivo. The conditional expression of shRNAs is achieved by a tetracycline-regulated mammalian expression system (T-REx™) that uses regulatory elements from the *Escherichia coli* Tn10-encoded Tet resistance operon (22, 23). Tetracycline regulation in the T-REx™ system is based on the binding of tetracycline to the Tet repressor and derepression of the promoter controlling expression of the shRNA or gene of interest (24) (Fig. 1).

We have divided the methodology in three parts. In the first one, we describe the procedure for the generation and characterization of a stable human cancer cell line that constitutively expresses sufficient levels of Tet repressor. In the second part, the steps are illustrated for the transfection of the Tet-repressor cell

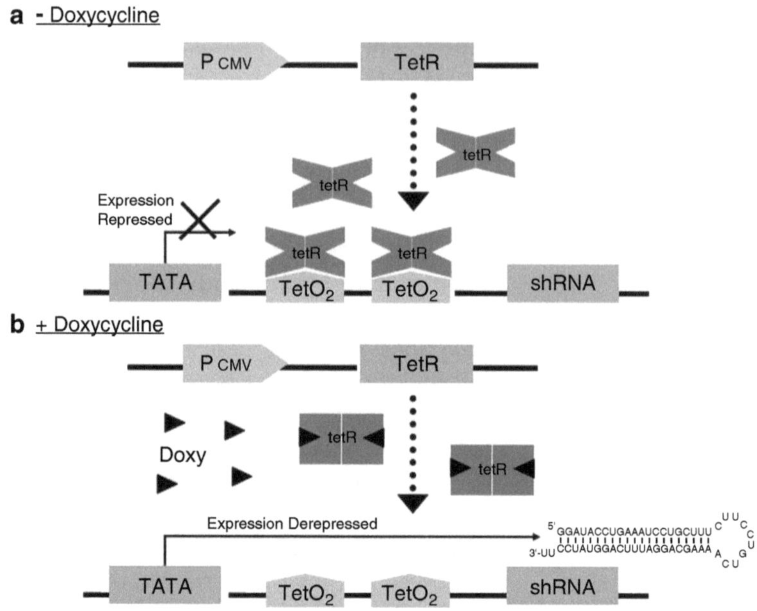

Fig. 1. Molecular mechanism of the tetracycline-regulated shRNA transcription. (**a**) In the absence of doxycycline, the constitutively expressed repressor homodimerizes and then binds to Tet operator 2 ($TetO_2$) sequences in the inducible expression vector preventing shRNA transcription. (**b**) When tetracycline is added to the medium, it causes a conformational change in the repressor, leading to its detachment from the Tet operator sequences and shRNA transcription induction.

line with the inducible shRNA plasmid (Fig. 2). Finally, in the third part, we describe the procedures needed to generate an shRNA-inducible xenograft mouse model.

2. Materials

2.1. Plasmids

1. pcDNA6/TR: a regulatory plasmid which encodes the Tet repressor (TetR) under the control of the human cytomegalovirus immediate-early (CMV) promoter (Invitrogen).
2. pSuperior retro: An inducible expression plasmid for shRNA expression under polymerase III H1 promoter. Vector sequence has been modified by inserting the tetracycline operator 2 ($TetO_2$) sites necessary for tetracycline-mediated expression.
3. pcDNA4/TO/LacZ: A control expression plasmid containing the lacZ reporter gene under a tetracycline-inducible promoter (Invitrogen).

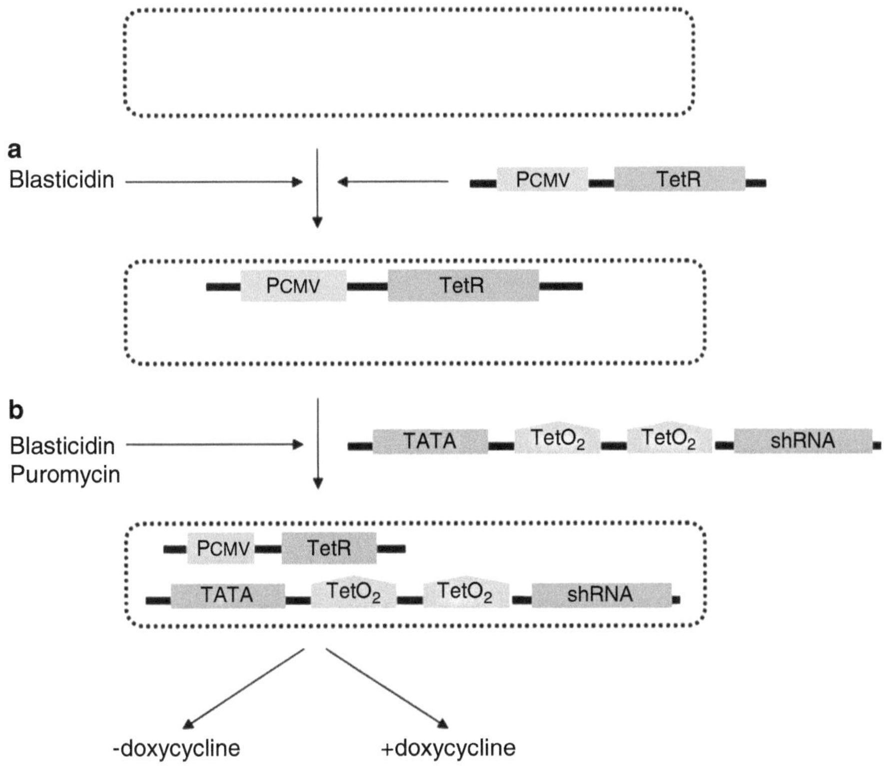

Fig. 2. Sequential procedure to generate a Tet-inducible shRNA-expressing human cancer cell line. (**a**) Delivery of the vector carrying the repressor under a constitutive promoter and the consequent selection with blasticidin allow to generate cells stably expressing the repressor (Subheading 3.1). (**b**) Only cells that integrate in their genome the vector carrying the shRNA cassette under the inducible promoter are able to survive the dual selection with blasticidin and puromycin and downregulate the target of interest after Tet induction (Subheading 3.2).

2.2. Reagents

1. Tet system approved fetal bovine serum (Clontech).
2. Selection antibiotic: Blasticidin and puromycin.
3. Lipofectamine 2000 (Invitrogen) as reagent for cell transfection.
4. Beta-Glo Assay (Promega) to screen clones for the expression of the repressor.
5. HCT116 cells and culture media (McCoy's + 2 mM L-glutamine + 10% FCS).
6. CD-1 Nude mice (Charles River).
7. Doxycycline and tetracycline.

3. Methods

To generate a stable human cancer cell line expressing an inducible shRNA, a sequential procedure should be followed (see Subheadings 3.1 and 3.2). The amount of reagents and the timelines indicated in this protocol are referred to the colon carcinoma HCT116 cell line and might need to be optimized for other cell lines.

The plasmid expressing Tet repressor (pcDNA6/TR) has a blasticidin-resistance cassette, while the plasmid used for the shRNA conditional expression (pSuperior) confers resistance to puromycin. Therefore, before starting the procedures, these two antibiotics need to be titrated on the selected cell line to find out the minimum inhibitory concentration necessary to kill completely untransfected cells.

The concentration of doxycycline should also be titrated at this stage using doses from 10 ng/ml to 10 μg/ml to rule out any unspecific toxic effect (see Note 1).

3.1. Generation of a Stable Cell Line Constitutively Expressing the Tet Repressor

1. Grow HCT116 cells in appropriate media. When cells have reached 50–60% confluence, transfect them with 20 μg of pcDNA6/TR and 40 μl of Lipofectamine 2000. Change medium 6 h later to prevent transfection reagent toxicity (see Note 2).
2. 24 h after transfection, split cells in several 100 mm petri dishes to obtain 20–25% confluence.
3. 48 h after transfection, start selection by adding blasticidin at a final concentration of 5 μg/ml. HCT116 cells without pcDNA6/TR integration will be counter selected. The selection media should be replaced every 3–4 days until single colonies appear (see Note 3).
4. Once colonies become visible, select viable colonies under the microscope and draw a circle on the bottom of the plate to mark them (see Note 4).
5. Aspirate growth medium and rinse the plate two times with PBS to remove any trace of medium and floating cells (see Note 5).
6. Pick up a cloning cylinder with a sterile forcep and press its bottom into sterile Vaseline. Remove it with a vertical movement and gently press it down on the previously drawn circle (see Note 6).
7. Fill cylinders with enough trypsin 0.25% EDTA to cover the plate surface inside the cylinder (normally, 40–50 μl are sufficient).
8. Repeat the procedure with each clone and leave the plate under the hood for some minutes until cells are detached (see Note 7).
9. Fill the cylinder with complete medium and slowly pipette up and down several times to detach the cells. Transfer each cell clone into a well of a 24-well plate previously filled with complete growth medium supplemented with blasticidin (see Note 8).
10. Repeat the process for all clones and incubate the plate in a tissue culture incubator (5% CO_2 at 37 °C) for several days, monitoring cell growth daily.
11. Once clones are grown, test them for the expression of the Tet repressor (see Note 9). Perform a functional assay by transient

transfection of a plasmid carrying a Tet-inducible reporter gene (Beta-Gal or GFP) (see Note 10).

12. For this functional assay we use the Beta-Glo assay (Promega), a luminescent assay able to quantify Beta-Gal expression. Other methods allowing a sensitive measurement of the reporter could be adopted (see Note 11).
13. Plate cells at the appropriate density in a 96-well white/black plate. Seed 4 wells for each clone (see Note 12).
14. 24 h later, transiently transfect 0.1 μg of reporter plasmid (pcDNA4/TO/LacZ) with 0.25 μl of Lipofectamine 2000. 24 h post-transfection induce half of the wells with doxycycline 2 μg/ml and assay reporter activity 24 or 48 h later.
15. Measure light emission (directly proportional to the amount of reporter produced) with a luminometer reader in induced versus not induced conditions, and make a ratio between values obtained to normalize them. Clones showing the highest induced/not induced ratio are the candidates to be used for the following step of integration of the inducible shRNA cassette. These are the clones that guarantee the highest level of repression in uninduced conditions. Positive clones should result in a tenfold ratio or more (see Note 13).
16. Clones have to be maintained in selective medium containing blasticidin (see Note 14).
17. Once the Tet-repressor clone has been obtained, it is helpful to compare its proliferation rate with the parental cell line to ensure a comparable duplication time. This will allow for a reliable comparison of data to that obtained from already performed drug treatments on the parental line or for the compliance with standardized protocols.

3.2. Transfection of the Tet-Inducible Target-Specific shRNA in the Tet-Repressor-Expressing Clone

1. Subclone shRNA sequences (see Note 15) directed against the gene of interest into the pSuperior vector (see Note 16).
2. Pretest shRNA activity to validate its efficacy by assessing target modulation through a transient transfection (see Note 17) in cells lacking the repressor (parental cell line) or in a TetR clone in the presence of doxycycline (see Note 18).
3. Transfect as previously described the Tet-repressor-expressing clone selected in subheading 3.1 point 15 with the Tet-driven shRNA construct that induces the desired knockdown levels.
4. The following day split cells in several 100 mm petri dishes to a confluence of 20–25%.
5. Select the clones with the appropriate concentration of the two antibiotics (blasticidin to maintain Tet-repressor construct and puromycin for the shRNA-expressing construct) changing medium every 3 or 4 days.

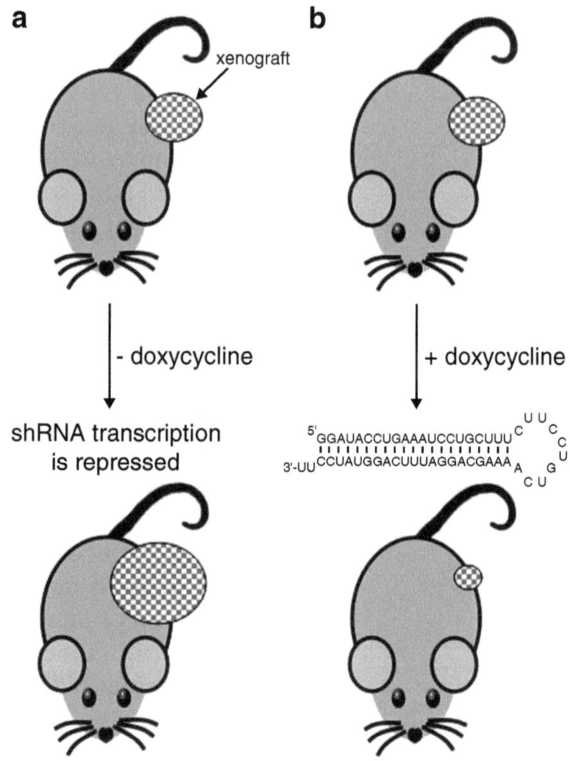

Fig. 3. Effects on the tetracycline-regulated shRNA expression in a mouse xenograft model. Xenografts of mice belonging to the control group, not receiving tetracycline, (**a**) cannot downregulate the oncogenic target and maintain their growth rate over time. Xenografts of mice receiving water supplemented with tetracycline (**b**) are able to express the shRNA against the target and hence to induce an arrest or a shrinkage of the tumoral mass.

6. Pick the colonies as indicated in subheading 3.1 from point 4 to point 9 and expand each clone in selection medium.
7. Once cells are grown, test them for shRNA expression and target modulation. Keep a fraction of the cells to maintain the lineage and use the others to perform the validation assay.
8. Seed cells and induce expression of the shRNA 24 h later by adding doxycycline (see Note 19). Harvest at different time points after induction and compare the expression of your target with uninduced cells (see Note 20).
9. Assess the downregulation of your target in the presence or absence of doxycycline by Western blot or quantitative PCR (qPCR) (see Note 21).
10. Maintain the generated cell lines always under dual selection pressure during the entire experiment (see Note 22).

3.3. Generation of the Human Xenograft Mouse Model

Validated Tet-inducible shRNA-expressing cell clones can then be used in nude mice.

Here we report the procedure for the in vivo validation of a HCT116 clone subcutaneously xenografted in nude mice (Fig. 3). Alternatively cells can be transplanted orthotopically.

1. Transplant 5×10^6 exponentially growing HCT116 cells subcutaneously into the flank of nude mice.
2. When tumors become measurable (about 100 mm^3), randomize mice in two homogenous groups one with and another without tetracycline, which is provided by dissolving it directly in drinking water (2 mg/ml). Replace tetracycline-supplemented drinking water every 2 days (see Note 23).
3. Divide each group of mice into two subgroups. Use one subgroup to evaluate tumor growth or other tumorigenesis-related readouts. The other group should be used for the evaluation of the dynamic expression profile of the target of interest after shRNA induction (see Note 24).
4. Measure tumor growth twice a week with a caliper, and calculate the estimates of tumor weight (assuming that 1 mg of tumor tissue has approximately a volume of 1 mm^3) as follows: (length (mm) x width $(mm)^2$)/2. Express activity parameters as T/C% (treated/control %) or tumor growth delay. Test the effective target modulation at different time points (e.g., every 24 h) by explanting the tumor from a Tet-induced mouse and assess target modulation by IHC or WB (see Note 25).

4. Notes

1. Doxycycline is normally used in a range of 1 or 2 μg/ml. In some cell lines this dose could be toxic and needs to be lowered (25).
2. It is recommended to linearize the plasmid before transfection to avoid aberrant integration of Tet-repressor coding sequence. Alternatively the construct can be delivered by viral-mediated infection. In this case use pLenti6/TR to express the repressor. It is recommended to optimize the transfection conditions by using an easily detectable reporter system such as green fluorescent protein (GFP).
3. Wait at least 24 h after transfection to allow for sufficient expression of the resistance gene before adding antibiotics. Include two plates as controls: one plate of transfected cells without antibiotic selection as positive control for cell viability and another plate of untransfected cells, seeded at the same density and treated with blasticidin to monitor cell resistance and antibiotic activity. All cells should die in this control under antibiotic selective pressure.

4. Avoid very close or large colonies which can be the result of a mixture of two different clones.
5. After the first wash, leave PBS a couple of minutes to promote the detachment of strongly adherent cells. Raise one side of the plate to remove any residual trace of PBS. Remaining PBS would loosen the interaction of the cloning cylinder with the plate surface.
6. Vaseline should be put as a thin layer in a glass petri dish and sterilized. The cylinder should be removed with a vertical movement to avoid the cylinder sliding, which would result in Vaseline filling the base of the cylinder. Be careful not to scratch the colony with the cylinder.
7. Cells detachment with trypsin is performed at room temperature and not at 37°C as usually indicated. This is needed to reduce the number of movements and hence to decrease the risk for the cylinder to detach from the surface with the consequent loss of the colony.
8. Avoid to culture the few cells forming the colony at too low density. This will often result in slow or absent growth. 24-well plates often provide a suitable surface for the growth of the small number of cells that are normally harvested with the cloning cylinder.
9. During the generation of the clone expressing the exogenous Tet repressor, every cell would integrate the cassette in a different position of the genome, with a wide associated variability in the amount of protein expressed. Lentiviral expression guarantees higher delivery efficiency. Nevertheless, a wide variability of construct expression is observed with both these approaches.
10. A Tet-inducible reporter gene is a gene whose expression is regulated by Tet operator sequences located within the promoter, and hence its expression is influenced by the amount of repressor inside the cell. Only clones that express high levels of repressor guarantee stringent regulation of the conditional expression. Positive clones should exhibit the highest fold induction of reporter expression after induction.
11. As an alternative, assess repressor expression by performing SDS-PAGE and probe the membrane with an anti-Tet-repressor antibody. This allows a direct visualization of the protein, but it is time consuming and does not provide any functional results.
12. Do not use clear plates for luminometric measurement because they allow light crossing between wells.
13. The highest levels of repressor expression are fundamental to obtain the best downregulation and prevent leakage in the absence of induction. Hence the accurate selection of few

selected single clones expressing the highest level of repressor is preferred instead of using the whole population of resistant cells. In fact some cells become resistant to antibiotic without expressing the repressor, probably following disruption of the repressor coding sequence during the integration event. If a mixed cell population is preferred in order to represent better the characteristics of the parental cell line, it is recommended to pool several clones which express good levels of repressor. Remember that the clone stably expressing the repressor could be also useful as a starting point to express in an inducible manner both shRNA and exogenous protein.

14. It is highly recommended to freeze and store some vials at early passage.
15. Because not all the shRNAs induce an appropriate downregulation of the target, it is recommended to test several sequences directed against different regions of the mRNA. shRNA sequences can be obtained by using bioinformatics tools or website such as (rnaidesigner.invitrogen.com).
16. Any alternative vector that allows a tetracycline-regulated expression of your shRNA can be used. Remember to consider the DNA delivery method (plasmid transfection or lentiviral transduction) to select the right vector. In the case of lentiviral infection, pLenti4/TO/V5-DEST (Invitrogen) could be another option.
17. Be sure to know the transfection efficiency. Poorly transfected cells could generate false-negative results. Set up a priori the optimal condition of transfection with a reporter gene.
18. Following transduction/transfection with the shRNA construct, cells have to be cultured in medium supplemented with FBS tested for the use with Tet-inducible vectors (Clontech). Cell culture serum often contains traces of tetracycline, which is enough to trigger shRNA expression. To prevent any possible unwanted expression of your shRNA, subculture Tet-expressing clones with Tet-free certified FBS for a couple of passages before performing shRNA construct delivery.
19. Doxycycline is more stable than tetracycline in culture medium and is also more potent, allowing it to be used at lower concentrations. Replace doxycycline-containing medium if a prolonged experiment is necessary (more than 72 h).
20. Make a time course from 24 to 96 h or more to evaluate target downregulation. Parameters influencing the target reduction rate are protein half-life, mRNA stability, and the absolute amount of protein in the cell.
21. Western blot provides a direct visualization of the residual amount of target expression; however, reliable results are dependent from antibody specificity. qPCR could be theoretically

adopted for every target, but it is less reliable when used to evaluate targets with very low expression. In some cases a partial target reduction is sufficient to observe an effect on cells, while in other cases a complete inhibition is necessary.

22. It is highly recommended to freeze and store some vials at early passage.
23. Tetracycline is less stable but cheaper than doxycycline. That is the main reason why it is preferred for its use in vivo, where a high amount of substance is necessary. Tetracycline reduced life-span can be compensated for by changing frequently the drinking water in order to reduce its degradation and the resulting variable induction over time.

    Considering that tetracycline is light sensitive, wrap drinking bottles with aluminum foil to preserve tetracycline integrity. Water can be also supplemented with 1–5% sucrose to improve its taste and make it more appealing for mice.
24. Each group should be made by at least 5–6 animals to follow tumor growth and 2–3 for each scheduled time point for the measurement of the residual expression of the target gene.
25. 1–4 days of induction are sufficient to test the in vivo modulatory effect of shRNA. 24 h can be considered the first standard point to assess target modulation, but this depends highly on target half-life.

References

1. Iorns E, Lord CJ, Turner N, Ashworth A (2007) Utilizing RNA interference to enhance cancer drug discovery. Nat Rev Drug Discov 6:556–568
2. Sandy P, Ventura A, Jacks T (2005) Mammalian RNAi: a practical guide. Biotechniques 39:215–224
3. Rankin EB, Fuh KC, Taylor TE et al (2010) AXL is an essential factor and therapeutic target for metastatic ovarian cancer. Cancer Res 70:7570–7579
4. Costa C, Hirsch E (2011) More than just kinases: the scaffolding function of PI3K. Curr Top Microbiol Immunol 346:171–181
5. Morgan-Lappe SE, Tucker LA, Huang X et al (2007) Identification of Ras-related nuclear protein, targeting protein for xenopus kinesin-like protein 2, and stearoyl-CoA desaturase 1 as promising cancer targets from an RNAi-based screen. Cancer Res 67:4390–4398
6. Hanahan D, Weinberg RA (2011) Hallmarks of cancer: the next generation. Cell 144:646–674
7. Singh M, Johnson L (2006) Using genetically engineered mouse models of cancer to aid drug development: an industry perspective. Clin Cancer Res 12:5312–5328
8. Elez R, Piiper A, Kronenberger B et al (2003) Tumor regression by combination antisense therapy against Plk1 and Bcl-2. Oncogene 22:69–80
9. Futami K, Kumagai E, Makino H et al (2008) Anticancer activity of RecQL1 helicase siRNA in mouse xenograft models. Cancer Sci 99:1227–1236
10. Oh BY, Lee RA, Kim KH (2011) siRNA targeting Livin decreases tumor in a xenograft model for colon cancer. World J Gastroenterol 17:2563–2571
11. Verma UN, Surabhi RM, Schmaltieg A et al (2003) Small interfering RNAs directed against beta-catenin inhibit the in vitro and in vivo growth of colon cancer cells. Clin Cancer Res 9:1291–1300
12. Oliveira S, Storm G, Schiffelers RM (2006) Targeted delivery of siRNA. J Biomed Biotechnol 2006:63675
13. Barres V, Ouellet V, Lafontaine J et al (2011) An essential role for Ran GTPase in epithelial ovarian cancer cell survival. Mol Cancer 9:272

14. Ganzinelli M, Carrassa L, Crippa F et al (2008) Checkpoint kinase 1 down-regulation by an inducible small interfering RNA expression system sensitized in vivo tumors to treatment with 5-fluorouracil. Clin Cancer Res 14:5131–5141
15. Lee-Hoeflich ST, Crocker L, Yao E et al (2008) A central role for HER3 in HER2-amplified breast cancer: implications for targeted therapy. Cancer Res 68:5878–5887
16. Sala G, Dituri F, Raimondi C et al (2008) Phospholipase Cgamma1 is required for metastasis development and progression. Cancer Res 68:10187–10196
17. Feng M, Zhang J, Anver M et al (2011) In vivo imaging of human malignant mesothelioma grown orthotopically in the peritoneal cavity of nude mice. J Cancer 2:123–131
18. Wang M, Gartel AL (2011) The suppression of FOXM1 and its targets in breast cancer xenograft tumors by siRNA. Oncotarget 2:1218–1226
19. Choy G, Choyke P, Libutti SK (2003) Current advances in molecular imaging: noninvasive in vivo bioluminescent and fluorescent optical imaging in cancer research. Mol Imaging 2:303–312
20. Luker GD, Luker KE (2008) Optical imaging: current applications and future directions. J Nucl Med 49:1–4
21. Richmond A, Su Y (2008) Mouse xenograft models vs GEM models for human cancer therapeutics. Dis Model Mech 1:78–82
22. Hillen W, Berens C (1994) Mechanisms underlying expression of Tn10 encoded tetracycline resistance. Annu Rev Microbiol 48:345–369
23. Hillen W, Gatz C, Altschmied L et al (1983) Control of expression of the Tn10-encoded tetracycline resistance genes. Equilibrium and kinetic investigation of the regulatory reaction. J Mol Biol 169:707–721
24. Yao F, Svensjo T, Winkler T et al (1998) Tetracycline repressor, tetR, rather than the tetR-mammalian cell transcription factor fusion derivatives, regulates inducible gene expression in mammalian cells. Hum Gene Ther 9:1939–1950
25. Marrazzo E, Marchini S, Previdi S et al (2006) Questioning the oncogenic role of DeltaNp73alpha in different cell lines expressing p53 or not. Cancer Biol Ther 5:794–803

Part VII

Translational Methods to Validate Biomarkers

Chapter 21

Bright-Field In Situ Hybridization Methods to Discover Gene Amplifications and Rearrangements in Clinical Samples

Hiroaki Nitta and Thomas M. Grogan

Abstract

Brightfield in situ hybridization (BISH) applications have significant advantages over traditional fluorescence in situ hybridization (FISH). BISH slides can be analyzed using a regular microscope while FISH slides require the use of a specialized fluorescence microscope. BISH slides allow observers for correlating the gene status (gene amplifications, gene rearrangements, and gene deletions) and tissue morphology better than FISH slides. Also, BISH slides are ideal for the permanent preservation of gene signals. Furthermore, BISH applications can be optimized using an automated tissue slide processing system. BISH applications are becoming a popular method for clinical examination of gene status for selecting cancer treatments.

Key words: Gene amplification, Gene rearrangement, *In situ* hybridization, Brightfield, Cancer, Personalized health care

1. Introduction

Fluorescence in situ hybridization (FISH) methods are used for the analyses of gene amplification, gene rearrangement, and gene deletion status of clinical samples. In general, FISH applications are labor-intensive manual protocols, and the observation of FISH slides requires a specialized fluorescence microscope in a dark room. Furthermore, additional hematoxylin and eosin (H&E) slides are necessary for correlating the tissue morphology and the gene signal of FISH slides. Recently, the technology of bright-field in situ hybridization (BISH) applications has advanced significantly, and a manual human epidermal growth factor receptor 2 (*HER2*) single-color BISH assay (Invitrogen Corporation, Camarillo, CA, USA) and an automated *HER2* dual-color BISH assay (Ventana Medical Systems, Inc., Tucson, AZ, USA) were approved as a clinical test for selecting breast cancer patient candidates for the humanized monoclonal antibody trastuzumab (Herceptin®, Genentech) by US

Jürgen Moll and Riccardo Colombo (eds.), *Target Identification and Validation in Drug Discovery: Methods and Protocols*, Methods in Molecular Biology, vol. 986, DOI 10.1007/978-1-62703-311-4_21,

Food and Drug Administration (FDA). For obtaining reproducible BISH assay results, because of a long assay protocol, the use of an automated slide processing system is recommended for avoiding human errors. Based on our successful assay development of BISH assays for gene amplifications (1–3) and gene rearrangements (4–6) using an automated slide processing system, we describe the importance of BISH assay optimization by using an automated instrument or for manual procedures in this chapter. However, because the preanalytical/tissue-processing steps have not been standardized, it is impossible to have one protocol that works on all clinical tissue samples. Therefore, this chapter should be used as a source of ideas for optimizing a BISH assay in a laboratory.

2. Materials

Formalin-fixed, paraffin-embedded xenograft tumors are ideal control samples for optimizing a BISH assay. Occasionally commercially available reagents are listed with specific vendors. However, this is not a complete list of commercial reagents, and alternative reagents can be found from other vendors. If reagents are prepared in a laboratory, distilled water and clean glassware should be used for making molecular-grade reagents.

2.1. Clinical Samples

1. Formalin-fixed, paraffin-embedded clinical samples (see Note 1).

2.2. Hapten-Labeled DNA Probes (Note 2)

1. ZytoDot2C SPEC probes (ZytoVision GmbH, Germany) (see Note 3).
2. FISH probes (Abnova, Taiwan, and Dako Denmark A/S, Denmark) (see Note 4).

   Select a probe set of your interest for a BISH assay from their lists of available probe sets.

2.3. Histology

1. Histology-grade xylene. Use for removing paraffin from paraffin tissue sections on glass slides.
2. Histology-grade alcohol. Use for hydrating or dehydrating tissue sections on glass slides.
3. Mayer's hematoxylin. Use for counterstaining tissue sections after completing BISH assays.
4. Histomount™ mounting solution (HS-103, National Diagnostics, Atlanta, GA, USA).

   Use for coverslipping BISH slides.

2.4. Pretreatment (See Note 5)

1. Heat pretreatment solution: Target Retrieval Solution (S1700, Dako Denmark A/S, Denmark).

Use as the first step of DNA retrieval procedure prior to the hybridization step.

2. SPoT-Light Tissue Pretreatment Kit (00-8401, Invitrogen Corporation, Camarillo, CA, USA).

   This kit contains a ready-to-use heat pretreatment solution and a ready-to-use enzyme reagent.

3. Proteases: Proteinase K, ready to use (PRK-004, Spring Bioscience, Inc., Pleasanton, CA, USA).

   Pepsin, ready to use (S-99003, 42 Life Science GmbH, Germany).

   Use proteinase K or pepsin as the second step of DNA retrieval procedure prior to the hybridization step.

2.5. Hybridization Buffer (See Note 6)

In situ hybridization buffer: 1.25× (ENZ-33808, Enzo Life Sciences, Farmingdale, NY, USA).

Use for diluting probe, adjusting the probe concentration, and/or inactivating alkaline phosphatase (AP) enzyme.

2.6. Stringency Wash Solution

SSC buffer: 20× concentrate (S6639-1L, Sigma-Aldrich, St. Louis, MO, USA).

Use for stringency wash steps after the hybridization step.

2.7. Immunodetection Reagents (See Note 7)

1. AP- and/or horseradish peroxidase (HRP)-conjugated antibodies for the haptens used for labeling probes (a direct detection method) or unconjugated antibodies for the haptens and AP- and/or HRP-labeled anti-species antibodies (an indirect detection method) (Vector Laboratories, Inc., Burlingame, CA, USA) (see Note 8).
2. Antibody diluent (ADS-125, Spring Bioscience).
3. Rinse solutions: PBS with Tween 20 (PBT-999, Spring Bioscience) or TBS with Tween 20 (TBT-999, Spring Bioscience).

   Choose one of rinse solutions for rinsing steps through the immunodetection for BISH signals.

4. Chromogen kits for AP- or HRP-based detection for BISH immunodetection reagents (see Note 9).

   Zyto*Dot*2C CISH Implementation Kit (C-3044-40, ZytoVision GmbH, Germany) (see Note 10).

   Zyto*Dot*2C SPEC HER2/CEN 17 Probe Kit (C-3022-40, ZytoVision) (see Note 11).

   SPoT-Light HER2 CISH Kit (84-0150, Invitrogen) (see Note 12).

   DuoCISH™ Kit (SK108, Dako) (see Note 13).

   Select appropriate detection kit for your purpose.

2.8. Hybridization Instruments (See Note 14)

1. StatSpin® ThermoBrite (TS01, IRIS Sample Processing, MA, USA).
2. Hybridizer (S2450, Dako).
3. SPoT-Light CISH Hybridizer (76-2000, Invitrogen).
4. PCR thermocycler with slide blocks.

   Use any of the aforementioned instruments for denaturation and hybridization of probes and DNA targets.

3. Methods

Dual-color BISH assays allow detecting two DNA targets on the same tissue sections. In general, manual dual-color BISH applications are labor-intensive 2-day protocols (Fig. 1). Paraffin sections must be treated for retrieving DNA targets so that probes can be hybridized effectively to the targets while tissue morphology is preserved as much as possible. The most effective way to unmask the DNA targets is a combination of heat pretreatment and protease digestion after removing paraffin from tissue sections. A pair of DNA probes labeled with two different haptens, such as digoxigenin (DIG) and 2,4 dinitrophenyl (DNP), and the haptens are immunologically visualized with HRP- and/or AP-based detection systems. Typically, a combination of HRP- and AP-based detections is used for dual-color BISH assays. However, a combination of two AP-based detections can be achieved by inactivating AP enzyme of the first BISH detection prior to the second AP-based BISH detection. A combination of two HRP-based detections can be achieved. However, there are more possible colors with AP detections than with HRP detections. A combination of two distinct colors should be selected for a dual-color BISH assay for easier analyses (Fig. 2), particularly for gene rearrangement ISH assays. Because of the lack of tissue-processing standardization, the optimization of BISH assays must be vigorously evaluated, especially for heat pretreatment and protease digestion steps for retrieving the DNA targets. In addition manufacturers' recommended protocols for dual-color BISH assays need to be modified for further assay optimization. This is one of the main reasons why the use of an automated slide staining system is highly recommended for improving the reproducibility of dual-color BISH assays.

3.1. Deparaffinization

1. Bake the glass slide with tissue sections approximately 5°C above the melting point temperature of paraffin in the oven for 15 min for improving the tissue section adhesive. The melting point temperature of paraffin depends on the specific paraffin product used.

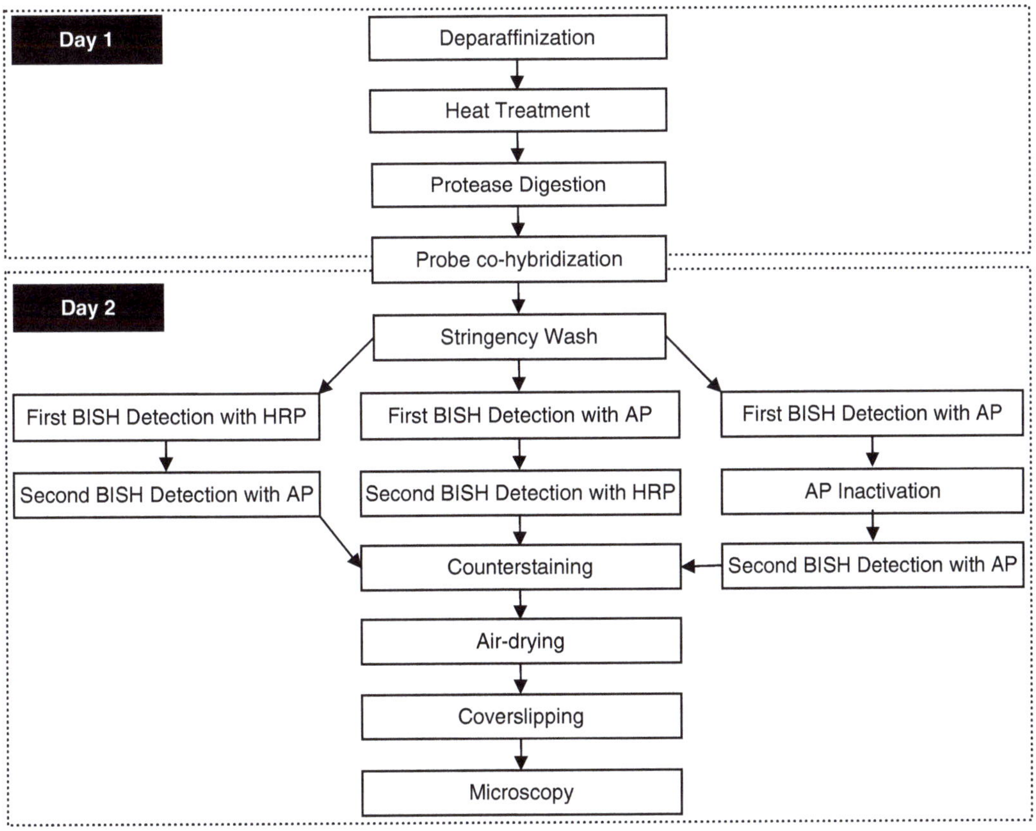

Fig. 1. A simplified flow chart for dual-color bright-field in situ hybridization (BISH) assays using horseradish peroxidase (HRP) and/or alkaline phosphatase (AP)-based immunological signal detections.

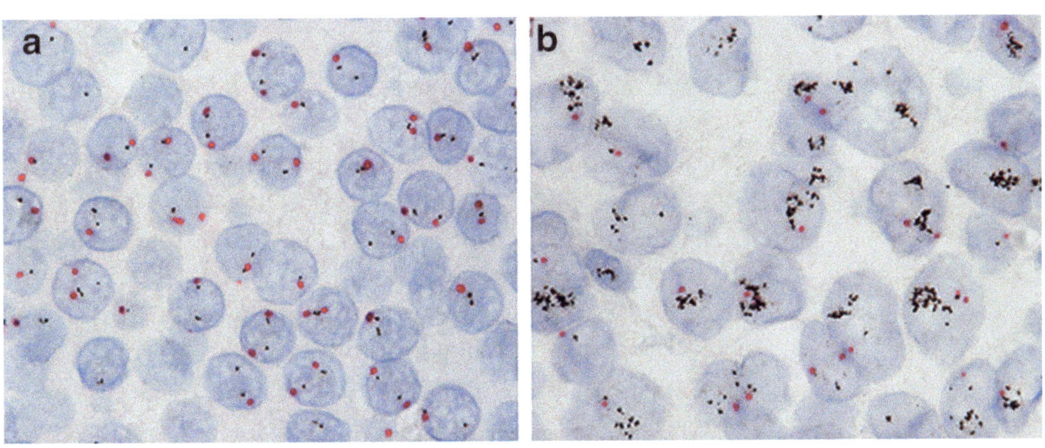

Fig. 2. Automated dual-color bright-field in situ hybridization (BISH) application for human epidermal growth factor receptor 2 (HER2) gene and chromosome 17 centromere (CEN17) of breast carcinomas. Normal HER2 gene (**a**) and amplified HER2 gene (**b**) cases are presented. HER2 gene was detected with horseradish peroxidase (HRP)-based silver detection (*black dots*), and CEN17 was detected with alkaline phosphatase (AP)-based fast red detection (*red dots*). 100× magnification.

2. Soak the slide in a Coplin jar of histology-grade xylene for 5 min at room temperature. Transfer the slide to fresh xylene and soak for 5 min.
3. Rinse the slide for removing xylene in a Coplin jar of 100% histology-grade alcohol for 2 min at room temperature. Transfer the slide to fresh 100% alcohol and soak for 2 min (see Note 15).
4. Transfer the slide to a Coplin jar of 80% alcohol and soak them for 2 min at room temperature.
5. Rinse the slide in a Coplin jar of distilled water for 2 min at room temperature. Transfer the slide to fresh distilled water and soak for 2 min.

3.2. Heat Pretreatment (See Note 16)

1. Place the slide in preheated heat pretreatment solution in a heat-resistant container according to the manufacturers' recommendations.
2. Cool the slide according to the manufacturers' recommendations.

3.3. Protease Digestion (See Note 17)

1. Apply a pepsin or proteinase K solution to the entire tissue sections and incubate for 3–5 min in a humidity-controlled chamber/box at room temperature.
2. Rinse the slide in a Coplin jar of distilled water for 2 min at room temperature.
3. Rinse the slide in a Coplin jar of 100% alcohol for 2 min at room temperature. Transfer the slide to fresh 100% alcohol and soak for 2 min.
4. Remove the slide and air-dry completely.

3.4. Hybridization (See Note 18)

1. Apply appropriate amount of a probe cocktail in a hybridization buffer onto tissue section.
2. Cover the tissue section with a coverslip by avoiding air bubbles under the coverslip.
3. Seal the edge of coverslip with rubber cement for avoiding the evaporation during denaturing and hybridization steps.
4. Dry the rubber cement for 15 min or until it dries completely at room temperature.
5. Place the covered slides on a PCR thermal cycler with slide blocks or a hybridization instrument.
6. Denature the probe and target at 95°C for 5 min.
7. Hybridize overnight at 37°C.

3.5. Stringency Wash

1. Remove the rubber cement from the slide.
2. Soak the slides in a Coplin jar of 2× SSC at room temperature until the coverslip to fall off from the slide.
3. Transfer the slide to a Coplin jar of 1× SSC for 5 min at 45°C. Repeat two additional washing in fresh 1× SSC for 5 min each.

4. Soak the slide in a Coplin jar of a rinse solution for 5 min at room temperature.

3.6. Signal Detection (See Note 19)

1. Remove the slide and apply an antibody for one of the haptens onto the slide and incubate for approximately 15 min in a humidity chamber/box (see Note 20) at room temperature. If the first hapten to visualize is DNP, anti-DNP antibody is applied. For a direct detection with the use of an enzyme-conjugated anti-hapten antibody, skip steps 2 and 3.
2. Rinse the slides twice with a rinse solution in a Coplin jar at room temperature.
3. Apply enzyme-conjugated anti-species antibody onto the slide and incubate for approximately 15 min in a humidity chamber/box at room temperature. If the primary antibody is a rabbit antibody, an enzyme-conjugated anti-rabbit antibody is applied.
4. Rinse the slides with rinse solution in a Coplin jar for three times at room temperature (see Note 21).
5. Apply a freshly prepared chromogen reagent mixture to the slide and incubate them in a humidity chamber/box at room temperature. Observe the color development under the light microscope and terminate the reaction by soaking in rinse solution when it is appropriate (see Note 22).
6. For a dual BISH assay with two AP-based detections, incubate the entire tissue section with hybridization buffer for 30 min in a humidity chamber/box at room temperature.
7. Rinse the slide very well with rinse buffer in a Coplin jar for 5 min for three times at room temperature.
8. Remove the side and apply an antibody for the other hapten onto the slide and incubate for approximately 15 min in a humidity chamber/box at room temperature. For a direct detection with the use of an enzyme-conjugated antibody, skip steps 9 and 10.
9. Rinse the slides twice with rinse solution in a Coplin jar at room temperature.
10. Apply enzyme-conjugated anti-species antibody onto the slide and incubate for approximately 15 min in a humidity chamber/box at room temperature. If the second anti-hapten antibody is a mouse antibody, an enzyme-conjugated anti-mouse antibody is applied.
11. Rinse the slide with a rinse solution in a Coplin jar for three times at room temperature.
12. Apply a freshly prepared chromogen reagent mixture to the slide and incubate it in a humidity chamber/box at room temperature. Observe the color development under the light microscope as needed and terminate the reaction by soaking in a rinse solution when it is appropriate.

3.7. Counterstaining and Mounting (See Note 23)

1. Transfer the slide to a Coplin jar of distilled water.
2. Soak the slide in a Coplin jar of Mayer's hematoxylin solution for 5 s at room temperature.
3. Rinse the slides with running tap water in a Coplin jar for 2 min.
4. Rinse the slides with distilled water in a Coplin jar for 2 min twice at room temperature.
5. Air-dry the slides on a bench until completely dried or dry in an oven at 65°C for 15 min.
6. Coverslip the slides with a permanent mounting solution, such as Histomount.

4. Notes

1. Any archived formalin-fixed, paraffin-embedded tissue blocks can be used for preparing tissue sections. However, there are cases that would never demonstrate successful BISH (or FISH) signals. Since, in general, information of how tissue samples were processed for paraffin embedding is very difficult to obtain, the exact cause of unsuccessful BISH assays cannot be clearly identified. The delay of placing tissue samples in a fixative is known to create difficulties of successful molecular histopathology assays. Tissue sections should be cut at 4–5 μm and placed onto SuperFrost® Plus slides (Erie Scientific Company, Portsmouth, NH, USA). The tissue sections can be stored in a slide box at room temperature.
2. Hapten-labeled DNA probes for FISH assays can be used for BISH assays and those probes can be purchased from various vendors. If commercial hapten-labeled DNA probes are not available, DNA probes can be labeled with haptens by nick translation or by PCR in a laboratory. Nick translation kit can be utilized for labeling DNA probes with hapten (10976776001, Roche Applied Science, Germany). New probes must be analyzed for the specificity by FISH or BISH assays using a comparative genomic hybridization metaphase control slide.
3. ZytoVision ZytoDot2C SPEC probes are labeled with DIG and DNP haptens and the probe hybridization sites can be visualized with a BISH detection kit including anti-DIG and anti-DNP antibodies.
4. Abnova FISH probes and Dako FISH probes are labeled with fluorescein isothiocyanate (FITC) and Texas Red (sulforhodamine 101 acid chloride) haptens, and the probe hybridization sites can be visualized with a BISH detection kit including anti-FITC and anti-Texas Red antibodies.

5. Using an automated slide processing system, we investigated tissue pretreatment conditions prior to the hybridization step for probes and DNA targets. We learned that the best pretreatment method is a combination of a heat pretreatment followed by a protease digestion with proteinase K or pepsin. Heat pretreatment alone or protease digestion alone is not adequate for retrieving DNA targets while preserving tissue morphology. Unfortunately, like tissue-processing protocols, the pretreatment conditions are not standardized. Because of unstandardized tissue-processing protocols, a robust pretreatment condition must be established in each laboratory individually. Also, the pretreatment protocol might need to be tweaked for different tissue samples.
6. In situ hybridization buffer is used for two purposes: (1) probe denaturing and hybridization and (2) AP enzyme inactivation between two sequential AP-based BISH detections. As we previously reported (4), we learned that AP enzyme can be inactivated by incubating the tissue section with an in situ hybridization buffer containing formamide for 30 min at room temperature without losing probe-target hybridization. Thus, after the first AP-based BISH detection, an in situ hybridization buffer should be applied for inactivating prior to the second AP-based BISH detection. Since formamide can denature proteins, the slides must be washed well with a rinse solution before the second immunodetection step.
7. You can choose to build your own BISH detection kit from individual reagents from various vendors or to purchase a commercial BISH kit. Selecting a right color combination with good color contrast is a key for successful BISH assay detection. Color blindness is an important factor for deciding a color combination (7, 8). Common color combinations are (1) blue (AP-based detection) and red (AP-based detection), (2) black (HRP-based detection) and red (AP-based detection), and (3) green (HRP-based detection) and red (AP-based detection).
8. Common haptens used for labeling DNA probes for BISH assays are biotin, DIG, DNP, FITC, and Texas Red. Based on the size of your DNA targets, you may choose from a direct detection or an indirect detection for BISH assays. In general, an indirect detection system can provide better sensitivity compared to a direct detection system. For an indirect detection, you need to select a combination of two antibodies raised with two different animal species, such as a mouse anti-DIG antibody and a rabbit anti-DNP antibody, so that enzyme-labeled anti-mouse antibody and anti-rabbit antibody can be applied for signal detection. If a direct BISH detection is going to be applied, anti-hapten antibodies raised in the same animal species that are labeled with either AP or HRP enzyme molecules

are used. Dilute antibodies according to manufacturer's recommendation with an antibody diluent of your choice.

9. AP- and HRP-based chromogen kits can be obtained from various histology companies. Vector Laboratories, Inc., Burlingame, CA, USA carries the most extensive product line for both AP- and HRP-based chromogen kits that can be used for BISH detection. 42 Life Science GmbH, Germany, also carries both AP- and HRP-based chromogen kits.
10. ZytoVision ZytoDot2C CISH Implementation Kit contains all necessary BISH detection reagents for the detection of DIG- and DNP-labeled DNA probes on formalin-fixed tissue sections. The probes can be purchased from ZytoVision or other vendors.
11. ZytoVision ZytoDot2C SPEC HER2/CEN 17 Probe Kit contains all necessary BISH detection reagents for the detection of HER2 and CEN17 DNA targets on formalin-fixed tissue sections. This is an ideal kit for manual dual-color BISH assay detection.
12. Invitrogen SPoT-Light HER2 CISH Kit contains all necessary reagents for performing a single-color BISH assay for DIG-labeled HER2 probe. One can modify the protocol by adding an AP-based BISH detection system (and replacing the probe with a probe cocktail) for dual-color BISH assays for two DNA targets.
13. Dako DuoCISH Kit contains HRP- and AP-based immunological detection reagents for Texas Red- and FITC-labeled probes. This is an ideal kit for manual dual-color BISH assay detection.
14. Those listed in situ hybridization instruments are for manual BISH applications, particularly for denaturing and hybridization steps. Moisture pads can be placed for creating humidity during hybridization, but we found the best way for avoiding the evaporation is to secure the sealing of coverslips with rubber cement. Automated slide processing systems that allow hybridization assays are also available from several manufactures, such as Ventana Medical Systems, Inc. (Tucson, AZ, USA) and Leica Microsystems GmbH (Germany). Ventana offers software and reagents for automated BISH assays.
15. Instead of dehydrating the tissue sections by using a series of alcohol solutions, one can simply air-dry the slides prior to a heat-treatment step.
16. Dako recommends heating target retrieval solution in a Coplin jar with a water bath (95–99°C without boiling) or with an autoclave (121°C) and cooling the jar to about 60°C. Invitrogen recommends heating their heat pretreatment solution by boiling on the hotplate and transferring the slides

to distilled water at room temperature. An important thing is to find an appropriate preheating step without destroying the tissue sections and to cool the slides prior to a protease digestion step.

17. Heat pretreatment alone is not adequate for retrieving DNA in the nuclei. Protease digestion is the most important step for successful BISH assays that allows observing the BISH signal and tissue morphology. Pepsin or proteinase K is commonly used for digesting the tissue sections for allowing the probes to hybridize the DNA targets. However, over-digestion of tissue sections will result in poor tissue morphology and less BISH signals. The use of nontumor cells near the tumor cell population as an internal assay performance control is the best way to evaluate the success of BISH assays. Nontumor cells should show 1–2 copies of a target DNA.

18. The Hybridization step is a technically challenging step in a manual protocol. The amount of probe solution needs to be adjusted based on the size of coverslip. Appropriate probe volumes are 10 μl for 22 × 22 mm coverslips, 20 μl for 22 × 30 mm coverslips, and 30 μl for 22 × 50 coverslips. Sealing the coverslip with rubber cement is not easy around the edge of coverslip with excess probe. If a larger size of coverslip is used, there are more chances to create air bubbles under the coverslip. In order to obtain consistent BISH signals, air bubbles must be avoided as much as possible. When the rubber cement did not seal the edge of coverslips, air bubbles will be created during the denaturing step. Air bubbles created during the denaturing step tend to disappear once the temperature goes down to the hybridization temperature. However, the area of tissue sections with air bubbles during hybridization will not have BISH signals.

19. H_2O_2-blocking step is often used for immunohistochemical assays, but we have not encountered the necessary of H_2O_2 blocking for BISH assays. Also, a protein-blocking step is incorporated prior to antibody incubation steps for immunohistochemical assays, but we found that it is not so important for BISH assays as long as the antibody concentrations are appropriate. However, the background staining caused by antibodies can be reduced or eliminated by adding casein in the antibody diluent as needed. In addition, protein blocking could also reduce the ISH signal.

20. Tissue sections should not be dried during immunological BISH detection. A humidity chamber/box can be assembled by placing wet paper towels in a tray and covering it with a plastic wrap sheet. PAP pen is often used for creating a hydrophobic barrier that is dissolved with xylene for immunohistochemistry. However, because BISH detection

with AP-based immunological detection(s) avoids the use of xylene for preserving BISH signals, it is not recommended using the PAP pen for BISH assays.

21. Washing steps prior to chromogen reaction is more important than washing steps between antibody incubations. Use plenty of a rinse solution for removing enzyme-conjugated antibodies for lowering the background staining. Gentle washing is important to preserve better tissue morphology.
22. Wear disposable gloves while handling slides covered with chromogen. Particularly, 3,3-diaminobenzidine (DAB) is a chromogen that is a well-known carcinogen.
23. The purpose of counterstaining for BISH slides is to visualize the nuclei so that an observer can correlate the tissue morphology and BISH signals. Very weak counterstaining should be achieved for avoiding obscuring BISH staining. Also, the use of bluing reagents might eliminate BISH staining dependent on which chromogen was used for immunodetection.

References

1. Nitta H, Hauss-Wegrzyniak B, Lehrkamp M et al (2008) Development of automated brightfield double *in situ* hybridization (BISH) application for *HER2* gene and chromosome 17 centromere (CEN 17) for breast carcinomas and an assay performance comparison to manual dual color *HER2* fluorescence *in situ* hybridization (FISH). Diagn Pathol 3:41
2. Stasik CJ, Nitta H, Zhang W et al (2010) Increased *MYC* gene copy number correlates with increased mRNA levels in diffuse large B-cell lymphoma. Haematologica 95: 597–603
3. Rodig SJ, Kutok JL, Paterson JC et al (2010) The pre-B-cell receptor associated protein VpreB3 is a useful diagnostic marker for identifying c-MYC translocated lymphomas. Haematologica 95:2056–2062
4. Nitta H, Zhang W, Kelly DB et al (2010) Automated brightfield break-apart in situ hybridization (ba-ISH) application: *ALK* and *MALT1* genes as models. Methods 52:352–358
5. Kim H, Choe J, Yoo S et al (2011) Detection of *ALK* gene rearrangement in non-small-cell lung cancer: a comparison of fluorescence *in situ* hybridization and chromogenic *in situ* hybridization with correlation of ALK protein expression. J Thorac Oncol 6:1359–1366
6. Yoshida A, Tsuta K, Nitta H et al (2011) Bright-field dual color chromogenic *in situ* hybridization for the diagnosis of *EML4-ALK*-postitive lung adenocarcinomas. J Thorac Oncol 6:1677–1686
7. Albrecht M (2010) Color blindness. Nat Methods 10:775
8. Landini G, Perryer DG (2011) More on color blindness. Nat Methods 8:891

Chapter 22

Combined MicroRNA In Situ Hybridization and Immunohistochemical Detection of Protein Markers

Boye Schnack Nielsen and Kim Holmstrøm

Abstract

MicroRNAs are short (18-23 nucleotides) non-coding RNAs involved in posttranscriptional regulation of gene expression through their specific binding to the 3'UTR of mRNAs. MicroRNAs can be detected in tissues using specific locked nucleic acid (LNA)-enhanced probes. The characterization of microRNA expression in tissues by in situ detection is often crucial following a microRNA biomarker discovery phase in order to validate the candidate microRNA biomarker and allow better interpretation of its molecular functions and derived cellular interactions. The in situ hybridization data provides information about contextual distribution and cellular origin of the microRNA. By combining microRNA in situ hybridization with immunohistochemical staining of protein markers, it is possible to precisely characterize the microRNA expressing cells and to identify the potential microRNA targets. This combined technology can also help to monitor changes in the level of potential microRNA targets in a therapeutic setting. In this chapter we present a fluorescence-based technology that allows the combination of microRNA in situ hybridization with immunohistochemistry exemplified by the in situ detection of miR-21 and miR-205 in combination with PDCD4 and smooth muscle α-actin.

Key words: MicroRNA, Locked nucleic acid, *In situ hybridization*, Immunohistochemistry, Fluorescence multiplexing

1. Introduction

MicroRNAs (miRNAs) constitute a group of short noncoding RNAs, 18–23 nucleotides in length. Since the discovery of the small regulatory noncoding RNA in C. elegans, Lin-4 (1–3), miRNAs have been found to direct essential mechanisms for regulation of gene expression in eukaryotic cells. In humans the number of annotated miRNAs has exceeded 1,500 (miRBase 18). Individual miRNAs contribute to the regulation of protein expression typically by binding to the 3'untranslated region (UTR) of mRNAs (4–6). MiRNAs are considered to affect translation of more than 30% of all

Jürgen Moll and Riccardo Colombo (eds.), *Target Identification and Validation in Drug Discovery: Methods and Protocols*, Methods in Molecular Biology, vol. 986, DOI 10.1007/978-1-62703-311-4_22,

mRNAs and thereby affect cellular processes such as stem cell fate, differentiation and proliferation (7–9), cell migration (10, 11), and adhesion (11). MiRNA genes are expressed as regular transcripts or as intronic transcripts containing regular short hairpin precursor structures. The sequential processing of the precursor transcript in the nucleus into mature single-stranded miRNAs in the cytoplasm may be unique for individual miRNAs; however, canonical pathways have been suggested (12, 13). The mature miRNA is loaded into the RNA-induced silencing complex (RISC), which binds complementary regions in the 3'UTR of mRNA for regulation of the protein expression activity. The unique nucleic acid sequence of individual miRNAs allows to some extent prediction of potential mRNA targets. Several databases based on target prediction algorithms, such as TargetScan (targetscan.org/vert_50/) and miRanda (14) (microrna.org/microrna/home.do) can be accessed online. The algorithms differ in particular in the criteria related to the pairing of the 3'-seed sequence resulting in different targets being identified. Critical use of these databases has recently been reviewed by Thomson et al. (15).

MiRNA can be identified in most biological samples including paraffin-embedded tissue biopsies and blood samples. A surprisingly high stability of miRNA compared to mRNA is probably related to the high stability of the RISC/miRNA complex. Isolation of miRNA from tissue samples and body fluids has led to miRNA profiling both at the level of microarray analysis (16, 17), PCR-based analyses (18), and deep sequencing (19, 20). These quantitative miRNA expression analysis platforms have been extensively used to identify novel biomarker candidates among the miRNAs, including profiles in differential diagnosis (21, 22). A common feature of these technologies is that the expression data represent a global average expression level in the tissue samples that is affected by the normalization procedure, which therefore must be chosen after detailed analysis and consideration. These technologies are strong quantitative tools with a high level of precision and reproducibility. However, to support and validate miRNA profiling as a biomarker discovery tool, more biological and molecular insight is needed to increase knowledge about the molecular mechanism regulated by the miRNA and to reach a better understanding based on the cellular origin of the miRNA. For instance, knowing the precise characteristics of a cell population under investigation will narrow down the repertoire of potential targets to the biologically most relevant.

Due to their functional involvement in specific molecular pathways, miRNAs have become interesting targets for therapeutic intervention. Examples of disease-related mechanisms affected by miRNAs include single-nucleotide polymorphisms (SNPs) in 3'UTR sequences (23, 24) or in miRNA genes (25, 26) that may affect normal protein expression and has been identified in association with heart failure (27) and cancer (25, 28). Chromosomal deletions and epigenetic changes involving miRNA also cause abnormal protein expression that is associated with malfunctions and diseases (29, 30). Aberrant protein expression processes may be normalized by therapeutic interference using either synthetic miRNA mimics (31), shRNAs (32–34), or miRNA inhibitors (anti-miR, antago-miR) (35–37). The effect of introducing such a therapeutic strategy may be followed by monitoring if the protein expression returns to its normal level after treatment. Thus, validating the uptake of the miRNA-modulating drug and monitoring the efficacy of the targeting molecule may require thorough histological localization analysis of the therapeutic nucleic acid construct as well as the targeted protein(s).

In situ hybridization (ISH) is a technology that enables characterization of the miRNA's cellular origin (38–40). For in situ detection of miRNAs, specific complementary probes consisting of DNA and locked nucleic acid (LNA) have been found to be a useful tool to achieve specific identification both in whole mount samples (41) and formalin-fixed and paraffin-embedded (FFPE) samples (38, 42, 43). The results gained from an ISH study provide specific information of contextual expression within a tissue; the miRNA may be focally expressed in particular tissue compartments, for example, necrotic areas. For some miRNAs, expression is cell specific, such as miR-126 and miR-145, which are specifically expressed in endothelial cells and vascular smooth muscle cells, respectively (38). In addition, simple histological analysis can often determine if a miRNA is expressed in the mesenchymal or epithelial compartment of a tissue. Specific identification of a cell group expressing a miRNA in cellular complex tissues like inflamed areas, tumor stroma, or regenerating tissues may require cell characterization using cell type-specific antibodies.

Here, we present a method that combines miRNA in situ hybridization using LNA-containing oligo probes with immunohistochemical detection of specific protein markers in order to precisely characterize the miRNA's cellular origin and monitor if a miRNA target gene is potentially valid by monitoring a parallel absence or downregulation of the specific downstream target protein.

2. Materials

2.1. Reagents

1. Double-labeled LNA™ probe (Exiqon). In this study we use miRCURY™ double-FAM-labeled LNA™ probes specific for miR-21 and miR-205.
2. In situ hybridization buffer (microRNA ISH buffer set (FFPE), Exiqon).
3. PBS 1x, RNase-free quality.
4. Tween-20 and a 10% solution in RNase-depleted water.
5. RNase-depleted water, for example, RNase-free Milli-Q water (Millipore).

6. 20× SSC buffer, RNase-free quality.
7. Sheep anti-FITC–POD (Roche). Primary antibodies: in this study we have used a mouse monoclonal antibody against smooth muscle α-actin (sm-α-actin, Dako) and rabbit polyclonal antibody against programmed cell death 4 (PDCD4, Rockland Immunochemicals).
8. Cy3-conjugated goat-anti-mouse and Cy3-conjugated goat-anti-rabbit antibodies (Jackson Immunoresearch).
9. Tyramine signal amplification (TSA) reagent, for example, TSA–FITC (PerkinElmer).
10. Anti-fade ProLong Gold with DAPI (Invitrogen).
11. SuperFrost® Plus (Thermo Fisher Scientific).
12. Hybridizer (Dako).
13. Shandon Sequenza Slide racks (Thermo Scientific).
14. Horizontal humidifying chamber.
15. For RNase-depleting working tools and surfaces: RNase ZAP, RNase Away, or similar.

2.2. Buffers

1. *Proteinase-K buffer*: To 900 ml Milli-Q water, add 5 ml of 1 M Tris–HCl (pH 7.4), 2 ml 0.5 M EDTA, and 0.2 ml 5 M NaCl. Adjust volume to 1,000 ml.
2. *Proteinase-K reagent*: To 10 ml proteinase-K buffer, add 7.5 μl proteinase-K stock of 20 mg/ml.
3. 0.1× SSC *buffer*: To 995 ml Milli-Q water, add 5 ml 20× SSC. The SSC buffer should be autoclaved.
4. *PBS–T*: to 1 l of PBS (pH 7.4), add 1 ml Tween-20.
5. *Blocking solution*: To 37.5 ml of Milli-Q water, add 5 ml 1 M Tris–HCl (pH 7.5), 1.5 ml 5 M NaCL, 5 ml fetal bovine serum (FBS), and 100 μl diluted Tween-20 (10% solution in water).

2.3. Glassware

All glassware, including coplin jars, glass-staining racks, and stacks of cover glass and bottles for buffers, should be heat treated in an oven at 180°C for 8 h. The items can be covered by aluminium foil before being placed in the oven in order to prevent contamination when removing the items afterward.

3. Methods

3.1. Tissue Sections

Before starting cutting sections from the tissue blocks, the whole workstation (bench top, microtome, blade holder, brushes, tweezers, cooling plate, water bath, etc.) should be cleaned with RNase ZAP. Set the cooling plate to −15°C and place the FFPE block on the plate.

Fill the warm water bath with RNase-free Milli–Q water and set temperature to 40–50°C depending on the type of paraffin used for embedding. Another water bath is prepared as an RNase-free coplin jar containing room-tempered RNase–free Milli–Q water.

Insert a new disposable blade in the knife carrier, and place the block in the cassette clamp. Trim the block, in order to avoid the first couple of sections, which might be contaminated during previous handling such as from tissue preparation or embedding. After trimming the block, cut 6 μm sections and place them into a dry sterilized jar with room–tempered RNase-free Milli–Q water, where any folding can be reversed. Transfer the slide to a heated (40–50°C) water bath, where the tissue is stretched to avoid folds and overlaps in the structure, and mount sections immediately on SuperFrost® Plus glass slides (glass slides taken from a new and untouched package are considered RNase-free). Mounting the section onto the slide should be done with care—allow water to slide away from in between the paraffin section and the glass slide—this process is important to avoid that sections fall off during deparaffination. Let the slides dry for 2 h at room temperature and store at 4°C in a dry box containing silica gel. Melt paraffin at 60°C on the day prior to the in situ hybridization experiment.

3.2. In Situ Hybridization

1. Deparaffinize slides in xylene and ethanol solutions in coplin jars ending up in PBS. In parallel, prepare a water bath and SSC buffer to be heated to 55°C (or the hybridization temperature). Place slides in xylene for 15 min (through 2–3 coplin jars) and then hydrate through ethanol solutions 99% (three coplin jars), 96% (two coplin jars), and 70% (two coplin jars) to PBS (two coplin jars). Each solution should include 5 min incubation.
2. Predigestion of tissue sections is done by applying 300 μl/slide of the proteinase-K reagent at 25 μg/ml (see Note 1) directly on the tissue and incubate for 40 min at 30°C in a horizontal humidifying chamber.
3. Discard the proteinase-K reagent and wash twice with PBS.
4. Apply 300 μl freshly prepared 3% hydrogen peroxide and incubate at room temperature for 10 min.
5. Discard the hydrogen peroxide and wash thrice with PBS.
6. Remove excess PBS and immediately apply 50 μl double-FAM LNA™ (see Note 2) probe solution and gently shield with cover glass (see Note 3). The probe solution is prepared as follows: denature LNA™ probe and dilute the probe in Exiqon ISH buffer. For example, for 2 ml hybridization mix containing 20 nM double-FAM-labeled miR-21 LNA™ probe (from 25 μM probe stock), transfer 4 μl into the bottom of a 2 ml nonstick RNase-free tube and place the tube at 90°C for 4 min. Spin down shortly using a tabletop centrifuge, and immediately

add 2 ml ISH buffer into the tube. LNA™ probes for other miRNAs may require optimization of the concentration (see Note 4).

7. Place the slides in the hybridizer and start a preset hybridization program for 1 h at 55°C.
8. Place slides into 55°C pre-warmed 0.1× SSC in a coplin jar. The cover slides will easily detach. Then transfer slides to another casket with 55°C pre-warmed 0.1× SSC. Wash slides thrice using 55°C pre-warmed 0.1× SSC.
9. Transfer slides to PBS–T and mount into Shandon Sequenza® slide racks. Avoid air bubbles during mounting.
10. Incubate with 200 μl blocking solution for 15 min at RT.
11. Detection of FAM probes is done by applying sheep anti-FAM–POD diluted 1:400 in blocking solution and incubated for 60 min at RT (preferably two times for 30 min). To 1,000 μl blocking solution, add 2.5 μl sheep anti-FAM–POD.
12. Wash each slide with 300 μl PBS–T thrice for 3 min.
13. Incubate with 150 μl freshly prepared TSA–FITC reagent 7–15 min at RT. Protect from light during development. For the preparation of 1,500 μl TSA–FITC substrate, add 30 μl TSA–FITC reagent into 1,500 μl diluent (see Note 5).
14. Wash each slide with 300 μl PBS–T thrice for 3 min.
15. Detection of protein markers is done by incubating 150 μl primary antibody (here anti-sm-α-actin diluted 1:200 and anti-PDCD4 diluted 1:300) (see Note 6) diluted in PBS containing 1% BSA for 60 min at room temperature.
16. Wash each slide with 300 μl PBS–T thrice for 3 min.
17. The protein marker antibodies are detected by incubating Cy3-conjugated detecting antibody (anti-mouse or anti-rabbit) diluted 1:400 in PBS for 30 min at room temperature.
18. Wash each slide with 300 μl PBS thrice for 3 min.
19. Mount slides directly with anti-fade ProLong Gold containing DAPI. Store slides dark and at 4°C. Evaluate staining result after 24 h.
20. Evaluate slides using a fluorescence microscope with filters allowing detection of FITC, Cy3, and DAPI emission.
21. The evaluation of miRNA ISH results requires considerations on the specificity of the signal. Specificity control analysis is an inherent part of molecular histology. For miRNA ISH analyses, there are several ways to address this question (40). Specific for double fluorescence analyses, one key point to address is preventing cross-reaction between the two parallel assays (see Note 7).

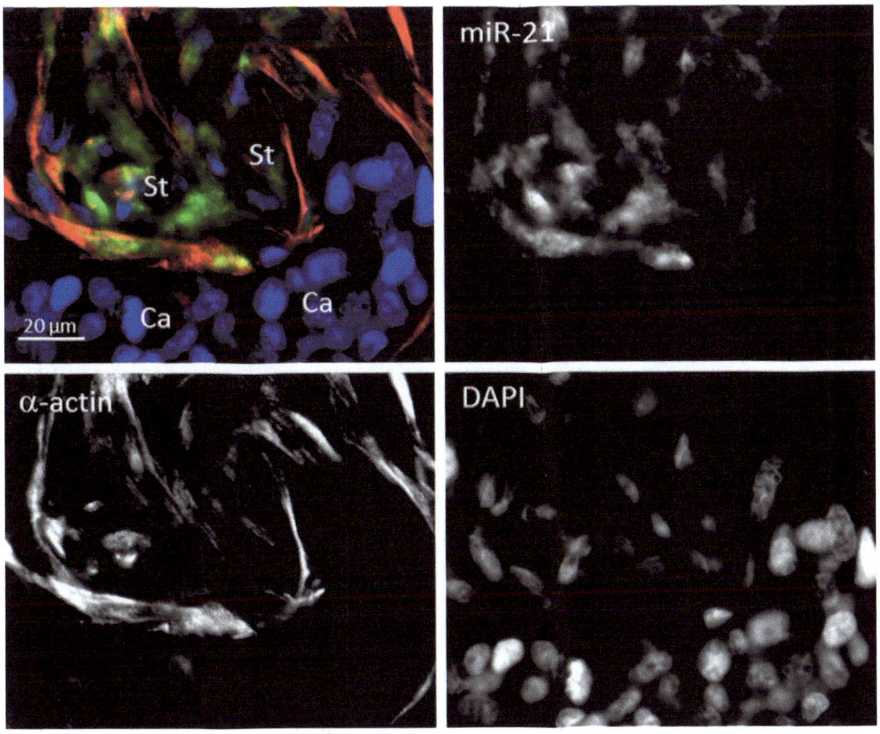

Fig. 1. Combined miR-21 ISH and sm-α-actin IHC for delineation of the cellular origin of miR-21 in breast cancer. MiR-21 was detected with TSA–FITC substrate (*green*) and sm-α-actin with Cy3 (*red*). Section was counter stained with DAPI (*blue*). The MiR-21 ISH signal is seen in sm-α-actin-positive myofibroblast located in the breast cancer stroma (St) surrounding clusters of miR-21 negative cancer cells (Ca). The individual fluorophores are shown in *black* and *white*.

3.3. Microscopy Observations: Some Examples

In this methodology chapter, we have combined (1) miR-21 ISH with sm-α-actin IHC, (2) miR-205 ISH with sm-α-actin IHC as examples to show a molecular characteristic of the miRNA positive cells, and (3) miR-21 ISH with PDCD4 IHC as an example to compare a miRNA (miR-21) expression with one of its validated targets (PDCD4). Expression of miR-21 and miR-205 in human breast lesions has been described by others (42, 44). Sm-α-actin is not expressed uniquely in one cell population in human breast tissue, but is expressed in different smooth muscle-differentiated cell populations: myoepithelial cells (the basal cells of the mammary gland and duct system), vascular smooth muscle cells (VSMC, surrounding the endothelial cells), and in myofibroblasts (myo-differentiated fibroblastic cells present in tumor stroma). These three cell populations can be distinguished by their morphological characteristics. In the first example, miR-21 ISH signal is seen in some of the sm-α-actin-positive tumor associated myofibroblasts (Fig. 1). In the second example, miR-205 ISH signal is seen in sm-α-actin-positive myoepithelial cells and is absent in VSMC and myoepithelial cells (Fig. 2).

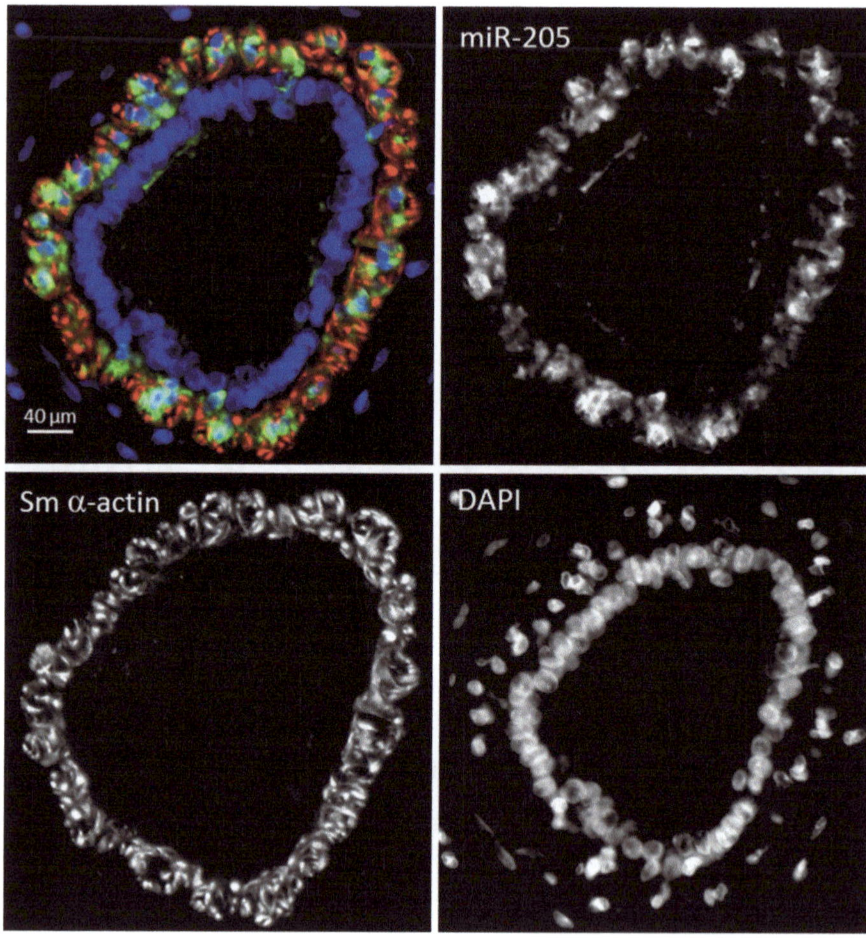

Fig. 2. Combined miR-205 ISH and sm-α-actin IHC for delineation of the cellular origin of miR-205 in normal breast gland. MiR-205 was detected with TSA–FITC substrate (*green*) and sm-α-actin with Cy3 (*red*). Section was counterstained with DAPI (*blue*). MiR-205 ISH signal is seen in sm-α-actin-positive myoepithelial cells surrounding the unstained luminal ductal epithelial cells.

The obvious differential expression of miR-205 and sm-α-actin is noteworthy. MiR-205 is likely not associated with smooth muscle differentiation in general since no expression is seen in the VSMC and myofibroblasts. MiR-205 is more likely related to the basal cell characteristics also seen in skin (20). The specific targets and role of miR-205 in differentiation of basal cells are not known (45). The differential expression of miR-21 and sm-α-actin is also interesting but is not as drastic as for miR-205. In addition to myofibroblasts, both smooth muscle cells in colon (46) and myoepithelial cells in breast (44) have been found to express miR-21.

PDCD4 has been found by several groups to be a direct target of miR-21 (47, 48). We combined miR-21 ISH and PDCD4 IHC in breast cancer samples and found that PDCD4 was more

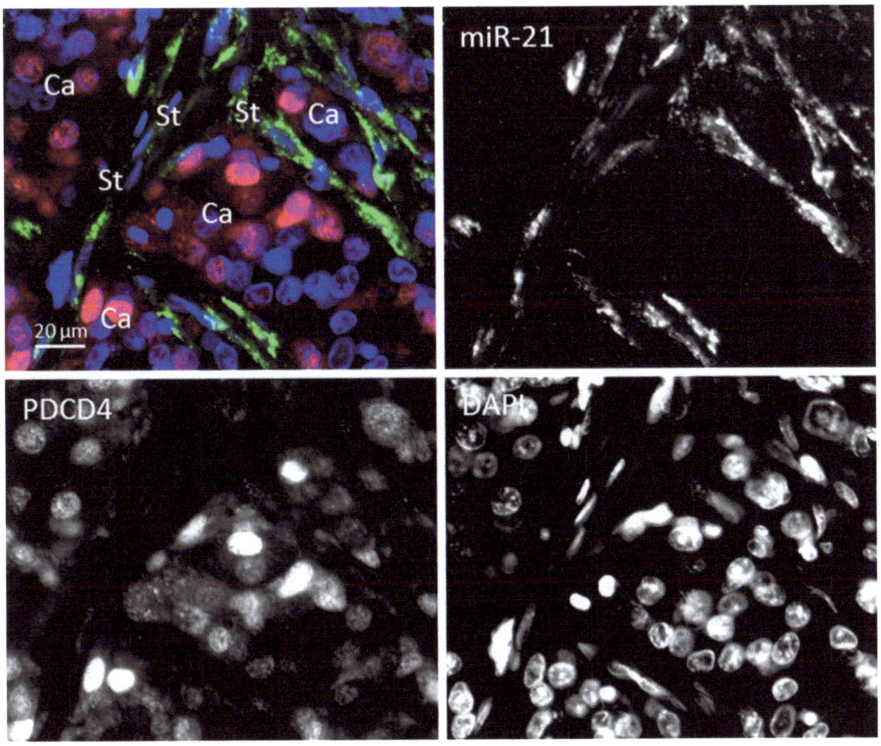

Fig. 3. Combined miR-21 ISH and PDCD4 IHC for delineation of regulatory targets of miR-21 in human breast cancer. MiR-21 was detected with TSA–FITC substrate (*green*) and PDCD4 with Cy3 (*red*). Section was counterstained with DAPI (*blue*). MiR-21 ISH signal is seen in the stroma (St) surrounding clusters of cancer cells (Ca). Intense PDCD4 signal can be seen in cancer cells, which are miR-21 negative, whereas a weaker PDCD4 signal is seen in miR-21-positive fibroblasts.

prevalent in the cancer cells than in adjacent stromal cells (Fig. 3). In contrast, miR-21 was most prevalent in breast cancer stromal cells. This differential expression should be expected if miR-21 downregulates its targets in the population where it is expressed, and the observation therefore strongly supports the many reports indicating that PDCD4 is a target of miR-21.

Here we report only one example of the many possibilities associated with combining miRNA ISH and protein marker IHC. In a therapeutic setting, where a miRNA or its target sequence is obstructed, this method has an obvious advantage in validating the efficacy of the RNA interference drug, for example, after interfering with the miR-21/PDCD4 pathway (47). Further fluorescence multiplexing by adding more markers into the combined assay is feasible and can be highly informative for clarification of the molecular pathway involved. Beyond this application, the technique might have a value in the future for differential diagnosis (49, 50).

4. Notes

1. Synchronous access to miRNAs and protein antigens requires that both probes (the LNA probe and the primary antibody) can bind to their targets. MiRNA ISH has been developed on a platform where tissue sections are predigested with proteolytic enzymes, such as proteinase-K. This pretreatment procedure also allows antigen detection by many primary antibodies; however, today heat-induced epitope retrieval (HIER) has become a frequent standard for many antibodies, whereas the effect of proteolytic treatment is often not reported. Therefore, note that novel antibodies on the market may not have been tested in a protocol based on proteolytic antigen retrieval. On the other hand, miRNA ISH may still find a way to its target in non-proteolytic pretreatment procedures. In this study, we performed proteinase-K treatment at 25 μg/ml at 30°C for 40 min. The optimal proteinase-K treatment is dependent on the tissue to be analyzed and the fixation protocol used. Routinely processed tissue samples, which are usually fixed overnight at room temperature in neutral buffered formalin, often show adequate performance. Denaturing fixatives should be avoided for use in ISH analyses. Proteinase-K treatment efficiency is dependent on three conditional parameters: the proteinase-K concentration, the incubation temperature, and the duration of incubation.
2. Probes can be differently labeled with hapten labels, for example carboxyfluorescein (6-FAM), digoxigenin (DIG) and biotin can be bound to LNA oligos. The choice of probe label depends on experimental design and the techniques available in the laboratory. The hapten label provides a template for crucial signal amplification since the FITC label on the oligo itself is not sufficient to allow detection in standard epifluorescence. In this study, the fluorescence signal was obtained with the TSA–FITC substrate, which allowed detection of miR-21 and miR-205.
3. For the hybridization step, we employed a Dako hybridizer; however, we also tested the protocol using a conventional hybridization oven and obtained similar ISH results. We applied 50 μl hybridization mix and sealed with a cover glass. In this experiment we did not seal the cover glass with rubber cement, which is proposed when incubating smaller hybridization reagent volumes (38, 40).
4. In this study we used the miR-21 probe at 10–20 nM. The probe concentration range may vary from 10 to 100 nM. Attention should be paid to unspecific binding and cross-hybridization when incubating probes at high-range probe concentrations. Use of a negative control probe (such as the scrambled

probe, Exiqon) often gives a good indication of unspecific probe binding, when incubated at the same concentration as the specific probe. Cross-hybridization cannot be evaluated using the scrambled probe and requires other specificity controls such as mismatch controls and positive controls (40).

5. The intensity of the ISH signal can be regulated by the incubation time of TSA–FITC substrate. We incubated the reagent for 5 min at room temperature, but the substrate reagent may be allowed to proceed for up to 15 min. A sufficient volume of the substrate reagent should be incubated to ensure that the reagent fully displaces the washing buffer in the slide chamber (at least 150 μl in the Shandon slide rack).
6. A series of antibody dilution tests prior to the combined ISH/IHC can give a good hint of optimal antibody dilution. Best fluorescence images are obtained when the intensity of the individual fluorophores are balanced.
7. The experiment must be designed without possibility of cross-reaction between the two parallel assays. For example, avoid using a goat primary antibody that needs detection with anti-sheep antibody since this may cross-react with the sheep anti-FAM antibody. Also, the anti-FAM antibody should not be used after applying TSA–FITC reagent.

Acknowledgment

We thank Trine Møller for excellent technical assistance and the Danish Ministry of Science, Innovation and Technology, for financial funding.

References

1. Ambros V (2001) microRNAs: tiny regulators with great potential. Cell 107:823–826
2. Lee RC, Feinbaum RL, Ambros V (1993) The C. elegans heterochronic gene lin-4 encodes small RNAs with antisense complementarity to lin-14. Cell 75:843–854
3. Lagos-Quintana M, Rauhut R, Lendeckel W et al (2001) Identification of novel genes coding for small expressed RNAs. Science 294:853–858
4. Bartel DP (2004) MicroRNAs: genomics, biogenesis, mechanism, and function. Cell 116: 281–297
5. Pillai RS, Bhattacharyya SN, Artus CG et al (2005) Inhibition of translational initiation by Let-7 MicroRNA in human cells. Science 309:1573–1576
6. Liu J, Rivas FV, Wohlschlegel J et al (2005) A role for the P-body component GW182 in microRNA function. Nat Cell Biol 7: 1261–1266
7. Ambros V (2011) MicroRNAs and developmental timing. Curr Opin Genet Dev 21: 511–517
8. Cordes KR, Sheehy NT, White MP et al (2009) miR-145 and miR-143 regulate smooth muscle cell fate and plasticity. Nature 460:705–710
9. Makeyev EV, Zhang J, Carrasco MA et al (2007) The MicroRNA miR-124 promotes neuronal differentiation by triggering brain-specific alternative pre-mRNA splicing. Mol Cell 27:435–448
10. Yao Q, Cao S, Li C et al (2011) Micro-RNA-21 regulates TGF-beta-induced myofibroblast

differentiation by targeting PDCD4 in tumor-stroma interaction. Int J Cancer 128: 1783–1792
11. Madhyastha R, Madhyastha H, Nakajima Y et al (2011) MicroRNA signature in diabetic wound healing: promotive role of miR-21 in fibroblast migration. Int Wound J 9:355–361
12. Medina PP, Slack FJ (2008) microRNAs and cancer: an overview. Cell Cycle 7:2485–2492
13. Winter J, Jung S, Keller S et al (2009) Many roads to maturity: microRNA biogenesis pathways and their regulation. Nat Cell Biol 11:228–234
14. Betel D, Wilson M, Gabow A et al (2008) The microRNA.org resource: targets and expression. Nucleic Acids Res 36:D149–D153
15. Thomson DW, Bracken CP, Goodall GJ (2011) Experimental strategies for microRNA target identification. Nucleic Acids Res 39:6845–6853
16. Ferdin J, Kunej T, Calin GA (2010) Non-coding RNAs: identification of cancer-associated microRNAs by gene profiling. Technol Cancer Res Treat 9:123–138
17. Sorensen KD, Orntoft TF (2010) Discovery of prostate cancer biomarkers by microarray gene expression profiling. Expert Rev Mol Diagn 10:49–64
18. Jensen SG, Lamy P, Rasmussen MH et al (2011) Evaluation of two commercial global miRNA expression profiling platforms for detection of less abundant miRNAs. BMC Genomics 12:435
19. Burnside J, Ouyang M, Anderson A et al (2008) Deep sequencing of chicken microRNAs. BMC Genomics 9:185
20. Joyce CE, Zhou X, Xia J et al (2011) Deep sequencing of small RNAs from human skin reveals major alterations in the psoriasis miRNAome. Hum Mol Genet 20:4025–4040
21. Schetter AJ, Leung SY, Sohn JJ et al (2008) MicroRNA expression profiles associated with prognosis and therapeutic outcome in colon adenocarcinoma. JAMA 299:425–436
22. Ralfkiaer U, Hagedorn PH, Bangsgaard N et al (2011) Diagnostic microRNA profiling in cutaneous T-cell lymphoma (CTCL). Blood 118:5891–5900
23. Clop A, Marcq F, Takeda H et al (2006) A mutation creating a potential illegitimate microRNA target site in the myostatin gene affects muscularity in sheep. Nat Genet 38:813–818
24. Richardson K, Lai CQ, Parnell LD et al (2011) A genome-wide survey for SNPs altering microRNA seed sites identifies functional candidates in GWAS. BMC Genomics 12:504
25. Permuth-Wey J, Thompson RC, Burton NL et al (2011) A functional polymorphism in the pre-miR-146a gene is associated with risk and prognosis in adult glioma. J Neurooncol 105:639–646
26. Lei B, Gao S, Luo LF et al (2011) A SNP in the miR-27a gene is associated with litter size in pigs. Mol Biol Rep 38:3725–3729
27. Nossent AY, Hansen JL, Doggen C et al (2011) SNPs in microRNA binding sites in 3′-UTRs of RAAS genes influence arterial blood pressure and risk of myocardial infarction. Am J Hypertens 24:999–1006
28. Zhang L, Liu Y, Song F et al (2011) Functional SNP in the microRNA-367 binding site in the 3′UTR of the calcium channel ryanodine receptor gene 3 (RYR3) affects breast cancer risk and calcification. Proc Natl Acad Sci USA 108:13653–13658
29. Lodygin D, Tarasov V, Epanchintsev A et al (2008) Inactivation of miR-34a by aberrant CpG methylation in multiple types of cancer. Cell Cycle 7:2591–2600
30. Calin GA, Ferracin M, Cimmino A et al (2005) A MicroRNA signature associated with prognosis and progression in chronic lymphocytic leukemia. N Engl J Med 353:1793–1801
31. Trang P, Wiggins JF, Daige CL et al (2011) Systemic delivery of tumor suppressor microRNA mimics using a neutral lipid emulsion inhibits lung tumors in mice. Mol Ther 19:1116–1122
32. Sioud M (2011) Promises and challenges in developing RNAi as a research tool and therapy. Methods Mol Biol 703:173–187
33. Wang Z, Rao DD, Senzer N et al (2011) RNA interference and cancer therapy. Pharm Res 28:2983–2995
34. Garofalo M, Croce CM (2011) microRNAs: master regulators as potential therapeutics in cancer. Annu Rev Pharmacol Toxicol 51:25–43
35. Gambari R, Fabbri E, Borgatti M et al (2011) Targeting microRNAs involved in human diseases: a novel approach for modification of gene expression and drug development. Biochem Pharmacol 82:1416–1429
36. Kasinski AL, Slack FJ (2010) Potential microRNA therapies targeting Ras, NFkappaB and p53 signaling. Curr Opin Mol Ther 12:147–157
37. Stenvang J, Silahtaroglu AN, Lindow M et al (2008) The utility of LNA in microRNA-based cancer diagnostics and therapeutics. Semin Cancer Biol 18:89–102
38. Jorgensen S, Baker A, Moller S et al (2010) Robust one-day in situ hybridization protocol for detection of microRNAs in paraffin samples using LNA probes. Methods 52:375–381

39. Soe MJ, Moller T, Dufva M et al (2011) A sensitive alternative for microRNA in situ hybridizations using probes of 2′-O-methyl RNA + LNA. J Histochem Cytochem 59:661–672
40. Nielsen BS (2012) MicroRNA in situ hybridization. Methods Mol Biol 822:67–84
41. Kloosterman WP, Wienholds E, de Brujin E et al (2006) In situ detection of miRNAs in animal embryos using LNA-modified oligonucleotide probes. Nat Methods 3:27–29
42. Sempere LF, Preis M, Yezefski T et al (2010) Fluorescence-based codetection with protein markers reveals distinct cellular compartments for altered MicroRNA expression in solid tumors. Clin Cancer Res 16:4246–4255
43. Nuovo GJ (2010) In situ detection of microRNAs in paraffin embedded, formalin fixed tissues and the co-localization of their putative targets. Methods 52:307–315
44. Rask L, Balslev E, Jorgensen S et al (2011) High expression of miR-21 in tumor stroma correlates with increased cancer cell proliferation in human breast cancer. APMIS 119:663–673
45. Greene SB, Herschkowitz JI, Rosen JM (2010) The ups and downs of miR-205: identifying the roles of miR-205 in mammary gland development and breast cancer. RNA Biol 7:300–304
46. Nielsen BS, Jorgensen S, Fog JU et al (2011) High levels of microRNA-21 in the stroma of colorectal cancers predict short disease-free survival in stage II colon cancer patients. Clin Exp Metastasis 28:27–38
47. Young MR, Santhanam AN, Yoshikawa N et al (2010) Have tumor suppressor PDCD4 and its counteragent oncogenic miR-21 gone rogue? Mol Interv 10:76–79
48. Ruan Q, Wang T, Kameswaran V et al (2011) The microRNA-21-PDCD4 axis prevents type 1 diabetes by blocking pancreatic beta cell death. Proc Natl Acad Sci USA 108:12030–12035
49. Sempere LF (2011) Integrating contextual miRNA and protein signatures for diagnostic and treatment decisions in cancer. Expert Rev Mol Diagn 11:813–827
50. Soe MJ, Okkels F, Sabourin D et al (2011) HistoFlex—a microfluidic device providing uniform flow conditions enabling highly sensitive, reproducible and quantitative in situ hybridizations. Lab Chip 11:3896–3907

Index

Jürgen Moll and Riccardo Colombo (eds.), *Target Identification and Validation in Drug Discovery: Methods and Protocols*, Methods in Molecular Biology, vol. 986, DOI 10.1007/978-1-62703-311-4,

H

I

K

L

M

N

MIX
Papier aus verantwortungsvollen Quellen
Paper from responsible sources
FSC® C105338

If you have any concerns about our products,
you can contact us on
ProductSafety@springernature.com

In case Publisher is established outside the EU,
the EU authorized representative is:
Springer Nature Customer Service Center GmbH
Europaplatz 3, 69115 Heidelberg, Germany

Printed by Libri Plureos GmbH
in Hamburg, Germany